PEAK POWER CONTROL IN MULTICARRIER COMMUNICATIONS

The implementation of multicarrier (MC) modulation in wireless and wireline communication systems, such as OFDM and DMT, is restricted by peak signal power, due to a sensitivity of the technique to distortions introduced by nonlinear devices. By controlling the peak power, the negative influence of signals with high peaks on the performance of the transmission system is greatly reduced. This book describes the tools necessary for analyzing and controlling the peak-to-average power ratio in MC systems, and how these techniques are applied in practical designs. The author starts with an overview of MC signals and basic tools and algorithms, before discussing properties of MC signals in detail: discrete and continuous maxima; statistical distribution of peak power, and codes with constant peak-to-average power ratio are all covered, concluding with methods to decrease peak power in MC systems. Current knowledge, problems, methods, and definitions are summarized using rigorous mathematics, with an overview of tools for the engineer. This book is aimed at graduate students and researchers in electrical engineering, computer science, and applied mathematics, as well as practitioners in the telecommunications industry. Further information on this title is available at www.cambridge.org/9780521855969.

SIMON LITSYN received his Ph.D. in Electrical Engineering from the Leningrad Electrotechnical Institute in 1982. He is currently a professor in the Department of Electrical Engineering Systems at the Tel Aviv University, where he has been since 1991. From 2000 to 2003 he served as an Associate Editor for Coding Theory in the IEEE Transactions on Information Theory. His research interests include coding theory, communications, and applications of discrete mathematics. He is the author of more than 150 journal articles.

PEAK POWER CONTROL IN
MULTICARRIER COMMUNICATIONS

SIMON LITSYN

School of Electrical Engineering, Tel Aviv University

CAMBRIDGE
UNIVERSITY PRESS

CAMBRIDGE UNIVERSITY PRESS
Cambridge, New York, Melbourne, Madrid, Cape Town, Singapore, São Paulo

Cambridge University Press
The Edinburgh Building, Cambridge CB2 2RU, UK

Published in the United States of America by Cambridge University Press, New York

www.cambridge.org
Information on this title: www.cambridge.org/9780521855969

First published 2007

Printed in the United Kingdom at the University Press, Cambridge

A catalog record for this publication is available from the British Library

ISBN-13   978-0-521-85596-9 hardback
ISBN-10   0-521-85596-9 hardback

To the memory of my mother.

Contents

Figures

Abbreviations

ACE	active constellation extension
ACI	adjacent channel interference
ACPR	adjacent channel power ratio
ADC	analog-to-digital converter
AM/PM	amplitude modulation/phase modulation
BCH	Bose–Chaudhuri–Hocquenghem (codes)
BER	bit error rate
(B)PSK	(binary) phase-shift keying
BS	block scaling
CCDF	complementary cumulative distribution function
CDMA	code division multiple access
CF	crest factor
CS	codes of strength
DAB	digital audio broadcasting
DAC	digital-to-analog converter
DC	direct current
(I)DFT	(inverse) discrete Fourier transform
DMT	discrete multitone
(A/H)DSL	(asymmetric/high speed) digital subscriber line
DVB	digital video broadcasting
EVM	error vector magnitude
(I)FFT	(inverse) fast Fourier transform
GI	guard interval
HIPERLAN	high performance radio local area network
HPA	high-power amplifier
IBO/OBO	input/output back-off
ICI	inter-carrier interference
ISI	inter-symbol interference

LDPC	low-density parity-check (codes)
LPF	low-pass filter
MC	multicarrier
MIMO	multiple-input multiple-output
OFDM	orthgonal frequency division multiplexing
OFDMA	orthgonal frequency division multiple access
PAPR	peak-to-average power ratio
PMEPR	peak-to-mean envelope power ratio
(Q)PSK	(quadrature) phase-shift keying
PRC	peak reduction carriers
PTS	partial transmit sequences
QAM	quadrature amplitude modulation
RM	Reed–Muller (codes)
RP	random phasor
RS	Reed–Solomon (codes)
SER	symbol error rate
SI	side information
SLM	selective mapping
SL	soft limiter
SNR	signal-to-noise ratio
SSPA	solid-state power amplifier
TI	tone injection
TS	trellis shaping
TWTA	traveling-wave tube amplifier
UWB	ultra wide band
WLAN	wireless local area network
WMAN	wireless metropolitan area network
WPAN	wireless personal area network

Notation

$\mathbb{Z}$	integer numbers
$\mathbb{N}$	natural numbers
$\mathbb{R}$	real numbers
$\mathbb{C}$	complex numbers
$\mathbb{F}$	finite field
$\Re(\cdot)$	real part
$\Im(\cdot)$	imaginary part
ι	$\sqrt{-1}$
a^*	complex conjugate of $a \in \mathbb{C}$
$|a|$	absolute value of $a \in \mathbb{C}$
$\arg(a)$	argument of a
$\mathbf{a}$	vector
$(\mathbf{a}, \mathbf{b})$	dot product of vectors $\mathbf{a}$ and $\mathbf{b}$
$\|\mathbf{a}\|$	norm of $\mathbf{a}$
A^t	transposed matrix A
E_{av}	average energy of constellation
E_{max}	maximum energy of a constellation point
f_0	carrier frequency
f_s	tone bandwidth
$\mathcal{M}_c$	continuous maximum
$\mathcal{M}_d$	discrete maximum
g.c.d.	greatest common divisor
i.i.d.	independent identically distributed
p.d.f.	probability density function
p.s.d.	power spectral density
deg	degree of a polynomial
sinh	hyperbolic sine
cosh	hyperbolic cosine
sign	sign

1

Introduction

In the mountains the shortest way is from peak to peak, but for that route thou must have long legs.

F. Nietzsche, *Thus Spake Zarathustra*

Multicarrier (MC) modulations such as orthogonal frequency division multiplexing (OFDM) and discrete multitone (DMT) are efficient technologies for the implementation of wireless and wireline communication systems. Advantages of MC systems over single-carrier ones explain their broad acceptance for various telecommunication standards (e.g., ADSL, VDSL, DAB, DVB, WLAN, WMAN). Yet many more appearances are envisioned for MC technology in the standards to come. A relatively simple implementation is possible for MC systems. Low complexity is due to the use of fast discrete Fourier transform (DFT), avoiding complicated equalization algorithms. Efficient performance of MC modulation is especially vivid in channels with frequency selective fading and multipath. Nonetheless, still a major barrier for implementing MC schemes in low-cost applications is its nonconstant signal envelope, making the transmission sensitive to nonlinear devices in the communication path. Amplifiers and digital-to-analog converters distort the transmit signals leading to increased symbol error rates, spectral regrowth, and reduced power efficiency compared with single carrier systems. Naturally, the transmit signals should be restricted to those that do not cause the undesired distortions. A reasonable measure of the relevance of the signals is the ratio between the peak power values to their average power (PAPR). Thus the goal of peak power control is to diminish the influence of transmit signals with high PAPR on the performance of the transmission system. Alternatives are either the complete exclusion of such signals or an essential decrease in the probability of their appearance. Neither of these goals can be achieved without a decrease in the efficient transmission rate or performance penalty.

In this monograph I describe methods of analysis and control of peak power effects on the performance of MC communication systems. This includes analysis of statistical properties of peak distributions in MC signals, descriptions of MC signals with low peaks, and approaches to decreasing high peaks in transmitted signals. Consequently, the organization of the book is as follows. In Chapter 2, I provide general definitions related to MC communication systems and MC signals, and introduce the main definitions related to peaks of MC signals. This is followed by a description of nonlinearities in power amplifiers and their influence on the performance. In Chapter 3, necessary mathematical tools are described. This is necessary since the mathematical arsenal of the peak power control research consists of many seemingly unrelated methods. Among them are harmonic analysis, probability, algebra, combinatorics, and coding theory. In Chapter 4, I explain how the continuous problem of peak estimation can be reduced to the discrete problem of analysis of maxima in the sampled MC signal. Chapter 5 deals with statistical distribution of peaks in MC signals. It is shown that the peak distribution is concentrated around a typical value and the proportion of signals that are essentially different from the typical maximum of the absolute value is small. Chapter 6 extends the analysis of the previous chapter to MC signals defined by coded information. In Chapter 7, I describe methods to construct MC signals with much smaller peaks than is typical. Finally, in Chapter 8, I analyze approaches for decreasing peaks in MC signals. Several algorithms are introduced and compared. Notes in the end of each chapter provide historical comments and attribute the results appearing in the chapter.

Several related topics are not treated in this monograph. For peak power control in CDMA see, e.g., [43, 64, 118, 228, 230, 231, 285, 299, 300, 308, 309, 324, 325, 326, 363, 421, 422, 423, 445] and references therein. Peak power reduction in MIMO systems is discussed in, e.g., [1, 66, 67, 154, 234, 235, 241, 278, 338, 389, 395, 411, 456]. For analysis of peak power control in OFDMA see, e.g., [154, 315, 427, 453]. Aspects of peak power reduction in radar systems are considered in [55, 236, 237, 238, 279, 430]. Peak power control in optical signals is considered in [371, 375].

In the process of writing the book I enjoyed advice, ideas and assistance from many friends and collaborators. Their expertise was crucial in determining the best ways of presenting the material and avoiding wrong concepts and mistakes. Here is a definitely incomplete (alphabetical) list of colleagues without whose kind support this book would definitely not have been written: Idan Alrod, Ella Barzani, Gregory Freiman, Masoud Sharif, Eran Sharon, Alexander Shpunt, Dov Wulich, Gerhard Wunder, and Alexander Yudin.

I also wish to thank the staff and associates of Cambridge University Press for their valuable assistance with production of this book. In particular I am grateful to editorial manager Dr. Philip Meyler, assistant editors Ms. Emily Yossarian and Ms. Anna Littlewood, production editor Ms. Dawn Preston, and copy editor Dr. Alison Lees.

2

Multicarrier signals

In this chapter, I introduce the main issues we will deal with in the book. In Section 2.1, I describe a multicarrier (MC) communication system. I introduce the main stages that the signals undergo in MC systems and summarize advantages and drawbacks of this technology. Section 2.2 deals with formal definitions of the main notions related to peak power: peak-to-average power ratio, peak-to-mean envelope power ratio, and crest factor. In Section 2.3, I quantify the efficiency of power amplifiers and its dependence on the power of processed MC signals. Section 2.4 introduces nonlinear characteristics of power amplifiers and describes their influence on the performance of communication systems.

2.1 Model of multicarrier communication system

The basic concept behind multicarrier (MC) transmission is in dividing the available spectrum into subchannels, assigning a carrier to each of them, and distributing the information stream between subcarriers. Each carrier is modulated separately, and the superposition of the modulated signals is transmitted. Such a scheme has several benefits: if the subcarrier spacing is small enough, each subchannel exhibits a flat frequency response, thus making frequency-domain equalization easier. Each substream has a low bit rate, which means that the symbol has a considerable duration; this makes it less sensitive to impulse noise. When the number of subcarriers increases for properly chosen modulating functions, the spectrum approaches a rectangular shape. The multicarrier scheme shows a good modularity. For instance, the subcarriers exhibiting a disadvantageous *signal-to-noise ratio* (SNR) can be discarded. Moreover, it is possible to choose the constellation size (bit loading) and energy for each subcarrier, thus approaching the theoretical capacity of the channel.

Figure 2.1 presents the structure of a MC transmitter. Let n be the number of subcarriers in this system. The following processing stages are employed to derive the transmit signal. Redundancy defined by an error-correcting code is appended to

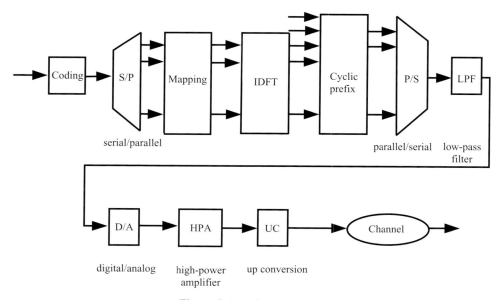

Figure 2.1 MC transmitter

the input information. The encoded data is converted to parallel form, and is mapped to n complex numbers defining points in the constellation used for modulation (e.g., QAM or PSK). These n complex numbers are inserted into an inverse discrete Fourier transform (IDFT) block, which outputs the time equidistributed samples of the baseband signal. The next block introduces a *guard interval* (GI) intended for diminishing the effect of the delay of the multipath propagation. The guard interval is usually implemented as a cyclic prefix (CP). Because of the CP, the transmit signal becomes periodic, and the effect of the time-dispersive multipath channel becomes equivalent to a cyclic convolution, discarding the GI at the receiver. Thus the effect of the multipath channel is limited to a pointwise multiplication of the transmitted data constellations by the channel transfer function, that is, the subcarriers remain orthogonal. Being converted back to the serial form, the samples are transformed by a low-pass filter (LPF) to give a continuous signal. This continuous signal is amplified by a high-power amplifier (HPA). Finally, if necessary, the baseband signal becomes passband by translation to a higher frequency. The reverse steps are performed by the receiver.

Implementation advantages of the MC communication system come from the simple structure of the DFT, which can be realized with a complexity proportional to $n \ln n$. Also, the equalization required for detecting the data constellations is an elementwise multiplication of the DFT output by the inverse of the estimated channel transfer function.

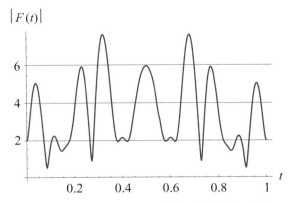

Figure 2.2 Envelope of a BPSK modulated MC signal for $n = 16$

However, several disadvantages arise with this concept, the most severe of which is the highly nonconstant envelope of the transmit signal (see Fig. 2.2), making MC modulation very sensitive to nonlinear components in the transmission path. A key component is the high-power amplifier (HPA). Owing to cost, design, and, most importantly, power efficiency considerations, the HPA cannot resolve the dynamics of the transmit signal and inevitably cuts off the signal at some point, causing additional in-band distortion and adjacent channel interference. The power efficiency penalty is certainly the major obstacle to implementing MC systems in low-cost applications. Moreover, in power-limited regimes determined by regulatory bodies, the average power is reduced in comparison to single-carrier systems, in turn reducing the range of transmission.

The main goal of peak power control is to diminish the influence of high peaks in transmit signals on the performance of the transmission system.

2.2 Peak power definitions

Let me give a more detailed description of the signals in the MC communication system. Denote by n the number of subcarriers (tones). The system receives at each time instant $0, \Upsilon, 2\Upsilon, \ldots$ a collection of n constellation symbols a_k, $k = 0, \ldots, n - 1$, where $a_k \in \mathbb{C}$, carrying the information to be transmitted. The subset $\mathcal{Q}$ of possible values of a_k depends on the type of the carrier modulation. The most popular complex constellations are BPSK, M-QAM, and M-PSK. We assume

$$\text{BPSK} = \{-1, 1\},$$
$$M\text{-QAM} = \left\{ A\left((2m_1 - 1) + \iota\left(2m_2 - 1\right)\right), m_1, m_2 \in \left\{ -\frac{m}{2} + 1, \ldots, \frac{m}{2} \right\} \right\},$$

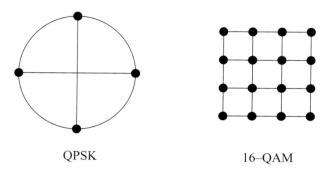

QPSK   16–QAM

Figure 2.3 Examples of standard constellations

for natural numbers $m > 1$, M such that $M = m^2$ and $A^2 = \frac{3}{2}(M-1)$, and

$$M\text{-PSK} = \left\{ 1, e^{\frac{2\pi \imath}{M}}, \ldots, e^{\frac{2\pi \imath (M-1)}{M}} \right\}$$

for $M > 2$ and $\imath = \sqrt{-1}$. With such normalization, the average energy of a constellation point is 1. Notice that the (envelope) power of all M-PSK signals is the same and equals n. Another example of signals with equal power is provided by *spherical constellations* for which the only imposed restriction is that $\sum_{k=0}^{n-1} |a_k|^2 = n$. However, e.g., for MC signals using M-QAM, with $M > 4$, there are signals having different power. In the case of constellation points of varying energy, we scale the signal in such a way that the average energy is normalized to 1,

$$E_{av} = E_{av}(\mathcal{Q}) = \frac{1}{|\mathcal{Q}|} \sum_{a \in \mathcal{Q}} |a|^2 = 1. \tag{2.1}$$

We denote the maximum energy of a constellation point by $E_{max} = E_{max}(\mathcal{Q})$, where

$$E_{max} = \max_{a \in \mathcal{Q}} |a|^2. \tag{2.2}$$

Let f_0 be the carrier frequency and f_s be the bandwidth of each tone. Ignoring the possibility of assigning a guard time (which is a common assumption) we set $f_s = \frac{1}{T}$. The transmitted signal on the interval $t = \left[0, \frac{1}{f_s}\right)$ is then represented by the real part of

$$S_{\mathbf{a}}(t) = \sum_{k=0}^{n-1} a_k e^{2\pi \imath (f_0 + k f_s) t}. \tag{2.3}$$

The *instantaneous power* of the transmit signal is $(\Re(S_{\mathbf{a}}(t)))^2$, while $|S_{\mathbf{a}}(t)|^2$ is called the *envelope power*. Denoting $\zeta = \frac{f_0}{f_s}$ and considering the signal on the interval $[0, 1)$, we have the following definition for the *peak-to-average power ratio* of $S_{\mathbf{a}}(t)$:

$$\text{PAPR}(\mathbf{a}) = \frac{1}{\sum_{k=0}^{n-1} |a_k|^2} \cdot \max_{t \in [0,1)} \left| \Re \left(\sum_{k=0}^{n-1} a_k e^{2\pi \imath (\zeta + k)t} \right) \right|^2. \tag{2.4}$$

It is straightforward that

$$\text{PAPR}(\mathbf{a}) \leq \text{PMEPR}(\mathbf{a}) = \frac{1}{\sum_{k=0}^{n-1} |a_k|^2} \cdot \max_{t \in [0,1)} |F_{\mathbf{a}}(t)|^2, \tag{2.5}$$

where

$$F_{\mathbf{a}}(t) = \sum_{k=0}^{n-1} a_k e^{2\pi \imath kt}, \tag{2.6}$$

and PMEPR stands for the *peak-to-mean-envelope-power ratio*. Another often considered parameter is the *crest factor* (CF) which is just the square root of PMEPR,

$$\text{CF}(\mathbf{a}) = \sqrt{\text{PMEPR}(\mathbf{a})}. \tag{2.7}$$

Although PMEPR provides an upper bound on PAPR, it is quite accurate for big values of ζ. Indeed, in the definition of PAPR we use $(\Re(S_{\mathbf{a}}(t)))^2$, while $|e^{2\pi \imath \zeta t} S_{\mathbf{a}}(t)|^2$ is used for PMEPR. If ζ is large, a tiny change in t drastically modifies the phase of $e^{2\pi \imath \zeta t} S_{\mathbf{a}}(t)$ while the value and phase of band-limited $S_{\mathbf{a}}(t)$ does not change significantly. Thus just in the very close vicinity of t_0 in which the maximum of $|S_{\mathbf{a}}(t)|$ is attained, it is possible to find t_1 such that $e^{2\pi \imath \zeta t_1} S_{\mathbf{a}}(t_1)$ is real and $|S_{\mathbf{a}}(t_1)|$ is very close to $|S_{\mathbf{a}}(t_0)|$. This analysis will be quantified later in Section 4.6.

In what follows, we will often consider a situation when the vector $\mathbf{a}$ belongs to a discrete set $\mathcal{C} \subset \mathbb{C}^n$, of size $|\mathcal{C}|$, called *code*. In this context, assuming that all the code words are equiprobable, we define the average power of an MC signal from code $\mathcal{C}$ as

$$P_{av}(\mathcal{C}) = \frac{1}{|\mathcal{C}|} \sum_{\mathbf{a} \in \mathcal{C}} \sum_{k=0}^{n-1} |a_k|^2. \tag{2.8}$$

Consequently,

$$\text{PAPR}(\mathcal{C}) = \frac{1}{P_{av}(\mathcal{C})} \cdot \min_{\mathbf{a} \in \mathcal{C}} \max_{t \in [0,1)} |\Re(S_{\mathbf{a}}(t))|^2, \tag{2.9}$$

where $S_{\mathbf{a}}(t)$ is defined in (2.3), and

$$\text{PMEPR}(\mathcal{C}) = \frac{1}{P_{av}(\mathcal{C})} \cdot \min_{\mathbf{a} \in \mathcal{C}} \max_{t \in [0,1)} |F_{\mathbf{a}}(t)|^2, \tag{2.10}$$

where $F_{\mathbf{a}}(t)$ is defined in (2.6). Finally,

$$\mathrm{CF}(\mathcal{C}) = \frac{1}{\sqrt{P_{av}(\mathcal{C})}} \cdot \min_{\mathbf{a} \in \mathcal{C}} \max_{t \in [0,1)} |F_{\mathbf{a}}(t)| . \qquad (2.11)$$

The expressions become especially simple if all the signals have the same energy n or when the coefficients are drawn independently from a constellation $\mathcal{Q}$ scaled such that the average energy of a constellation point is 1. Then

$$\mathrm{PAPR}(\mathcal{C}) = \frac{1}{n} \cdot \min_{\mathbf{a} \in \mathcal{C}} \mathrm{PAPR}(\mathbf{a}),$$

$$\mathrm{PMEPR}(\mathcal{C}) = \frac{1}{n} \cdot \min_{\mathbf{a} \in \mathcal{C}} \mathrm{PMEPR}(\mathbf{a}),$$

$$\mathrm{CF}(\mathcal{C}) = \frac{1}{\sqrt{n}} \cdot \min_{\mathbf{a} \in \mathcal{C}} \mathrm{CF}(\mathbf{a}).$$

We measure PAPR and PMEPR in decibels (dB), using $10 \lg \mathrm{PAPR}(\mathbf{a})$ or $10 \lg \mathrm{PMEPR}(\mathbf{a})$ as an indication of the quality of a signal $\mathbf{a}$. Correspondingly, for CF we will replace the factor of 10 by 20. Notice that the maximum of an MC signal in an n-carrier system is n, thus the maximum of PAPR and PMEPR is n while the maximum CF is $\sqrt{n}$, which corresponds to $10 \lg n$.

Example 2.1   Let $n = 4$, and $\mathbf{a} = (1, -1, -1, 1)$. Then

$$F_{\mathbf{a}}(t) = 1 - e^{2\pi \imath t} - e^{4\pi \imath t} + e^{6\pi \imath t}.$$

The maximum of $|F_{\mathbf{a}}(t)|^2$ can be determined as follows. Indeed,

$$|F_{\mathbf{a}}(t)|^2 = (1 - \cos 2\pi t - \cos 4\pi t + \cos 6\pi t)^2$$
$$+ (- \sin 2\pi t - \sin 4\pi t + \sin 6\pi t)^2.$$

Differentiating and equating the result to zero, after simple trigonometric manipulations, we reduce the problem to finding a solution to

$$1 + 2 \cos 2\pi t - 3 \cos^2 2\pi t = 0.$$

Thus the maximum of $|F_{\mathbf{a}}(t)|^2$ occurs when

$$t = \frac{1}{2\pi} \arccos \left(-\frac{1}{3} \right) = 0.304086,$$

and equals 9.48148. This corresponds to

$$10 \lg \frac{9.48148}{4} = 3.74816 \, \mathrm{dB}.$$

$\square$

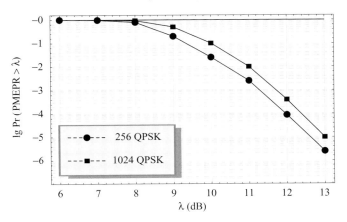

Figure 2.4 CCDFs of PMEPR of MC signal with 256 and 1024 QPSK modulated subcarriers

Another widely used characterization of MC signals deals with the probabilistic distribution of peak power values. Namely, we will be using the complementary cumulative distribution function (CCDF) of PAPR, PMEPR or CF. For instance, in the case of PMEPR, CCDF is just the probability that PMEPR of a randomly chosen MC signal exceeds a predefined threshold λ, $\Pr(\text{PMEPR} > \lambda)$. An example of such CCDF for a QPSK modulated MC system with 256 and 1024 subcarriers is presented in Fig. 2.4.

Example 2.2 For $n = 4$ and BPSK modulation, Table 2.1 contains the list of PMEPR values for all possible sequences. In this situation (in contrast with large values of n) we are able to find the explicit distribution of probabilities of PMEPR,

PMEPR dB	2.48	3.75	6.02
Probability	1/2	1/4	1/4

$\square$

There are several possible goals of the peak power control:

- Restriction of the set of used MC signals to those with peaks not exceeding a prescribed level;
- Restriction of the set of used MC signals in such a way that the probability of having a peak exceeding a prescribed level is much smaller than in the unrestricted set;
- Modification of the used MC signals in such a way that the probability of errors in the reconstructed coefficient vector is small, while at the same time the peaks are bounded with high probability.

Which of the goals is to be addressed depends on the system requirements and regulations.

Table 2.1 *PMEPRs for n = 4 and BPSK modulation*

Vector **a**				PMEPR(**a**) (dB)
−1	−1	−1	−1	6.02
1	−1	−1	−1	2.48
−1	1	−1	−1	2.48
1	1	−1	−1	3.75
−1	−1	1	−1	2.48
1	−1	1	−1	6.02
−1	1	1	−1	3.75
1	1	1	−1	2.48
−1	−1	−1	1	2.48
1	−1	−1	1	3.75
−1	1	−1	1	6.02
1	1	−1	1	2.48
−1	−1	1	1	3.75
1	−1	1	1	2.48
−1	1	1	1	2.48
1	1	1	1	6.02

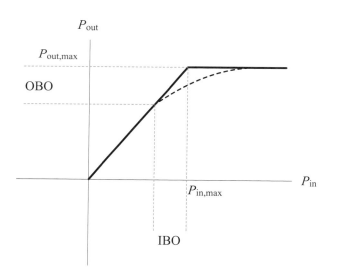

Figure 2.5 Input–output power characteristic of a HPA

2.3 Efficiency of power amplifiers

The perfectly linear ideal memoryless amplifier produces an output that is a multiple of the input. In reality, there is no amplifier able to provide unlimited output. The amplifier output is always limited to some value, called *saturation*. A typical characteristic of a HPA is presented in Fig. 2.5.

Instantaneous input back-off (IBO) and output back-off (OBO) are measured in decibels,

$$\mathrm{IBO} = 10 \lg \frac{P_{\mathrm{in,max}}}{P_{\mathrm{in}}},$$

and

$$\mathrm{OBO} = 10 \lg \frac{P_{\mathrm{out,max}}}{P_{\mathrm{out}}},$$

where P_{in} and P_{out} are the instantaneous powers of the signal before and after the HPA. The back-off definitions could easily be generalized to nonconstant envelope signals by averaging P_{in} and P_{out} over the signal's period.

Power amplifier efficiency is a significant factor for the efficiency of most wireless systems. Poor efficiency of the last power amplifier stage leads to large energy loss, not only deteriorating system efficiency, but also exacerbating thermal issues with the devices. Two measurements of the efficiency of HPAs are generally used. The first one is the power supply (or DC) efficiency η_{DC} defined as the ratio between output power P_{out} and the HPA power consumption P_{DC},

$$\eta_{DC} = \frac{P_{\mathrm{out}}}{P_{DC}}. \tag{2.12}$$

As the expression for η_{DC} does not depend on the HPA input power, it can give, in the case of a HPA with low gain, too optimistic a view of the HPA's efficiency. This is why the *power-added efficiency* is defined,

$$\eta_{\mathrm{PAE}} = \frac{P_{\mathrm{out}} - P_{\mathrm{in}}}{P_{DC}}, \tag{2.13}$$

with P_{in} being the power at HPA input.

Equations (2.12) and (2.13) characterize the efficiency in the case of constant-envelope signals at the HPA input. For nonconstant-envelope signals defined on $[0, 1)$ the time-average efficiency is introduced,

$$\overline{\eta} = \int_0^1 \eta(t)\,\mathrm{d}t. \tag{2.14}$$

Power amplifiers are commonly classified into two main groups: linear and nonlinear HPAs. The term "linear amplifier" does not mean generally that such an amplifier is perfectly linear, but rather that an effort is made at linear amplification of the input signal. Within each of these groups, the amplifiers can be further classified to a number of classes. The main types of linear amplifiers, see, e.g., [200], are A, AB, and B. Nonlinear amplifiers belong to C, D, E, F, G, H, and S classes. I will deal throughout with linear amplifiers.

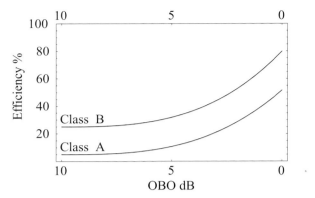

Figure 2.6 Efficiency as a function of output power back-off

The A, AB, and B classes are defined by the so-called conduction angle, where the conduction angle of 2π corresponds to A class, π defines B class, and AB is characterized by intermediate values of the angle. The efficiency grows with reducing of the conduction angle. However, when the conduction angle decreases, the achievable output power falls down. Figure 2.6 depicts HPA efficiency as a function of the output power back-off for different HPA classes. As seen from the picture, the maximum efficiency is achieved at the operation point at saturation.

In [270] the following relation between $\overline{\eta}$ and PMEPR of a MC signal **a** is given:

$$\overline{\eta} = G \cdot e^{-g \cdot \mathrm{PMEPR}(\mathbf{a})}, \tag{2.15}$$

where the PMEPR is in dB, and $G = 0.587, g = 0.1247$ for class A, and $G = 0.907$, $g = 0.1202$ for class B.

2.4 Models of HPA nonlinearities

Commonly used models for the nonlinearity of high power amplifiers (HPA) are given by their amplitude dependent amplitude distortion and amplitude dependent phase distortion conversion characteristics, e.g., [347, 351]. If the characteristics are memoryless, the result of the conversion depends only on the current value of the signal. Let the input signal be presented in polar coordinates,

$$x = \rho e^{\iota \psi},$$

and the complex envelope of the output signal be

$$g(x) = F(\rho)e^{\iota(\psi + \Phi(\rho))}.$$

Here, $F(\rho)$ and $\Phi(\rho)$ are, respectively, the AM/AM and AM/PM characteristics.

The simplest way to characterize the nonlinear behavior of a HPA is the 1dB compression point. It is the point where the gain of the amplifier is reduced by 1dB with respect to the gain in its linear region.

- *Soft limiter (SL)*
  The SL's AM/AM and AM/PM are

$$F(\rho) = \begin{cases} \rho, & \rho \le A, \\ A, & \rho > A, \end{cases}$$

$$\Phi(\rho) = 0.$$

It is argued in [5, 6, 181, 314] that SL is an adequate model for the nonlinearity if the HPA is linearized by a suitable predistorter.

For MC signals we introduce the *clipping ratio* γ, defined as

$$\gamma = \frac{A}{\sqrt{P_{\text{in}}}}, \tag{2.16}$$

where P_{in} is the average input power of the MC signal before clipping. When $\gamma = \infty$ the signal is not distorted, while when γ tends to 0, as will be shown, the output signal converges to the one with constant envelope.

- *Rapp model*
  The model [337] describes accurately the conversion transform of solid-state power amplifiers (SSPA). The AM/AM and AM/PM characteristics are:

$$F(\rho) = \frac{\rho}{\left(1 + \left(\frac{\rho}{A}\right)^{2p}\right)^{\frac{1}{2p}}},$$

$$\Phi(\rho) = 0.$$

The parameter p reflects the smoothness of the transition from the linear region to the limiting saturation region. When p grows the SSPA model converges to SL.

- *Saleh model*
  This simple model involving two-parameter functions was suggested in [349] for describing traveling-wave tube amplifiers (TWTA). Appropriate selections for the amplitude and phase coefficients also provide an accurate model for SSPA.
  The AM/AM and AM/PM characteristics of the model are:

$$F(\rho) = \frac{\alpha_1 \rho}{1 + \alpha_2 \rho^2},$$

$$\Phi(\rho) = \frac{\beta_1 \rho^2}{1 + \beta_2 \rho^2}.$$

As an example, the set of parameters that closely matches the TWTA data given in [349] is

$$\alpha_1 = 2.1587, \quad \alpha_2 = 1.1517, \quad \beta_1 = 4.0330, \quad \beta_2 = 9.1040.$$

The coefficients in the model can be made frequency dependent.

- *Ghorbani model*
  This model is similar to Saleh's one. It uses four-parameter functions and is claimed to fit better the AM/AM characteristics of SSPA. The AM/AM and AM/PM characteristics of the model are:

$$F(\rho) = \frac{\alpha_1 \rho^{\alpha_2}}{1 + \alpha_3 \rho^{\alpha_2}} + \alpha_4 \rho,$$

$$\Phi(\rho) = \frac{\beta_1 \rho^{\beta_2}}{1 + \beta_3 \rho^{\beta_2}} + \beta_4 \rho.$$

As an example, the set of parameters that closely matches GaAs FET SSPA data [129] is

$$\alpha_1 = 8.1081, \quad \alpha_2 = 1.5413, \quad \alpha_3 = 6.5202, \quad \alpha_4 = -0.0718$$
$$\alpha_1 = 4.6645, \quad \alpha_2 = 2.0965, \quad \alpha_3 = 10.8800, \quad \alpha_4 = -0.0030.$$

- *Polynomial model*
  This model uses the Taylor series with complex coefficients. The HPA output is expressed as

$$g(x) = x \sum_{j=0}^{m} b_j |x|^j,$$

for some prescribed order m.

As a result of the nonlinearity in a HPA the transmit signal alters. The main effects on the constellation diagram are rotation, attenuation, offset, warping, and cloud-like shape of constellation points. Moreover, nonlinear processing of the signal in a HPA results in out-of-band radiation.

To characterize the effect of nonlinearity in a HPA on the performance of the communication system several parameters are used. The basic ones are the error vector magnitude (EVM), adjacent channel power ratio (ACPR), and symbol error rate (SER) or bit error rate (BER).

The EVM evaluates the effects of nonlinearities on the constellation diagram. For a single-carrier signal using constellation $\mathcal{Q}$ with $E_{av} = 1$ it is defined as:

$$\text{EVM}(\mathcal{Q}) = \sqrt{\frac{1}{|\mathcal{Q}|} \sum_{a \in \mathcal{Q}} |g(a) - a|^2},$$

where $g(x)$ is the function describing the HPA. Let an MC signal be defined by the coefficient vector $\mathbf{a}$ belonging to $\mathcal{C} \subseteq \mathcal{Q}^n$, and $\hat{\mathbf{a}} \in \mathbb{C}^n$ be the vector reconstructed from the signal after the HPA. Then the EVM can be defined as:

$$\text{EVM}(\mathcal{C}) = \frac{\sqrt{\frac{1}{|\mathcal{C}|} \sum_{\mathbf{a} \in \mathcal{C}} \sum_{k=0}^{n-1} |a_k - \hat{a}_k|^2}}{\sqrt{\frac{1}{|\mathcal{C}|} \sum_{\mathbf{a} \in \mathcal{C}} \sum_{k=0}^{n-1} |a_k|^2}}.$$

The ACPR characterizes nondesirable spectral regrowth due to nonlinearities in a HPA. The ACPR is defined as the ratio between power transmitted to an adjacent channel and power transmitted to the main channel.

The SER characterizes the probability of an MC coefficient vector to be wrong due to the nonlinearity in a HPA. For an $\hat{a} \in \mathcal{C}$, let $\tilde{a}$ be the closest to its constellation point,

$$\tilde{a} = \arg \min_{a \in \mathcal{Q}} |\hat{a} - a|.$$

Let $\mathbf{a}$ be a coefficient vector picked from a code $\mathcal{C} \subseteq \mathcal{Q}^n$, and $\hat{\mathbf{a}} \in \mathbb{C}^n$ be the coefficient vector reconstructed from the signal after HPA. Define $\tilde{\mathbf{a}} \in \mathcal{Q}^n$ as the vector consisting of the constellation points closest to the components of $\hat{\mathbf{a}}$. Then

$$\text{SER} = \frac{1}{|\mathcal{C}|} \sum_{\mathbf{a} \in \mathcal{C}} \chi(\mathbf{a}, \tilde{\mathbf{a}}),$$

where $\chi(\mathbf{a}, \tilde{\mathbf{a}})$ is 1 if $\mathbf{a} \neq \tilde{\mathbf{a}}$, and 0 otherwise.

Analogously we define **BER**. Let bin(**a**) be the binary vector of length m corresponding to the coefficient vector $\mathbf{a}$. Then

$$\text{SER} = \frac{1}{m|\mathcal{C}|} \sum_{\mathbf{a} \in \mathcal{C}} d(\text{bin}(\mathbf{a}), \text{bin}(\tilde{\mathbf{a}})),$$

where $d(\cdot)$ is the number of bits in which two vectors differ.

2.5 Notes

Section 2.1 The following sources provide much more detailed treatment of multicarrier communication systems: Hanzo *et al.* [155], Bingham [26] and Ebert and Weinstein [99] on general MC communications, Bahai *et al.* [17], Prasad [334], van Nee and Prasad [286], Wang and Giannakis [428] on OFDM, Bingham [27], Starr *et al.* [391] on DSL. These are just samples of many relevant texts.

There is an extra reason for avoiding high peaks in MC signals related to the digital processing of the signals on the intermediate stages in the transmitter and receiver. For instance, the number of bits necessary for representation of a signal at the output of IDFT is a derivative of the signal's dynamic range, and thus increases if there are signals with high peaks. This in turn yields a complexity increase of the Fourier and digital-to-analog transform blocks. The same occurs at the receiver side where the signals should be processed with increased complexity due to their excessive dynamic range.

Section 2.3 To learn more about models of amplifiers one may read, e.g., Kennington [200] and Cripps [81, 82].

Section 2.4   The memoryless models exhibit good accuracy for low powers. However, when the band of the input signal increases, more memory effects are encountered. There are two main categories of HPA models including memory – physically-based transistor models (heating effects) and behavioral models describing the HPA as a nonlinear function without explicit correspondence to physical phenomena. The following are samples of models used to describe HPAs with memory:

- Saleh model with memory, where the coefficients of the Saleh memoryless model are dependent on the frequency;
- Volterra model, where the relationship between the input and output is expressed in the form of complex Volterra series, see [61];
- Wiener–Hammerstein model, based on the concatenation of linear dynamic system and static nonlinearity, see [80];
- Quadrature polynomial model [92];
- Neural network model [171];
- Augmented behavioral model [13].

3

Basic tools and algorithms

In this chapter I collect the basic mathematical tools which are used throughout the book. Most of them are given with rigorous proofs. I mainly concentrate here on results and methods that do not appear in the standard engineering textbooks and omit those that happen to be common technical knowledge. On the other hand, I have included some material that is not directly used in further arguments, but I feel that it might prove useful in further research on peak power control problems. It should be advised that the chapter is mainly for reference purposes and may be omitted in the first reading.

The chapter is organized as follows. Section 3.1 deals with harmonic analysis. In Section 3.1.1 I describe the Parseval equality and its generalizations. Section 3.1.2 introduces some useful trigonometric relations. Chebyshev polynomials and interpolation are described in Section 3.1.3. Finally, in Section 3.1.4, I prove Bernstein's inequality relating the maximum of the absolute value of a trigonometric polynomial and its derivative. In Section 3.2 I deal with some notions related to probability. I prove the Chernoff bound on the probability of deviations of values of random variables. In Section 3.3, I introduce tools from algebra. In Section 3.3.1 groups, rings, and fields are defined. Section 3.3.2 describes exponential sums in finite fields and rings. A short account of results from coding theory is presented in Section 3.4. Section 3.4.1 deals with properties of the Hamming space. In Section 3.4.2, definitions related to error-correction codes are introduced. Section 3.4.3 deals with the distance distributions of codes. In Section 3.4.4, I analyze properties of Krawtchouk polynomials playing an important role in the MacWilliams transform of the distance distributions. Section 3.4.5 lists main families of error-correcting codes. Finally, in Section 3.5, I describe a fast algorithm for estimating the maximum element of DFT with linear complexity.

3.1 Elements of harmonic analysis

We start with some very basic relations which will be used for analysis of MC signals. Throughout we use notation ι for $\sqrt{-1}$, a^* for the complex conjugates of complex a, and assume that j, k, ℓ, m, and n are integers.

Theorem 3.1   *(orthogonality)*

$$\sum_{k=0}^{n-1} e^{2\pi \iota k \frac{j}{n}} = \begin{cases} 0 & \text{if } j \equiv 0 \bmod n, \\ n & \text{otherwise.} \end{cases}$$

Proof   If $j \equiv 0 \bmod n$,

$$\sum_{k=0}^{n-1} e^{2\pi \iota k \frac{j}{n}} = \sum_{k=0}^{n-1} 1 = n.$$

Assume $j \not\equiv 0 \bmod n$. Using the sum of geometric progression we obtain

$$\sum_{k=0}^{n-1} e^{2\pi \iota k \frac{j}{n}} = \frac{e^{2\pi \iota j} - 1}{e^{2\pi \iota \frac{j}{n}} - 1} = 0.$$

$\square$

Theorem 3.2   *(discrete Fourier transform (DFT)) Let*

$$F\left(\frac{j}{n}\right) = \sum_{k=0}^{n-1} a_k e^{2\pi \iota k \frac{j}{n}}. \tag{3.1}$$

Then

$$a_k = \frac{1}{n} \sum_{\ell=0}^{n-1} F\left(\frac{\ell}{n}\right) e^{-2\pi \iota \ell \frac{k}{n}}. \tag{3.2}$$

Proof   Indeed,

$$\frac{1}{n} \sum_{\ell=0}^{n-1} F\left(\frac{\ell}{n}\right) e^{-2\pi \iota \ell \frac{k}{n}} = \frac{1}{n} \sum_{\ell=0}^{n-1} \sum_{j=0}^{n-1} a_j e^{2\pi \iota j \frac{\ell}{n}} e^{-2\pi \iota \ell \frac{k}{n}}$$

$$= \frac{1}{n} \sum_{j=0}^{n-1} a_j \sum_{\ell=0}^{n-1} e^{2\pi \iota \ell \frac{j-k}{n}} = a_k.$$

In the last equation we used Theorem 3.2.

$\square$

The vector $(\sqrt{n}a_0, \sqrt{n}a_1, \ldots, \sqrt{n}a_{n-1})$ in what follows will be addressed as the result of DFT of $\left(F\left(\frac{0}{n}\right), F\left(\frac{1}{n}\right), \ldots, F\left(\frac{n-1}{n}\right)\right)$. Analogously, the vector $\left(\frac{1}{\sqrt{n}}F\left(\frac{0}{n}\right), \frac{1}{\sqrt{n}}F\left(\frac{1}{n}\right), \ldots, \frac{1}{\sqrt{n}}F\left(\frac{n-1}{n}\right)\right)$ is the result of inverse DFT (IDFT) of $(a_0, a_1, \ldots, a_{n-1})$.

3.1.1 Parseval identity and its generalizations

Theorem 3.3   *(Parseval identity) Let*

$$F\left(\frac{j}{n}\right) = \sum_{k=0}^{n-1} a_k e^{2\pi \iota k \frac{j}{n}}.$$

Then

$$\sum_{j=0}^{n-1} \left| F\left(\frac{j}{n}\right) \right|^2 = n \sum_{k=0}^{n-1} |a_k|^2.$$

Proof   Indeed,

$$\sum_{j=0}^{n-1} \left| F\left(\frac{j}{n}\right) \right|^2 = \sum_{j=0}^{n-1} F\left(\frac{j}{n}\right) F^*\left(\frac{j}{n}\right)$$

$$= \sum_{j=0}^{n-1} \left(\sum_{k=0}^{n-1} a_k e^{2\pi \iota k \frac{j}{n}}\right) \left(\sum_{\ell=0}^{n-1} a_\ell^* e^{-2\pi \iota \ell \frac{j}{n}}\right)$$

$$= \sum_{k=0}^{n-1} \sum_{\ell=0}^{n-1} a_k a_\ell^* \sum_{j=0}^{n-1} e^{2\pi \iota j \frac{k-\ell}{n}}$$

$$= n \sum_{k=0}^{n-1} a_k a_k^* = n \sum_{k=0}^{n-1} |a_k|^2.$$

$\square$

For two complex-valued vectors $\mathbf{a} = (a_0, a_1, \ldots, a_{n-1})$ and $\mathbf{b} = (b_0, b_1, \ldots, b_{n-1})$ we denote by $(\mathbf{a}, \mathbf{b})$ their dot product $\sum_{j=0}^{n-1} a_j b_j^*$, and use $\|\mathbf{a}\|$ for the norm of $\mathbf{a}$,

$$\|\mathbf{a}\| = \sqrt{(\mathbf{a}, \mathbf{a})} = \sqrt{\sum_{j=0}^{n-1} a_j a_j^*} = \sqrt{\sum_{j=0}^{n-1} |a_j|^2}.$$

Then denoting

$$\mathbf{e}(j) = \left(1, e^{-2\pi \iota \frac{j}{n}}, e^{-2\pi \iota \frac{2j}{n}}, \ldots, e^{-2\pi \iota \frac{(n-1)j}{n}}\right),$$

we derive an equivalent form of the Parseval identity:

$$\sum_{j=0}^{n-1} |(\mathbf{a}, \mathbf{e}(j))|^2 = n \|\mathbf{a}\|^2. \tag{3.3}$$

There exist many generalizations of the Parseval identity. Let me present several useful relations when instead of mutually orthogonal vectors $\mathbf{e}(j)$ one uses an arbitrary set of M not necessarily orthogonal n-dimensional complex-valued vectors. We start with proving a basic Cauchy–Schwartz inequality.

Theorem 3.4 *(Cauchy–Schwartz inequality) Let* **a** *and* **b** *be arbitrary n-dimensional complex-valued vectors. Then*

$$|(\mathbf{a}, \mathbf{b})| \leq \|\mathbf{a}\| \cdot \|\mathbf{b}\|. \tag{3.4}$$

Proof If $(\mathbf{a}, \mathbf{b}) = 0$, the inequality trivially holds since the right-hand side of (3.4) is nonnegative. Assume $(\mathbf{a}, \mathbf{b}) \neq 0$. Then define an auxiliary nonnegative function of a real λ,

$$\varphi(\lambda) = \|\mathbf{a} + \lambda(\mathbf{a}, \mathbf{b})\mathbf{b}\|^2 = (\mathbf{a} + \lambda(\mathbf{a}, \mathbf{b})\mathbf{b}, \mathbf{a} + \lambda(\mathbf{a}, \mathbf{b})\mathbf{b}).$$

Furthermore,

$$\varphi(\lambda) = (\mathbf{a}, \mathbf{a}) + \lambda(\mathbf{a}, \mathbf{b})^*(\mathbf{a}, \mathbf{b}) + \lambda(\mathbf{a}, \mathbf{b})(\mathbf{b}, \mathbf{a}) + \lambda^2|(\mathbf{a}, \mathbf{b})|^2(\mathbf{b}, \mathbf{b})$$
$$= \lambda^2|(\mathbf{a}, \mathbf{b})|^2(\mathbf{b}, \mathbf{b}) + 2\lambda|(\mathbf{a}, \mathbf{b})|^2 + (\mathbf{a}, \mathbf{a}) \geq 0.$$

For the last inequality to be true for all real λ it is necessary that the discriminant of the quadratic equation be nonpositive. This is equivalent to

$$|(\mathbf{a}, \mathbf{b})|^4 - |(\mathbf{a}, \mathbf{b})|^2(\mathbf{a}, \mathbf{a})(\mathbf{b}, \mathbf{b}),$$

which reduces to

$$|(\mathbf{a}, \mathbf{b})|^2 \leq (\mathbf{a}, \mathbf{a})(\mathbf{b}, \mathbf{b}),$$

and we are done. $\square$

The following result is from Bombieri [34].

Theorem 3.5 *Let* **a**, **c** *and* $\mathbf{f}(j)$, $j = 0, 1, \ldots, M - 1$, *be arbitrary n-dimensional complex-valued vectors. Then*

$$\left|\sum_{j=0}^{M-1} c_j(\mathbf{a}, \mathbf{f}(j))\right| \leq \|\mathbf{c}\| \cdot \|\mathbf{a}\| \cdot \left(\max_{0 \leq j \leq M-1} \sum_{\ell=0}^{M-1} |(\mathbf{f}(j), \mathbf{f}(\ell))|\right)^{\frac{1}{2}}. \tag{3.5}$$

Proof We have

$$\sum_{j=0}^{M-1} c_j(\mathbf{a}, \mathbf{f}(j)) = \left(\mathbf{a}, \sum_{j=0}^{R-1} c_j^* \mathbf{f}(j)\right).$$

By the Cauchy–Schwartz inequality, it implies

$$\left|\sum_{j=0}^{M-1} c_j(\mathbf{a}, \mathbf{f}(j))\right| \leq \|\mathbf{a}\| \cdot \left\|\sum_{j=0}^{M-1} c_j^* \mathbf{f}(j)\right\|. \tag{3.6}$$

Using the inequality

$$|c_k^* c_\ell| \leq \frac{1}{2}\left(|c_k|^2 + |c_\ell^2|^2\right)$$

we proceed with the second product term in the right-hand side of (3.6) as follows:

$$\left\|\sum_{j=0}^{M-1} c_j^* \mathbf{f}(j)\right\|^2 = \sum_{k=0}^{M-1}\sum_{\ell=0}^{M-1} c_k^* c_\ell (\mathbf{f}(k), \mathbf{f}(\ell))$$

$$\leq \sum_{k=0}^{M-1} |c_k|^2 \sum_{\ell=0}^{M-1} |(\mathbf{f}(k), \mathbf{f}(\ell))|$$

$$\leq \left(\sum_{k=0}^{M-1} |c_k|^2\right) \cdot \max_{0 \leq j \leq M-1} \sum_{\ell=0}^{M-1} |(\mathbf{f}(k), \mathbf{f}(\ell))|. \qquad (3.7)$$

Substituting this result into (3.6) we obtain the claim.   □

Corollary 3.6   *Let* $\mathbf{a}$ *and* $\mathbf{f}(j)$, $j = 0, 1, \ldots, M-1$, *be arbitrary n-dimensional complex-valued vectors. Then*

$$\sum_{j=0}^{M-1} |(\mathbf{a}, \mathbf{f}(j))|^2 \leq \|\mathbf{a}\|^2 \max_{0 \leq j \leq M-1} \sum_{\ell=0}^{M-1} |(\mathbf{f}(j), \mathbf{f}(\ell))|. \qquad (3.8)$$

Proof   Set $c_j = (\mathbf{a}, \mathbf{f}(j))^*$ in the previous theorem, and the claimed result follows after division by $\left(\sum_{j=0}^{M-1} |(\mathbf{a}, \mathbf{f}(j))|^2\right)^{\frac{1}{2}}$ and squaring both sides of the inequality.   □

A similar inequality was obtained independently by Boas [30] and Bellman [24].

Corollary 3.7   *Let* $\mathbf{a}$ *and* $\mathbf{f}(j)$, $j = 0, 1, \ldots, M-1$, *be arbitrary n-dimensional complex-valued vectors. Then*

$$\sum_{j=0}^{M-1} |(\mathbf{a}, \mathbf{f}(j))|^2 \leq \|\mathbf{a}\|^2 \left(\max_{0 \leq j \leq M-1} \|\mathbf{f}(j)\| + \left(\sum_{j,\ell=0;\ j \neq \ell}^{M-1} |(\mathbf{f}(j), \mathbf{f}(\ell))|^2\right)^{\frac{1}{2}}\right). \qquad (3.9)$$

□

Another estimate is given by Halász [147].

Corollary 3.8   *Let* $\mathbf{a}$ *and* $\mathbf{f}(j)$, $j = 0, 1, \ldots, M-1$, *be arbitrary n-dimensional complex-valued vectors. Then*

$$\sum_{j=0}^{M-1} |(\mathbf{a}, \mathbf{f}(j))| \leq \|\mathbf{a}\| \left(\sum_{j=0}^{M-1}\sum_{\ell=1}^{M-1} |(\mathbf{f}(j), \mathbf{f}(\ell))|\right)^{\frac{1}{2}}. \qquad (3.10)$$

Proof   Choose in Theorem 3.5

$$c_j = e^{-\iota \arg(\mathbf{a}, \mathbf{f}(j))}, \quad j = 0, 1, \ldots, M-1.$$

In this case for the right-hand side of (3.6) we derive a bound

$$\sum_{j=0}^{M-1}\sum_{\ell=0}^{M-1} c_j^* c_\ell (\mathbf{f}(j), \mathbf{f}(\ell)) \le \sum_{j=0}^{M-1}\sum_{\ell=1}^{M-1} |(\mathbf{f}(j), \mathbf{f}(\ell))|,$$

and we obtain the result by substituting this into (3.6).   □

Yet another estimate is attributed to Selberg, see [275].

Theorem 3.9   *Let* $\mathbf{a}$ *and* $\mathbf{f}(j)$, $j = 0, 1, \ldots, M-1$, *be arbitrary n-dimensional complex-valued vectors. Then*

$$\sum_{j=0}^{M-1} |(\mathbf{a}, \mathbf{f}(j))|^2 \frac{1}{\sum_{\ell=0}^{M-1} |(\mathbf{f}(j), \mathbf{f}(\ell))|} \le \|\mathbf{a}\|^2. \tag{3.11}$$

Proof   For any complex numbers c_j we have

$$\left\| \mathbf{a} - \sum_{j=0}^{M-1} c_j \mathbf{f}(j) \right\|^2 \ge 0.$$

Therefore,

$$\|\mathbf{a}\|^2 - 2\Re\left(\sum_{j=0}^{M-1} c_j^* (\mathbf{a}, \mathbf{f}(j))\right) + \sum_{j=0}^{M-1}\sum_{\ell=0}^{M-1} c_j^* c_\ell (\mathbf{f}(j), \mathbf{f}(\ell)) \ge 0.$$

From (3.7) we obtain

$$2\Re\left(\sum_{j=0}^{M-1} c_j^* (\mathbf{a}, \mathbf{f}(j))\right) \le \|\mathbf{a}\|^2 + \sum_{j=0}^{M-1} |c_j|^2 \sum_{\ell=0}^{M-1} |(\mathbf{f}(j), \mathbf{f}(\ell))|,$$

and choosing

$$c_j = (\mathbf{a}, \mathbf{f}(j)) \frac{1}{\sum_{\ell=0}^{M-1} |(\mathbf{f}(j), \mathbf{f}(\ell))|},$$

we obtain the sought result.   □

Properties of DFT dealing with changing the order of summation will be useful in some further considerations.

Theorem 3.10   *If* g.c.d.$(\ell, n) = 1$ *and* f *is an arbitrary real-valued function, then*

$$\sum_{k=0}^{n-1} e^{2\pi \imath f(k)} = \sum_{k=0}^{n-1} e^{2\pi \imath f(k\ell+m)}.$$

Proof   Under the conditions of the theorem when k goes from 0 to $n-1$, $(k\ell + m) \bmod n$ takes on each value from the residue system modulo n exactly once.   □

Another useful property is related to cyclic rotation of the transformed vector.

Theorem 3.11   *For an* $\mathbf{a} = (a_0, a_1, \ldots, a_{n-1})$, *and any* $j = 0, 1, \ldots, n - 1$, $s = 0, 1, \ldots, n - 1$,

$$\left| \sum_{k=0}^{n-1} a_k e^{2\pi \iota k \frac{j}{n}} \right| = \left| \sum_{k=0}^{n-1} a_{(k+s) \bmod n} \, e^{2\pi \iota k \frac{j}{n}} \right|. \tag{3.12}$$

Proof   Indeed,

$$\sum_{k=0}^{n-1} a_{(k+s) \bmod n} \, e^{2\pi \iota k \frac{j}{n}} = e^{-2\pi \iota j \frac{s}{n}} \cdot \sum_{k=0}^{n-1} a_k e^{2\pi \iota k \frac{j}{n}}.$$

$\square$

3.1.2 Useful relations

In this section I collect several relations used throughout.

Let us start with inequalities related to the absolute value. We will use, on many occasions, the following straightforward inequality valid for any complex a, b,

$$|a + b| \leq |a| + |b|. \tag{3.13}$$

For estimation of the maximum of a sum the following result is useful.

Theorem 3.12   *Let* $(a_0, a_1, \ldots, a_{n-1})$ *and* $(b_0, b_1, \ldots, b_{n-1})$ *be two complex-valued vectors. Then*

$$\left| \sum_{k=0}^{n-1} a_k b_k \right| \leq \left(\max_{k=0,1,\ldots,n-1} |b_k| \right) \cdot \sum_{k=0}^{n-1} |a_k|.$$

Proof   Indeed, by (3.13),

$$\left| \sum_{k=0}^{n-1} a_k b_k \right| \leq \sum_{k=0}^{n-1} |a_k b_k| = \sum_{k=0}^{n-1} |a_k||b_k|.$$

$\square$

Theorem 3.13   *Let* $\mathbf{a} = (a_0, a_1, \ldots, a_{n-1})$ *and* $\mathbf{b} = (b_0, b_1, \ldots, b_{n-1})$ *be two real-valued vectors with nonnegative components. Moreover, let the components of the vectors be sorted in the nonincreasing order, i.e.,*

$$a_0 \geq a_1 \geq \ldots \geq a_{n-1}, \quad b_0 \geq b_1 \geq \ldots \geq b_{n-1}.$$

Then

$$\max_{\pi} \sum_{k=0}^{n-1} a_k b_{\pi(k)} = \sum_{k=0}^{n-1} a_k b_k,$$

where the maximum is taken over all possible permutations π of the components of $\mathbf{b}$.

Proof Indirectly assume that in a permutation π delivering a maximum to $S = \sum_{k=0}^{n-1} a_k b_{\pi(k)}$ for some index j, $0 \le j \le n - 2$, we have $b_{\pi(j)} < b_{\pi(j+1)}$. Then by exchanging the places of $b_{\pi(j+1)}$ and $b_{\pi(j)}$ we will not decrease S. Indeed,

$$0 \le (a_j - a_{j+1})(b_{\pi(j+1)} - b_{\pi(j)})$$
$$= (a_j b_{\pi(j+1)} + a_{j+1} b_{\pi(j)}) - (a_j b_{\pi(j)} + a_{j+1} b_{\pi(j+1)}).$$

$\square$

Now I list several useful trigonometric relations.

Theorem 3.14 *For any $a, b \in \mathbb{R}$,*

$$\sum_{k=0}^{n-1} \sin(ka + b) = \frac{\sin \frac{(n-1)a+2b}{2} \cdot \sin \frac{na}{2}}{\sin \frac{a}{2}}, \tag{3.14}$$

$$\sum_{k=0}^{n-1} \cos(ka + b) = \frac{\cos \frac{(n-1)a+2b}{2} \cdot \sin \frac{na}{2}}{\sin \frac{a}{2}}. \tag{3.15}$$

Proof We will make use of the following product-to-sum and sum-to-product trigonometric identities:

$$\sin x \cdot \sin y = \frac{1}{2}(\cos(x - y) - \cos(x + y)), \tag{3.16}$$

$$\cos x - \cos y = -2 \sin \frac{1}{2}(x + y) \cdot \sin \frac{1}{2}(x - y). \tag{3.17}$$

Let $S = \sin b + \sin(a + b) + \sin(2a + b) + \ldots + \sin((n - 1)a + b)$. Then by (3.16) and (3.17),

$$S \cdot \sin \frac{a}{2} = \frac{1}{2}\left(\cos \frac{-a + 2b}{2} - \cos \frac{a + 2b}{2} + \cos \frac{a + 2b}{2} - \cos \frac{3a + 2b}{2} \right.$$
$$+ \cos \frac{3a + 2b}{2} - \cos \frac{5a + 2b}{2} + \ldots$$
$$\left. + \cos \frac{(2n - 3)a + 2b}{2} - \cos \frac{(2n - 1)a + 2b}{2} \right)$$
$$= \frac{1}{2}\left(\cos \frac{-a + 2b}{2} - \cos \frac{(2n - 1)a + 2b}{2} \right)$$
$$= \sin \frac{(n - 1)a + 2b}{2} \cdot \sin \frac{na}{2}.$$

For cosines we act analogously but using

$$\sin x \cdot \cos y = \frac{1}{2}(\sin(x + y) + \sin(x - y)),$$

$$\sin x - \sin y = 2\cos\frac{1}{2}(x + y) \cdot \sin\frac{1}{2}(x - y).$$

$\square$

Theorem 3.15 *For any $a, b \in \mathbb{R}$,*

$$\sum_{k=0}^{n-1} \sin^2(ka + b) = \frac{n}{2} - \frac{\cos((n-1)a + 2b) \cdot \sin(na)}{2\sin a}, \qquad (3.18)$$

$$\sum_{k=0}^{n-1} \cos^2(ka + b) = \frac{n}{2} + \frac{\cos((n-1)a + 2b) \cdot \sin(na)}{2\sin a}. \qquad (3.19)$$

Proof We will prove the second inequality. Indeed,

$$\sum_{k=0}^{n-1} \cos^2(ka + b) = \frac{1}{2}\sum_{k=0}^{n-1}(1 + \cos 2(ka + b))$$

$$= \frac{n}{2} + \frac{1}{2}\sum_{k=0}^{n-1} \cos(k \cdot 2a + 2b).$$

Now the result follows from (3.15). The second inequality follows from $\sin^2(ka + b) = 1 - \cos^2(ka + b)$.

$\square$

An analogous integral form is presented in the next theorem.

Theorem 3.16

$$\int \sin^2 x \, dx = \frac{x}{2} - \frac{1}{4}\sin 2x, \qquad (3.20)$$

$$\int \cos^2 x \, dx = \frac{x}{2} + \frac{1}{4}\sin 2x. \qquad (3.21)$$

Proof Use $\sin^2 x = \frac{1}{2} - \frac{1}{2}\cos 2x$, $\cos^2 x = \frac{1}{2} + \frac{1}{2}\cos 2x$.

$\square$

Theorem 3.17 *For $|t| \le \frac{1}{2}$,*

$$|\sin \pi t - \pi t| \le \frac{\pi^3 t^3}{6} \qquad (3.22)$$

and

$$|\sin \pi t| \ge 2|t|. \qquad (3.23)$$

Proof The first inequality follows from the sign-alternating Taylor series for sine. The second one is due to the observation that $\sin \pi t$ is convex in the interval $\left[0, \frac{1}{2}\right]$, and thus the line connecting its extreme points lies beyond its graph. □

Theorem 3.18 *For any t and positive integer n,*

$$\frac{\sin \pi n t}{\sin \pi t} \leq n.$$

Proof We will use induction. Clearly, the inequality is correct for $n = 1$. Assume it holds for $(n - 1)$. Then,

$$
\begin{aligned}
\sin \pi n t &= \sin \pi t \cdot \cos \pi (n - 1)t + \cos \pi t \cdot \sin \pi (n - 1)t \\
&\leq |\sin \pi t| + |\sin \pi (n - 1)t| \\
&\leq n |\sin \pi t|.
\end{aligned}
$$

□

We will make extensive use of Euler's formula, which states that, for any real t,

$$e^{\iota t} = \cos t + \iota \sin t. \tag{3.24}$$

It follows that

$$\cos t = \frac{1}{2}(e^{\iota t} + e^{-\iota t}), \tag{3.25}$$

and

$$\iota \sin t = \frac{1}{2}(e^{\iota t} - e^{-\iota t}). \tag{3.26}$$

Theorem 3.19 *For any t,*

$$|e^{\iota t} - 1| \leq t, \tag{3.27}$$

$$|e^{\iota t} - 1 - \iota t| \leq \frac{t^2}{2}. \tag{3.28}$$

Proof Since

$$\iota \int_0^t e^{\iota x} dx = e^{\iota x} \Big|_0^t = e^{\iota t} - 1,$$

and $|\iota e^{\iota x}| = 1$, we conclude that

$$|e^{\iota t} - 1| \leq \left| \int_0^t 1 dx \right| = t.$$

Furthermore,

$$\iota \int_0^t (e^{\iota x} - 1) = e^{\iota t} - 1 - \iota t,$$

and

$$\left| \int_0^t (e^{ix} - 1) \right| \leq \left| \int_0^t x \, dx \right| = \frac{t^2}{2},$$

and we are done.   □

3.1.3 Chebyshev polynomials and interpolation

Let θ be defined by $x = \cos\theta$, then the Chebyshev polynomials of the first and second kind are defined for positive integer n and $x \in [-1, 1]$ as

$$T_n(x) = \cos n\theta, \quad U_n(x) = \frac{1}{n+1} T'_{n+1}(x) = \frac{\sin(n+1)\theta}{\sin\theta}.$$

It is straightforward to show that $T_n(x)$ and $U_n(x)$ are indeed polynomials in x of degree n.

Lemma 3.20 *For $-1 \leq x \leq 1$, and any positive integer n,*

$$|T_n(x)| \leq 1, \quad |U_n(x)| \leq n+1. \tag{3.29}$$

Proof The first inequality trivially follows from $\cos n\theta \leq 1$. The second one follows from

$$\frac{\sin(n+1)\theta}{\sin\theta} = \cos n\theta + \cos\theta \frac{\sin n\theta}{\sin\theta}$$

$$= \cos n\theta + \cos\theta \cos(n-1)\theta + \cos^2\theta \frac{\sin(n-1)\theta}{\sin\theta}$$

$$= \sum_{j=0}^{n} \cos^i \theta \cos(n-i)\theta.$$

The last sum contains $n+1$ summands, each of them not exceeding 1.   □

By direct substitution we check that $x_j = \cos\frac{(2j-1)\pi}{2n}$, $j = 1, 2, \ldots, n$, are zeros of $T_n(x)$, and these are all zeros since $T_n(x)$ has degree n. Let us find the values of the derivative of $T_n(x)$, $T'_n(x)$, in the zeros of $T_n(x)$. Indeed, differentiating $T_n(x) = \cos n \arccos x$, we derive

$$T'_n(x_j) = (-1)^{j-1} \frac{n+1}{\sqrt{1 - x_j^2}}, \quad j = 1, 2, \ldots, n. \tag{3.30}$$

We aim at interpolating functions from the values in zeros of Chebyshev polynomials. Let a polynomial of degree at most $n-1$ be defined by its values $P(x_1), P(x_2), \ldots, P(x_n)$. Then it can be presented in the form

$$P(x) = P(x_1) f_1(x) + P(x_2) f_2(x) + \ldots + P(x_n) f_n(x), \tag{3.31}$$

where $f_j(x)$ are the fundamental interpolation polynomials,

$$f_j(x) = \frac{1}{T_n'(x_j)} \frac{T_n(x)}{x - x_j}, \quad j = 1, 2, \ldots, n. \tag{3.32}$$

Formula (3.31) follows from the easily checked fact that

$$f_j(x_k) = \begin{cases} 0 & \text{if } j \neq k \\ 1 & \text{otherwise.} \end{cases}$$

Theorem 3.21 *(Lagrange interpolation) Every polynomial $P(x)$ of degree $(n-1)$ can be presented as*

$$P(x) = \frac{1}{n} \sum_{j=1}^{n} (-1)^{j-1} \sqrt{1 - x_j^2}\, P(x_j) \frac{T_n(x)}{x - x_j}. \tag{3.33}$$

where $x_j = \cos(2j - 1)\frac{\pi}{2n}, \ j = 1, 2, \ldots, n$.

Proof   This follows from (3.30), (3.32), and (3.31).   □

The following lemma provides a useful restriction on the maximum of polynomials.

Lemma 3.22   *Let $P(x)$ be a polynomial of degree $n - 1$, satisfying for $x \in [-1, 1]$,*

$$\sqrt{1 - x^2}\, |P(x)| \leq 1.$$

Then for any $x \in [-1, 1]$,

$$|P(x)| \leq n.$$

Proof   For $x \in [x_1, 1]$, using Theorem 3.21, (3.29), and the nonnegativity of $T_n(x)$ in the considered interval, we deduce that

$$|P(x)| \leq \frac{1}{n} \sum_{j=1}^{n} \frac{T_n(x)}{x - x_j} = \frac{T_n'(x)}{n} = U_{n-1}(x) \leq n.$$

Analogous   arguments   are   valid   for $x \in [-1, x_n]$.   In   the   case $x \in \left[x_n = -\cos\frac{\pi}{2n}, x_1 = \cos\frac{\pi}{2n} \right]$ we have

$$\sqrt{1 - x^2} \geq \sin\frac{\pi}{2n} > \frac{2}{\pi} \cdot \frac{\pi}{2n} = \frac{1}{n}.$$

In the last inequality we used (3.23).

3.1.4 Trigonometric polynomials and Bernstein's inequality

A trigonometric polynomial of order m is

$$g(\theta) = \lambda_0 + \lambda_1 \cos\theta + \mu_1 \sin\theta + \lambda_2 \cos 2\theta$$
$$+ \mu_2 \sin 2\theta + \ldots + \lambda_m \cos m\theta + \mu_m \sin m\theta,$$

with the complex coefficients λ_0 and $\lambda_j, \mu_j,\ j = 1, 2, \ldots, m$. The polynomials with all real coefficients are called real trigonometric polynomials.

Lemma 3.23 *For $a_k \in \mathbb{C},\ k = 0, 1, \ldots, n-1,\ a_{n-1} \neq 0$,*

$$F(\theta) = \sum_{k=0}^{n-1} a_k e^{\iota k \theta}$$

is a trigonometric polynomial of order $(n-1)$, and

$$F_R(\theta) = \Re \left(\sum_{k=0}^{n-1} a_k e^{\iota k \theta} \right)$$

is a real trigonometric polynomial of order $(n-1)$.

Proof Using

$$e^{\iota k \theta} = \cos k\theta + \iota \sin k\theta,$$

we have

$$F(\theta) = a_0 + \sum_{k=1}^{n-1} (a_k \cos k\theta + \iota a_k \sin k\theta),$$

$$F_R(\theta) = \Re(a_0) + \sum_{k=1}^{n-1} (\Re(a_k) \cos k\theta + \Re(\iota a_k) \sin k\theta).$$

$\square$

Lemma 3.24 *Let $a_k \in \mathbb{C},\ k = 0, 1, \ldots, n-1,\ a_{n-1} \neq 0$, and*

$$F(\theta) = \sum_{k=0}^{n-1} a_k e^{\iota k \theta}.$$

Then $|F(\theta)|^2$ is a real trigonometric polynomial of order $(n-1)$.

Proof Indeed,

$$|F(\theta)|^2 = F(\theta) F^*(\theta) = \sum_{k=0}^{n-1} \sum_{j=0}^{n-1} a_j a_k^* e^{\iota(j-k)\theta}.$$

Singling out the terms with $j = k$ and taking into account that for every complex a and b,

$$ab^* + a^*b = 2\Re\{ab^*\},$$

we have

$$|F(\theta)|^2 = \sum_{k=0}^{n-1} |a_k|^2 + 2\Re \left\{ \sum_{k=0}^{n-1} \sum_{j=k+1}^{n-1} a_j a_k^* e^{\iota(j-k)\theta} \right\}$$

$$= \Re \left\{ \left(\sum_{k=0}^{n-1} |a_k|^2 \right) + \sum_{j=1}^{n-1} \left(2 \sum_{k=0}^{n-j-1} a_{k+j} a_k^* \right) e^{\iota j \theta} \right\},$$

and the result follows from the second claim of the previous lemma.   □

The following theorem from Bernstein [25] relates values of a real trigonometric polynomial and its derivative. The proof is attributed to Fejér (see [331]).

Theorem 3.25   *(Bernstein) Let $g(\theta)$ be a real trigonometric polynomial of order $(n-1)$. Then for every θ,*

$$|g'(\theta)| \le (n-1) \max_{\theta \in [0,2\pi)} |g(\theta)|.$$

Proof   Let $g(\theta) = \lambda_0 + \sum_{k=0}^{n-1}(\lambda_k \cos k\theta + \mu_k \sin k\theta)$, $\lambda_0, \lambda_k, \mu_k \in \mathbb{R}$ for $k = 1, 2, \ldots, n-1$. Assume that for every θ, $|g(\theta)| \le 1$. Consider the function

$$s(\Delta) = \frac{g(\theta + \Delta) - g(\theta - \Delta)}{2}. \tag{3.34}$$

Using the trigonometric identities

$$\cos(\alpha + \beta) - \cos(\alpha - \beta) = -2\sin\alpha\sin\beta,$$
$$\sin(\alpha + \beta) - \sin(\alpha - \beta) = 2\cos\alpha\sin\beta,$$

we have

$$s(\Delta) = \sum_{k=1}^{n-1} \xi_k \sin k\Delta, \tag{3.35}$$

where

$$\xi_k = -\lambda_k \sin k\theta + \mu_k \cos k\theta.$$

We will now prove by induction that for $k \ge 1$,

$$\sin k\Delta = \sin\Delta \cdot P_{k-1}(\cos\Delta), \quad \deg P_{k-1} = k-1, \tag{3.36}$$
$$\cos k\Delta = G_k(\cos\Delta), \quad \deg G_k = k, \tag{3.37}$$

where P and G are polynomials, $P_0 = 1$, $P_1 = 2\cos\Delta$, $G_1 = \cos\Delta$. It is clearly correct for $k = 1, 2$. For $k > 2$,

$$\cos k\Delta = \cos\Delta \cdot \cos(k-1)\Delta - \sin\Delta \cdot \sin(k-1)\Delta$$
$$= \cos\Delta \cdot G_{k-1}(\cos\Delta) - \sin^2\Delta \cdot P_{k-2}(\cos\Delta)$$
$$= G_k(\cos\Delta).$$

Furthermore,

$$\begin{aligned}
\sin k\Delta &= \sin \Delta \cdot \cos(k-1)\Delta + \cos \Delta \cdot \sin(k-1)\Delta \\
&= \sin \Delta \cdot G_{k-1}(\cos \Delta) + \cos \Delta \cdot \sin \Delta \cdot P_{k-2}(\cos \Delta) \\
&= \sin \Delta \cdot P_{k-1}(\cos \Delta),
\end{aligned}$$

and (3.36) and (3.37) are proven.

Returning to (3.35) we conclude that

$$s(\Delta) = \sin \Delta \cdot \sum_{k=1}^{n-1} \xi_k P_{k-1}(\cos \Delta) = \sin \Delta \cdot P(\cos \Delta), \qquad (3.38)$$

where $\deg P = n - 2$. Note that, by (3.34),

$$g'(\theta) = \lim_{\Delta \to 0} \frac{s(\Delta)}{\Delta} = \lim_{\Delta \to 0} \frac{s(\Delta)}{\sin \Delta} = \lim_{\Delta \to 0} P(\cos \Delta). \qquad (3.39)$$

Substituting $x = \cos \Delta$, $x \in [-1, 1]$, into (3.38) we obtain

$$s(\Delta) = \sqrt{1 - x^2} \, P(x).$$

On the other hand, by (3.34), $|s(\Delta)| \le 1$, and application of Lemma 3.22 accomplishes the proof. □

Now we are ready to state the main result of the section.

Theorem 3.26 *(Bernstein's inequality for complex trigonometric polynomials)*
Let $a_k \in \mathbb{C}$, $k = 0, 1, \ldots, n-1$, $a_{n-1} \ne 0$, and

$$F(t) = \sum_{k=0}^{n-1} a_k e^{2\pi \imath k t}.$$

Then

$$\max_{t \in [0,1)} |F'(t)| \le 2\pi (n-1) \max_{t \in [0,1)} |F(t)|. \qquad (3.40)$$

Proof Let $\theta = 2\pi t$, then $F(\theta) = F\left(\frac{t}{2\pi}\right)$. Notice that $F(\theta)$ is periodic with period 2π. Let $|F'(\theta)|$ attain its maximum at $\theta = \tau$. Pick real α such that $e^{\imath\alpha} F'(\tau)$ is real. Set $F_R(\theta) = \Re(e^{\imath\alpha} F(\theta))$. By Lemma 3.23, $F_R(\theta)$ is a real trigonometric polynomial. Then

$$\begin{aligned}
\frac{1}{2\pi} \max_t |F'(t)| &= \max_\theta |F'(\theta)| = e^{\imath\alpha} \cdot F'(\tau) = F_R'(\tau) \\
&\le (n-1) \max_t |F_R(t)| \le (n-1) \max_\theta |F(\theta)| \\
&= (n-1) \max_t |F(t)|.
\end{aligned}$$

The inequality before the last is given by Theorem 3.25. □

3.2 Elements of probability

We start with estimates on deviations of random variables. Let X be a random variable with expectation $E(X)$ and second central moment $\text{var}(X) = E((X - E(X))^2)$. Let X take only nonnegative values.

Theorem 3.27 *(Markov's inequality) For any nonnegative random variable, X,*

$$\Pr(X \geq \lambda) \leq \frac{E(X)}{\lambda}.$$

Proof This is a proof for discrete random variables; generalization to the continuous case is straightforward. Since X takes only nonnegative values,

$$E(X) \geq \sum_{x \geq \lambda} x \cdot \Pr(X = x) \geq \sum_{x \geq \lambda} \lambda \cdot \Pr(X = x)$$
$$= \lambda \sum_{x \geq \lambda} \Pr(X = x) = \lambda \, \Pr(X \geq \lambda).$$

$\square$

Note that we can substitute any positive function f for X,

$$\Pr(f(X) \geq f(\lambda)) \leq \frac{E(f(X))}{f(\lambda)}.$$

If, moreover, f is a nondecreasing function, we get

$$\Pr(X \geq \lambda) = \Pr(f(X) \geq f(\lambda)) \leq \frac{E(f(X))}{f(\lambda)}.$$

Theorem 3.28 *(Chebyshev's inequality) For any random variable X,*

$$\Pr(|X - E(X)| \geq \lambda) \leq \frac{\text{var}(X)}{\lambda^2}.$$

Proof We choose $f(X) = X^2$, and have

$$\Pr(|X - E(X)| \geq \lambda) \leq \Pr((X - E(X))^2 \geq \lambda^2)$$
$$\leq \frac{E((X - E(X))^2)}{\lambda^2} = \frac{\text{var}(X)}{\lambda^2}.$$

$\square$

Let X now be a sum of n independent random variables $X_j, j = 0, 1, \ldots, n - 1$. We denote by μ the expectation of X, and have

$$\mu = E(X) = E\left(\sum_{j=0}^{n-1} E(X_j)\right).$$

Theorem 3.29 *(Chernoff bound)*

$$\Pr(X \geq (1 + \delta)\mu) \leq \min_{\varepsilon > 0} e^{-\varepsilon(1+\delta)\mu} \cdot \prod_{j=0}^{n-1} E(e^{\varepsilon X_j}).$$

Proof We pick $f(X) = e^{\varepsilon X}$, and compute

$$\Pr(X \geq (1 + \delta)\mu) = \Pr\left(e^{\varepsilon X} \geq e^{\varepsilon(1+\delta)\mu}\right) \leq \frac{E(e^{\varepsilon X})}{e^{\varepsilon(1+\delta)\mu}}.$$

Now

$$E(e^{\varepsilon X}) = E\left(e^{\varepsilon \sum_{j=0}^{n-1} X_j}\right) = E\left(\prod_{j=0}^{n-1} e^{\varepsilon X_j}\right) = \prod_{j=0}^{n-1} E(e^{\varepsilon X_j}),$$

and we are done. $\qquad\square$

In the above we essentially used the multiplicative property of the function $e^{\varepsilon x}$, namely that

$$e^{\varepsilon(x+y)} = e^{\varepsilon x} \cdot e^{\varepsilon y}.$$

The exponential function of the purely imaginary argument,

$$e^{\imath \varepsilon x} = \cos \varepsilon x + \imath \sin \varepsilon x,$$

also possesses this property.

Let X be a random variable having probability density function (p.d.f.), $f(x)$. The characteristic function φ of X is defined as

$$\varphi(\zeta) = \int_{-\infty}^{\infty} e^{\imath \zeta x} f(x)\, dx = u(\zeta) + \imath v(\zeta), \tag{3.41}$$

where

$$u(\zeta) = \int_{-\infty}^{\infty} \cos \zeta x \cdot f(x)\, dx, \quad v(\zeta) = \int_{-\infty}^{\infty} \sin \zeta x \cdot f(x)\, dx.$$

In harmonic analysis, φ is conventionally addressed as the *Fourier–Stiltjes transform*. The characteristic function is continuous, $\varphi(0) = 1$ and $|\varphi(\zeta)| \leq 1$ for all ζ.

Let $\varphi(\zeta)$ be such that there exists $\int_{-\infty}^{\infty} |\varphi(\zeta)|\, d\zeta$. Then the p.d.f. can be computed (inverse Fourier–Stiltjes transform) as

$$f(x) = \frac{1}{2\pi} \int_{-\infty}^{\infty} e^{-\imath \zeta x} \varphi(\zeta)\, d\zeta. \tag{3.42}$$

3.3 Elements of algebra

In this section I only briefly go through some basic definitions and results about finite fields and rings that are required later.

3.3.1 Main algebraic structures

Let G be a nonempty set and $\circ$ be a binary operation defined on G. The pair $(G, \circ)$ is a *group* if the following three properties hold:

(i) $(a \circ b) \circ c = a \circ (b \circ c)$ for all $a, b, c \in G$.
(ii) There is an identity element e such that $e \circ a = a \circ e = a$ for all $a \in G$.
(iii) For every $a \in G$ there exists an inverse $a^{-1} \in G$ such that $a \circ a^{-1} = a^{-1} \circ a = e$.

If, furthermore,

(iv) $a \circ b = b \circ a$ for all $a, b \in G$,

the group is called abelian or commutative.

We often use the notation of ordinary multiplication or addition for the group operation. Using the multiplicative notation, we write $a^n = a \cdot a \cdot \ldots \cdot a$ (n factors a), and in the additive notation $na = a + a + \ldots + a$ (n summands a). If n is negative, we define $a^n = (a^{-1})^{-n}$ and $na = (-n)(-a)$ respectively.

A group $(G, \cdot)$ is called cyclic if there is an element $a \in G$ such that every $b \in G$ is of the form a^j for some $j \in \mathbb{Z}$. Such an element is called a *generator* of the group.

Let R be a nonempty set and $+$ and $\cdot$ be two binary operations defined on R. The triple $(R, +, \cdot)$ is a *ring* if

(i) $(R, +)$ is an abelian group,
(ii) $(a \cdot b) \cdot c = a \cdot (b \cdot c)$ for all $a, b, c \in R$,
(iii) there is an element $1 \in R$ such that $1 \cdot a = a \cdot 1 = a$ for all $a \in R$,
(iv) $a \cdot (b + c) = a \cdot b + a \cdot c$ and $(a + b) \cdot c = a \cdot c + b \cdot c$ for all $a, b, c \in R$.

The ring $(R, +, \cdot)$ is called commutative if

(v) $a \cdot b = b \cdot a$ for all $a, b \in R$.

The identity element of $(R, +)$ is denoted by 0.

A ring $(F, +, \cdot)$ is called a *field* if the pair $(F \setminus \{0\}, \cdot)$ is an abelian group. We denote $F^* = F \setminus \{0\}$. As with the usual multiplication, we often omit the symbol $\cdot$ and simply write ab instead of $a \cdot b$. The ring of integers modulo n is denoted by $\mathbb{Z}_n$. When n is a prime, $\mathbb{Z}_n$ is a field.

The smallest integer p such that $p1 = 1 + 1 + \ldots + 1 = 0$ is called the *char-acteristic* of the field. The characteristic of a finite field is always a prime. In a field

F with characteristic p

$$(a + b)^p = a^p + b^p$$

for all $a, b \in F$, because all the binomial coefficients $\binom{p}{i}$, $0 < i < p$, are divisible by p.

If $(F, +, \cdot)$ is a field, the set of polynomials over F defined by

$$F[x] = \{a_0 + a_1 x + \ldots + a_m x^m : m = 0, 1, \ldots; a_i \in F, 0 \leq i \leq m\},$$

together with the addition

$$\left(\sum a_i x^i\right) + \left(\sum b_i x^i\right) = \sum (a_i + b_i) x^i$$

and multiplication

$$\left(\sum a_i x^i\right) \left(\sum b_j x^j\right) = \sum \left(\sum_{i+j=k} a_i b_j\right) x^k$$

forms a ring $(F[x], +, \cdot)$ called the polynomial ring over F.

A polynomial $a_0 + a_1 x + \ldots + a_m x^m \in F[x]$ with $m \geq 1$ and $a_m \neq 0$ is called *irreducible* if it cannot be written as a product of two polynomials in $F[x]$ both of degree less than m.

Theorem 3.30 *If p is a prime, then for every $m \geq 1$ there exists an irreducible polynomial $g(x) \in \mathbb{Z}_p[x]$ of degree m.* □

Theorem 3.31 *If p is a prime and $g(x)$ is an irreducible polynomial of degree m in $\mathbb{Z}_p[x]$, then the residue classes*

$$a(x) + <g(x)> = \{a(x) + s(x)g(x) : s(x) \in \mathbb{Z}_p[x]\},$$

where $a(x) \in \mathbb{Z}_p[x]$, together with the usual addition

$$(a(x) + <g(x)>) + (b(x) + <g(x)>) = \big(a(x) + b(x)\big) + <g(x)>$$

and multiplication

$$(a(x) + <g(x)>)(b(x) + <g(x)>) = a(x)b(x) + <g(x)>$$

form a finite field with p^m elements. □

The characteristic of this field is p. Consequently, for every prime, p, and every integer, $m \geq 1$, there exists a finite field with $q = p^m$ elements, which is denoted by $\mathbb{F}_q$. It can be shown that for each such q there is a unique field with q elements (up to isomorphism). The field $\mathbb{F}_q$ is a vector space over $\mathbb{F}_p$. If we choose a basis for $\mathbb{F}_q$ over $\mathbb{F}_p$, every element of $\mathbb{F}_q$ can be written uniquely as an $\mathbb{F}_p$-linear combination of the basis elements.

Table 3.1 *The field of 16 elements*

0	$= 0$	$= 0000$
$1 = \alpha^{15}$	$= 1$	$= 0001$
α	$= \alpha$	$= 0010$
α^2	$= \alpha^2$	$= 0100$
α^3	$= \alpha^3$	$= 1000$
α^4	$= \alpha + 1$	$= 0011$
α^5	$= \alpha^2 + \alpha$	$= 0110$
α^6	$= \alpha^3 + \alpha^2$	$= 1100$
α^7	$= \alpha^3 + \alpha + 1$	$= 1011$
α^8	$= \alpha^2 + 1$	$= 0101$
α^9	$= \alpha^3 + \alpha$	$= 1010$
α^{10}	$= \alpha^2 + \alpha + 1$	$= 0111$
α^{11}	$= \alpha^3 + \alpha^2 + \alpha$	$= 1110$
α^{12}	$= \alpha^3 + \alpha^2 + \alpha + 1$	$= 1111$
α^{13}	$= \alpha^3 + \alpha^2 + 1$	$= 1101$
α^{14}	$= \alpha^3 + 1$	$= 1001$

Example 3.1   The polynomial $g(x) = x^4 + x + 1 \in \mathbb{Z}_2[x]$ is irreducible, and can be used to construct a finite field of 16 elements. The residue classes $a(x) + <g(x)>$ are represented by polynomials in $\mathbb{Z}_2[x]$ of degree less than four. If we denote the residue class $x + <g(x)>$ by α, then all the elements of the field are 0, 1, α, $\alpha + 1$, α^2, $\alpha^2 + 1$, $\alpha^2 + \alpha$, $\alpha^2 + \alpha + 1$, α^3, $\alpha^3 + 1$, $\alpha^3 + \alpha$, $\alpha^3 + \alpha + 1$, $\alpha^3 + \alpha^2$, $\alpha^3 + \alpha^2 + 1$, $\alpha^3 + \alpha^2 + \alpha$, $\alpha^3 + \alpha^2 + \alpha + 1$. Addition in the field is easy. For example,

$$(\alpha^3 + \alpha^2 + 1) + (\alpha^2 + \alpha + 1) = \alpha^3 + \alpha,$$

because the characteristic is two. Multiplication of any two elements is almost as easy. For instance,

$$(\alpha^3 + \alpha^2 + 1)(\alpha^2 + \alpha + 1) = \alpha^5 + \alpha + 1.$$

To see which of the sixteen elements listed above this is, we use the fact that $\alpha^4 = \alpha + 1$ and hence $\alpha^5 = \alpha(\alpha + 1) = \alpha^2 + \alpha$. Consequently $\alpha^5 + \alpha + 1 = \alpha^2 + 1$. In this way it is easy to verify that all the nonzero elements of the field are powers of α, as shown in Table 3.1.

Using this table the multiplication is even easier: to multiply $\alpha^3 + \alpha^2 + 1$ and $\alpha^2 + \alpha + 1$, we check from the table that $\alpha^3 + \alpha^2 + 1 = \alpha^{13}$ and $\alpha^2 + \alpha + 1 = \alpha^{10}$, and hence their product is $\alpha^{23} = \alpha^8 = \alpha^2 + 1$.   □

Theorem 3.32   *The multiplicative group $(\mathbb{F}_q^*, \cdot)$ of $\mathbb{F}_q$ is cyclic.*   □

A generator of the multiplicative group of $\mathbb{F}_q$ is called a *primitive element* of the field. If α is a primitive element of the field $\mathbb{F}_q$, then $\alpha^{q-1} = 1$. Consequently, the elements of the finite field $\mathbb{F}_q$ are the q roots of the equation $x^q = x$. If $k > 0$ is not divisible by $q - 1$, then

$$\sum_{a \in \mathbb{F}_q} a^k = \sum_{i=0}^{q-2} \alpha^{ik} = \frac{\alpha^{k(q-1)} - 1}{\alpha^k - 1} = 0. \tag{3.43}$$

We will be using the following result about prime fields.

Theorem 3.33 *(Euler's criterion)* Let p be an odd prime and $\beta \in \mathbb{F}_p^*$. Then

$$x^2 = \beta \bmod p \tag{3.44}$$

has a solution if and only if

$$\beta^{\frac{p-1}{2}} = 1 \bmod p. \tag{3.45}$$

Proof Let α be a primitive element in $\mathbb{F}_p$. Assuming that $x = \alpha^r$ satisfies (3.44) for some r, we have $\beta = \alpha^{2r} \bmod p$. Then

$$\beta^{\frac{p-1}{2}} = \alpha^{2r \frac{p-1}{2}} = \alpha^{r(p-1)} = 1 \bmod p.$$

Conversely, suppose that (3.45) holds. We have $\beta = \alpha^r \bmod p$, for some integer r. Thus $\alpha^{r \frac{p-1}{2}} = 1 \bmod p$, and so $(p - 1)$ divides $r \frac{p-1}{2}$, which implies that r is even. Thus $\beta = \left(\alpha^{\frac{r}{2}} \right)^2 \bmod p$, so β is congruent to a square modulo p. $\square$

Corollary 3.34 *If for some c and odd prime p*

$$c^2 = -1 \bmod p,$$

then $p = 1 \bmod 4$.

Proof By the previous theorem the condition is equivalent to

$$(-1)^{\frac{p-1}{2}} = 1 \bmod p.$$

If $p = 4m + 3$ then $(-1)^{2m+1} = -1 \bmod p$, a contradiction. $\square$

3.3.2 Exponential sums over fields and rings

Let $(G, \cdot)$ be a finite abelian group and 1 the identity element of G. A *character* of G is a mapping φ from G to the set of complex numbers with norm 1 such that

$$\varphi(ab) = \varphi(a)\varphi(b) \text{ for all } a \in G, b \in G.$$

The mapping φ such that $\varphi(a) = 1$ for all $a \in G$ is called the *trivial character*. For the trivial character, the sum $\sum_{a \in G} \varphi(a)$ equals the cardinality of the group.

Theorem 3.35   *If φ is a nontrivial character of an abelian group G, then*

$$\sum_{a \in G} \varphi(a) = 0.$$

Proof   There is an element b of G such that $\varphi(b) \neq 1$. When a runs through G, so does ba. Hence

$$(1 - \varphi(b)) \sum_{a \in G} \varphi(a) = \sum_{a \in G} \varphi(a) - \sum_{a \in G} \varphi(ba) = 0.$$

$\square$

The characters of the additive and multiplicative groups of a finite field are called *additive* and *multiplicative characters*, respectively.

In the field $\mathbb{F}_q$, where $q = p^m$, the *trace function* Tr is defined by

$$\mathrm{Tr}(x) = x + x^p + x^{p^2} + \ldots + x^{p^{m-1}} \text{ for all } x \in \mathbb{F}_q.$$

The trace function satisfies the property

$$\mathrm{Tr}(a + b) = \mathrm{Tr}(a) + \mathrm{Tr}(b) \text{ for all } a, b \in \mathbb{F}_q$$

and it is a surjective mapping from $\mathbb{F}_q$ to $\mathbb{F}_p$. Let

$$e(x) = e^{2\pi i \, \mathrm{Tr}(x)/p} \text{ for all } x \in \mathbb{F}_q.$$

For every $a \in \mathbb{F}_q$, the function

$$\psi_a(x) = e(ax) \text{ for all } x \in \mathbb{F}_q$$

is an additive character of $\mathbb{F}_q$. When a runs through all the nonzero elements of $\mathbb{F}_q$, the functions ψ_a run through all the $q - 1$ different nontrivial additive characters of the field $\mathbb{F}_q$.

Let $q = 2^m$. The additive characters of $\mathbb{F}_q$ and multiplicative characters of $\mathbb{F}_q^*$ are correspondingly

$$\psi_a(x) = (-1)^{\mathrm{Tr}(ax)}, \quad a, x \in \mathbb{F}_q,$$
$$\chi_j(x) = e^{2\pi i \frac{j}{2^m - 1}}, \quad j = 0, 1, \ldots, 2^m - 2; \quad x \in \mathbb{F}_q^*.$$

In particular, ψ_0 and χ_0 are the trivial characters. The order of a multiplicative character χ is the least positive integer d such that $\chi^d = \chi_0$.

Let $f(x) \in \mathbb{F}_q[x]$ and suppose f is not expressible in the form $g^2(x) + g(x) + b$ where $g(x) \in \mathbb{F}_q[x]$ and $b \in \mathbb{F}_q$. Then f is *nondegenerate*. A sufficient condition for f to be nondegenerate is that f has odd degree. I now state bounds on exponential sums.

Theorem 3.36 *(Weil–Carlitz–Uchiyama bound) Let ψ be a nontrivial additive character of $\mathbb{F}_q$ and let $f(x) \in \mathbb{F}_q[x]$ be a nondegenerate polynomial of degree r. Then*

$$\left| \sum_{x \in \mathbb{F}_q} \psi(f(x)) \right| \le (r - 1) 2^{\frac{m}{2}}.$$

□

Theorem 3.37 *Let ψ be a nontrivial additive character of $\mathbb{F}_q$. Let χ be a non-trivial multiplicative character of $\mathbb{F}_q^*$ of order d with $d | (q - 1)$. Let $f(x) \in \mathbb{F}_q[x]$ have degree r, where r is odd. Suppose $g(x) \in \mathbb{F}_q[x]$ has s distinct roots and that g.c.d.$(d, \deg g) = 1$. Then*

$$\left| \sum_{x \in \mathbb{F}_q^*} \psi(f(x)) \chi(g(x)) \right| \le (r + s - 1) 2^{\frac{m}{2}}.$$

□

Now we pass to Galois rings. In what follows, $R_{e,m}$ denotes the ring of characteristic 2^e and degree m. This ring contains 2^{em} elements, has characteristic 2^e, and can be shown to be isomorphic to the factor ring $\mathbb{Z}_{2^e}[x]/(f(x))$ where f is a monic basic irreducible of degree m.

The units $R_{e,m}^*$ in $R_{e,m}$ contain a cyclic subgroup $\mathcal{T}_{e,m}^*$ of order $2^m - 1$. We denote by β a generator of this set. We write

$$\mathcal{T}_{e,m} = \mathcal{T}_{e,m}^* \cup \{0\} = \{\beta^k, k = 0, 1, \ldots, 2^m - 2\} \cup \{0\},$$

and call $\mathcal{T}_{e,m}$ the *Teichmuller set* in $R_{e,m}$. Every element $x \in R_{e,m}$ has a 2-adic expansion,

$$x = x_0 + 2x_1 + \ldots + 2^{e-1} x_{e-1}, \quad x_k \in \mathcal{T}_{e,m}, \quad k = 0, 1, \ldots, e - 1.$$

We define the *Frobenius automorphism*, σ, on $R_{e,m}$ by

$$\sigma(x) = x_0^2 + 2x_1^2 + \ldots + 2^{e-1} x_{e-1}^2,$$

and by analogy with the finite-field case, the absolute trace function Tr on $R_{e,m}$ by

$$\mathrm{Tr}(x) = \sum_{k=0}^{m-1} \sigma^k(x).$$

We also define characters for the ring $R_{e,m}$. For odd a, $a = 1, 3, \ldots, 2^e - 1$, let $\psi_a : R_{e,m} \to \mathbb{C}$ denote the additive character of $R_{e,m}$ defined by

$$\psi_a(x) = e^{2\pi \imath a \frac{\mathrm{Tr}(x)}{2^e}}, \quad x \in R_{e,m}.$$

For each integer j, $j = 0, 1, \ldots, 2^m - 2$, let $\chi_j : R_{e,m}^* \to \mathbb{C}$ denote the multiplicative character defined by

$$\chi_j(x) = e^{2\pi i \frac{jk}{2^m - 1}}, \quad x \in R_{e,m}^*, \quad x = \beta^k \bmod 2, \ 0 \le k < 2^m - 1.$$

The characters ψ_0 and χ_0 are the trivial characters for the Galois ring.

Let $f(x) \in R_{e,m}[x]$ and suppose f is not expressible in the form

$$\sigma(g(x)) - g(x) + b,$$

for any $g(x) \in R_{e,m}[x]$, and any $b \in R_{e,m}$. Here,

$$\sigma\left(\sum_k g_k x^k\right) = \sum_k \sigma(g_k) x^{2k}.$$

Then we say that f is nondegenerate. An easily verified condition for f of degree at least 1 to be nondegenerate is that f contains no terms of even degree.

Now let f be a polynomial with 2-adic expansion

$$f(x) = F_0[x] + 2F_1[x] + \ldots + 2^{e-1}F_{e-1}[x],$$
$$F_k[x] \in \mathcal{T}_{e,m}[x], \ k = 0, 1, \ldots, e - 1.$$

We define the *weighted degree* of f to be

$$D_f = \max\{2^{e-1}d_0, 2^{e-2}d_1, \ldots, d_e\},$$

where d_k is the degree of F_k. The following results (see [156, 224]) allow bounding exponential sums over Galois rings.

Theorem 3.38 *Let ψ be a nontrivial additive character of $R_{e,m}$. Let $f(x) \in R_{e,m}[x]$ be nondegenerate and of weighted degree D_f. Then*

$$\left| \sum_{x \in \mathcal{T}_{e,m}} \psi(f(x)) \right| \le (D_f - 1)2^{\frac{m}{2}}.$$

$\square$

Theorem 3.39 *Let ψ be a nontrivial additive character of $R_{e,m}$. Let $f(x) \in R_{e,m}[x]$ be nondegenerate and of weighted degree D_f. Let χ be a nontrivial multiplicative character of $R_{e,m}$. Then*

$$\left| \sum_{x \in \mathcal{T}_{e,m}^*} \psi(f(x))\chi(x) \right| \le D_f 2^{\frac{m}{2}}.$$

$\square$

3.4 Elements of coding theory

In this section I survey the theory of error-correcting codes.

3.4.1 Hamming space

A binary code of length n is simply a nonempty set of binary vectors of length n. More generally, we have the following definition. Let Q be a finite set with q elements. A nonempty subset C of $Q^n = Q \times Q \times \ldots \times Q$ is called a q-ary code of length n.

The vectors belonging to a code are called *code words*. A code with only one code word is called trivial. Whenever convenient, codes are assumed to have at least two code words. The set Q is called the *alphabet*. We use the term *vector* for an n-tuple over an arbitrary alphabet, not only in the case when Q is a field. The elements of Q^n are also called *points* or *words*. The set Q^n is called the (q-ary) *Hamming space*.

The *Hamming distance* between two vectors $\mathbf{x} = (x_0, x_1, \ldots, x_{n-1})$, $\mathbf{y} = (y_0, y_1, \ldots, y_{n-1})$ in Q^n is the number of coordinates in which they differ, i.e.,

$$d(\mathbf{x}, \mathbf{y}) = |\{j : x_j \neq y_j\}|.$$

The Hamming distance satisfies the *triangle inequality*,

$$d(\mathbf{x}, \mathbf{y}) + d(\mathbf{y}, \mathbf{z}) \geq d(\mathbf{x}, \mathbf{z}),$$

for all $\mathbf{x}, \mathbf{y}, \mathbf{z} \in Q^n$, and is a metric. If $V \subseteq Q^n$, then we denote

$$d(\mathbf{x}, V) = \min_{\mathbf{v} \in V} d(\mathbf{x}, \mathbf{v}).$$

Assume that $0 \in Q$. The *Hamming weight* $w(\mathbf{x})$ of a vector $\mathbf{x} = (x_0, x_1, \ldots, x_{n-1}) \in Q^n$ is defined by

$$w(\mathbf{x}) = d(\mathbf{x}, \mathbf{0}), \tag{3.46}$$

where $\mathbf{0} = 0^n = (0, 0, \ldots, 0)$. A vector with even (odd) weight is called *even (odd)*. The set of even (odd) code words of a code C is called the *even weight (odd weight) subcode* of C and is denoted by C_e (C_o). The *support* of a vector $\mathbf{x} \in Q^n$ is the set $\{j : x_j \neq 0\}$ and is denoted by supp($\mathbf{x}$). In the case of a binary alphabet, vectors can be identified with their supports, and for two binary vectors, $\mathbf{x}$ and $\mathbf{y}$, of the same length, $\mathbf{x} \subset \mathbf{y}$, $\mathbf{x} \cup \mathbf{y}$, $\mathbf{x} \cap \mathbf{y}$ and $\mathbf{x} \setminus \mathbf{y}$ refer to the supports of $\mathbf{x}$ and $\mathbf{y}$.

The *Hamming sphere* (or *ball*) $B_r(\mathbf{x})$ of radius r centred at the vector $\mathbf{x} \in Q^n$ is defined by

$$B_r(\mathbf{x}) = \{\mathbf{y} \in Q^n : d(\mathbf{y}, \mathbf{x}) \leq r\},$$

and its cardinality is

$$V_q(n, r) = \sum_{j=0}^{r} \binom{n}{j}(q - 1)^j.$$

The subscript q is usually omitted if $q = 2$. More generally, if $V \subseteq Q^n$, we denote

$$B_r(V) = \bigcup_{\mathbf{x} \in V} B_r(\mathbf{x}).$$

It is also convenient to have a notation for the *layers* (or *shells*) of the Hamming sphere, and we therefore denote

$$S_r(\mathbf{x}) = \{\mathbf{y} \in Q^n : d(\mathbf{y}, \mathbf{x}) = r\},$$

and

$$S_r = \{\mathbf{y} \in Q^n : w(\mathbf{y}) = r\}.$$

The (binary) *entropy function* $H(x)$ is defined by

$$H(x) = -x \log_2 x - (1 - x) \log_2(1 - x)$$

where $0 \le x \le 1$. The following two lemmas can be obtained using Stirling's formula

$$\sqrt{2\pi}\, n^{n+\frac{1}{2}} e^{-n+\frac{1}{12n}-\frac{1}{360n^3}} < n! < \sqrt{2\pi}\, n^{n+\frac{1}{2}} e^{-n+\frac{1}{12n}}. \tag{3.47}$$

Lemma 3.40 *Suppose that $0 < \lambda < 1$ and λn is an integer. Then*

$$\frac{2^{nH(\lambda)}}{\sqrt{8n\lambda(1 - \lambda)}} \le \binom{n}{\lambda n} \le \frac{2^{nH(\lambda)}}{\sqrt{2\pi n\lambda(1 - \lambda)}}. \tag{3.48}$$

$\square$

Lemma 3.41 *Suppose that $0 < \lambda < \frac{1}{2}$ and λn is an integer. Then*

$$\frac{2^{nH(\lambda)}}{\sqrt{8n\lambda(1 - \lambda)}} \le V(n, \lambda n) \le 2^{nH(\lambda)}. \tag{3.49}$$

$\square$

The behavior of the binomial coefficient $\binom{n}{k}$ in the vicinity of $k = \frac{n}{2}$ is also of interest. The following results can be obtained using Taylor series of the logarithmic function.

Lemma 3.42 *Let*

$$k = \frac{n + c\sqrt{n}}{2},$$

where $c = c(n) = o\left(n^{\frac{1}{6}}\right)$. Then

$$\binom{n}{k} = \binom{n}{n/2} \cdot e^{-\frac{c^2}{2}} \cdot (1 + o(1)).$$

$\square$

Another range includes small k.

Lemma 3.43 *If $k = o\left(n^{\frac{2}{3}}\right)$ then*

$$\binom{n}{k} = e^{-\frac{k^2}{2n}} \cdot \frac{n^k}{k!}(1 + o(1)).$$

If $k = o\left(n^{\frac{3}{4}}\right)$ then

$$\binom{n}{k} = e^{-\frac{k^2}{2n} - \frac{k^3}{6n^2}} \cdot \frac{n^k}{k!}(1 + o(1)).$$

$\square$

Along with the Hamming distance, sometimes it is of interest to consider the *Lee distance*. Let the alphabet Q be identified with the set of integers $\{0, 1, \ldots, q - 1\}$. Then the Lee distance between two vectors $\mathbf{x}$ and $\mathbf{y}$ is

$$d_L(\mathbf{x}, \mathbf{y}) = \sum_{j=0}^{n-1} \min(|x_j - y_j|, q - |x_j - y_j|). \tag{3.50}$$

Although there exist some essential differences, the geometries of the Hamming and Euclidean spaces are, in many respects, similar. One of the most crucial properties is that the sphere is the set of maximal size among the sets of given diameter (Theorem 3.46). The *diameter* of a set $S \subseteq \mathbb{F}^n$ is defined by

$$\mathrm{diam}(S) = \max_{\mathbf{x}, \mathbf{y} \in S} d(\mathbf{x}, \mathbf{y}).$$

A set $S \subseteq \mathbb{F}^n$ is said to be *hereditary* if

$$\mathbf{x} \in S \text{ and } \mathrm{supp}(\mathbf{y}) \subseteq \mathrm{supp}(\mathbf{x}) \Rightarrow \mathbf{y} \in S.$$

A simple property on the diameter of a hereditary set is the following:

Lemma 3.44 *For any two elements, $\mathbf{s}_1$ and $\mathbf{s}_2$, of a hereditary set S,*

$$|\mathrm{supp}(\mathbf{s}_1) \cup \mathrm{supp}(\mathbf{s}_2)| \leq \mathrm{diam}(S).$$

Proof The vector $\mathbf{s}_1 \setminus \mathbf{s}_2$, with support $\mathrm{supp}(\mathbf{s}_1) \setminus \mathrm{supp}(\mathbf{s}_2)$, belongs to S by heredity, and has disjoint support with $\mathbf{s}_2$; thus $|\mathrm{supp}(\mathbf{s}_1) \cup \mathrm{supp}(\mathbf{s}_2)| = d(\mathbf{s}_1 \setminus \mathbf{s}_2, \mathbf{s}_2)$ is upper bounded by the diameter of S. $\square$

Let us split the elements of a set S into two sets $S_0^{(j)}$ and $S_1^{(j)}$, according to whether their jth component is 0 or 1. Define a set of transformations τ_j by

$$\tau_j(\mathbf{s}) = \begin{cases} \mathbf{s} + \mathbf{e}_j, & \text{if } \mathbf{s} \in S_1^{(j)} \text{ and } \mathbf{s} + \mathbf{e}_j \notin S_0^{(j)} \\ \mathbf{s}, & \text{otherwise}, \end{cases}$$

where $\mathbf{e}_j$ is the vector having 1 in the jth coordinate and zeros elsewhere. Notice that performing τ_j can only decrease the weight.

Lemma 3.45   *A set S is one-to-one mapped by τ_j onto a set with smaller or equal diameter. If S is stable under all values of τ_j, then S is hereditary.*

Proof   We first check injectivity: let $\mathbf{s}_1$ and $\mathbf{s}_2$ be two distinct elements of S; $\mathbf{s}_1'$ and $\mathbf{s}_2'$ their respective images under τ_j. There is something to prove only if exactly one element, say $\mathbf{s}_2$, moves; thus $\mathbf{s}_1' = \mathbf{s}_1$ and $\mathbf{s}_2' = \mathbf{s}_2 + \mathbf{e}_j$. If $\mathbf{s}_2' = \mathbf{s}_1'$, then $\mathbf{s}_2 + \mathbf{e}_j = \mathbf{s}_1$, and $\mathbf{s}_2$ should not have moved. Thus $\mathbf{s}_2' \neq \mathbf{s}_1'$.

Now $d(\mathbf{s}_1', \mathbf{s}_2') = d(\mathbf{s}_1, \mathbf{s}_2 + \mathbf{e}_j) = d(\mathbf{s}_1, \mathbf{s}_2) - 1$, if $\mathbf{s}_1 \in S_0^{(j)}$. If $\mathbf{s}_1 \in S_1^{(j)}$, then $\mathbf{s}_1^* = \mathbf{s}_1 + \mathbf{e}_j \in S_0^{(j)}$ (otherwise, $\mathbf{s}_1$ would have moved). But then, $d(\mathbf{s}_1', \mathbf{s}_2') = d(\mathbf{s}_1^*, \mathbf{s}_2) \leq \text{diam}(S)$. Thus the diameter cannot increase. Heredity can be rephrased as

$$\mathbf{s} \in S_1^{(j)} \Rightarrow \mathbf{s} + \mathbf{e}_j \in S_0^{(j)}, \text{ for all } j.$$

This property is clearly satisfied if S is stable under every τ_j.   □

The following theorem is from Kleitman [205].

Theorem 3.46 *(Kleitman)*   *Let $B \subseteq \mathbb{F}^n$; $n \geq 2r + 1$ be a set of diameter $2r$. Then $|B| \leq V(n, r)$.*

Proof   Let B be a set of diameter $2r$ in $\mathbb{F}^n$. Apply the transformations τ_i coordinatewise until B stabilizes to some B^*. Combining Lemmas 3.45 and 3.44, B^* is a hereditary set of size $|B|$, with diameter at most $2r$ and

$$|\text{supp}(\mathbf{a}) \cup \text{supp}(\mathbf{b})| \leq 2r, \tag{3.51}$$

for all $\mathbf{a}, \mathbf{b} \in B^*$. In particular, $\mathbf{0} \in B^*$ and any element in B^* has weight at most $2r$.

This is already sufficient for an asymptotic version of Theorem 3.46.

Theorem 3.47   *For n large enough with respect to r, the largest subset of $\mathbb{F}^n$ with diameter $2r$ is realized by a sphere of radius r.*

Proof   Let S be such a set, and S^* its stabilized version under τ_j. Take an element $\mathbf{s}_1 \in S^*$ of maximal weight $r + k$. Assume $k > 0$ (otherwise there is nothing to prove). For any $\mathbf{s}_2 \in S^*$, $\text{supp}(\mathbf{s}_2)$ is the disjoint union of $\text{supp}(\mathbf{s}_2) \cap \text{supp}(\mathbf{s}_1)$ and supp(s_2)\supp(s_1), the latter of size at most $r - k$ by (3.51) applied to $s_1, s_2 \setminus s_1 \in S^*$. Thus the number of such s_2s is upperbounded by

$$2^{r+k} \sum_{i=0}^{r-k} \binom{n-r-k}{i} = o(V(n,r)).$$

$\square$

We have now accomplished the proof of Theorem 3.46. We apply another set of transformations σ_j to B^*, with the goal of "pushing 1s to the left". Again split the elements of B^* in two: the set $B_{01}^{(j)}$ of vectors with a 0 in position $j - 1$ and 1 in position j, and its complement in B^*. Define σ_i by

$$\sigma_j(\mathbf{a}) = \begin{cases} \mathbf{a} + \mathbf{e}_{j-1} + \mathbf{e}_j, & \text{if } \mathbf{a} \in B_{01}^{(j)} \text{ and } \mathbf{a} + \mathbf{e}_{j-1} + \mathbf{e}_j \notin B^* \\ \mathbf{a}, & \text{otherwise.} \end{cases}$$

Apply these transformations coordinatewise until B^* stabilizes to some B''. Analogously to the previous case one easily checks that B'' is a hereditary set (it remains so after each σ_j) of size $|B|$ and diameter at most $2r$. Moreover, along with any vector $\mathbf{a}$ the set B'' contains all the vectors obtained from $\mathbf{a}$ by interchanging any number of 1s with 0s, each 0 lying to the left of the corresponding 1 in $\mathbf{a}$.

For $\mathbf{a} \in B''$, assume that $w(\mathbf{a}) = r + k$ where $k > 0$. Then the following property (P) holds: at most $r - k$ of these 1s can be associated to distinct 0s in $\mathbf{a}$, each 0 lying to the left of the corresponding 1. Indeed, if there were $r - k + 1$ such 1s, the vector $\mathbf{b}$ obtained by interchanging the 1s with the 0s would also belong to B'', and $d(\mathbf{a}, \mathbf{b}) = 2(r - k + 1)$. Then

$$|\text{supp}(\mathbf{a}) \cup \text{supp}(\mathbf{b})| = d(\mathbf{a}, \mathbf{b}) + |\text{supp}(\mathbf{a} * \mathbf{b})|$$
$$= 2(r - k + 1) + w(\mathbf{a}) - (r - k + 1) > 2r$$

would hold, contradicting (3.51).

We now define a last set of transformations: split B'' into $B_l'' = \{\mathbf{a} \in B'' : w(\mathbf{a}) \leq r\}$ and $B_h'' = \{\mathbf{a} \in B'' : w(\mathbf{a}) = r + k, 0 < k \leq r\}$ (for "light" and "heavy"). At the end of the following procedure, B_l'' is unchanged, and B_h'' is transformed into A', with

$$|A'| = |B_h''|, \quad A' \cap B_l'' = \emptyset, \quad A' \subseteq B_r(\mathbf{0}).$$

To every $\mathbf{a}$ in B_h'', associate the corresponding vectors $\mathbf{a}'$ and $\mathbf{a}''$ as follows. The vector $\mathbf{a}'$ has 0s in all positions where $\mathbf{a}$ has 1s. To fill the 1s in $\mathbf{a}'$ we start from the rightmost 1 in $\mathbf{a}$ and set 1 in $\mathbf{a}'$ in the first position to the left of this 1 where $\mathbf{a}$ has 0. If there is no such 0 we assume that the positions with indices n and less are to the left of the first coordinate. We proceed in the same manner with the next $r - k$ (to the left) nonzero positions of $\mathbf{a}$, adding in each step a new 1 to $\mathbf{a}'$. Clearly

$w(\mathbf{a}') = r - k + 1$. Since $\mathbf{a}$ and $\mathbf{a}'$ have disjoint supports, $d(\mathbf{a}, \mathbf{a}') = 2r + 1$, and thus $\mathbf{a}' \notin B''$.

Let i be the number of steps for constructing $\mathbf{a}'$ when the inserted 1 has index less than the index of the corresponding 1 in $\mathbf{a}$, and let $\mathbf{a}''$ be the vector we get after the ith step. By the property (P), $\mathbf{a}''$ is always defined; evidently, $w(\mathbf{a}'') = i$. Notice that i may actually be deduced from $\mathbf{a}'$ since i is the maximal number of 1s in $\mathbf{a}'$ which can be paired with distinct 0s of $\mathbf{a}'$, each 0 lying to the right of the corresponding 1; and, given $\mathbf{a}'$, we may determine $\mathbf{a}''$. From $\mathbf{a}''$, moreover, we may recover $\mathbf{a}$, by inserting i 1s in the places as close as possible to the right of the 1s in $\mathbf{a}''$; the remaining 1s in $\mathbf{a}$ are inserted in the first $r + k - i$ left-hand places which are not 1s of $\mathbf{a}''$ or already 1s in $\mathbf{a}$.

So, we have constructed a one-to-one mapping from $\mathbf{a}$ to $\mathbf{a}'$ and, therefore, a one-to-one mapping from B into a subset of $B_r(\mathbf{0})$.   □

Example 3.2   Let $n = 11, r = 4, k = 1$, and consider how the vector $\mathbf{a} \in B_h''$ given below is transformed:

$$
\begin{array}{ll}
\mathbf{a} \;\; 10110000011 & \\
0*00*****00 & \text{Step 0} \\
0*00****100 & \text{Step 1} \\
0*00***1100 & \text{Step 2} \\
0100***1100 & \text{Step 3} \\
0100**11100 & \text{Step 4} \\
\mathbf{a}' \;\; 01000011100 &
\end{array}
$$

Here $i = 3$ and $\mathbf{a}'' = 01000001100$ can be deduced from $\mathbf{a}'$. Now $\mathbf{a}$ can be recovered from $\mathbf{a}''$ by putting three 1s immediately to the right of the 1s in $\mathbf{a}''$: $\mathbf{a} = *01****0011$, then 1s in the two leftmost positions where $\mathbf{a}$ and $\mathbf{a}''$ have no 1s: $\mathbf{a} = 10110000011$.   □

Note that for q large enough, Theorem 3.46 does not hold: choose $B = \mathbb{F}_q^{2r} \oplus 0^{n-2r}$. Then, B clearly has diameter $2r$, and as a function of q:

$$|B| = q^{2r} > V_q(n, r) = O(q^r).$$

3.4.2 Parameters of codes

The *minimum distance* d of a code $C \subseteq Q^n$ is the smallest of the distances between different code words of C, i.e.,

$$d = d(C) = \min_{\mathbf{a}, \mathbf{b} \in C, \mathbf{a} \neq \mathbf{b}} d(\mathbf{a}, \mathbf{b}).$$

Two codes $C_1 \subseteq Q^n$ and $C_2 \subseteq Q^n$ are called *equivalent* if C_2 is obtained from C_1 by applying to all the code words of C_1 a fixed permutation of the coordinates and to each coordinate a permutation of the symbols in the alphabet (which may vary with the coordinates). A vector $\mathbf{x}$ is said to be *r-covered* (or simply *covered* if r is clear from the context) by a vector $\mathbf{y}$ if $d(\mathbf{x}, \mathbf{y}) \leq r$, i.e., $\mathbf{x} \in B_r(\mathbf{y})$ or equivalently $\mathbf{y} \in B_r(\mathbf{x})$. A vector $\mathbf{x}$ is *r*-covered by a set V if it is *r*-covered by at least one element of V.

The *covering radius* of a code $C \subseteq Q^n$ is the smallest integer R such that every vector $\mathbf{x} \in Q^n$ is R-covered by at least one code word of C, i.e.,

$$R = R(C) = \max_{\mathbf{x} \in Q^n} d(\mathbf{x}, C) = \max_{\mathbf{x} \in Q^n} \min_{\mathbf{c} \in C} d(\mathbf{x}, \mathbf{c}). \tag{3.52}$$

In other words, the covering radius measures the distance between the code and the farthest-off vectors in the space. The covering radius is also the smallest integer R such that the union of the Hamming spheres of radius R centred at the code words is the whole space.

For a code C with minimum distance d, the integer

$$e = \lfloor (d-1)/2 \rfloor$$

is called the *packing radius* or the *error-correcting capability* of the code C. Notice that e is the largest integer such that the Hamming spheres of radius e centered at the code words are disjoint. Therefore,

$$|C| \leq \frac{q^n}{V_q(n, e)},$$

which is called the *sphere-packing bound* or *Hamming bound*.

Usually it is convenient to choose the alphabet Q to possess a certain algebraic structure. If our alphabet is $\mathbb{Z}_q$ or $\mathbb{F}_q$, we can define the sum and difference of two vectors $\mathbf{x} = (x_0, x_1, \ldots, x_{n-1})$ and $\mathbf{y} = (y_0, y_1, \ldots, y_{n-1})$ as

$$\mathbf{x} + \mathbf{y} = (x_0 + y_0, x_1 + y_1, \ldots, x_{n-1} + y_{n-1})$$

and

$$\mathbf{x} - \mathbf{y} = (x_0 - y_0, x_1 - y_1, \ldots, x_{n-1} - y_{n-1}),$$

and their *componentwise product* as

$$\mathbf{x} * \mathbf{y} = (x_0 y_0, x_1 y_1, \ldots, x_{n-1} y_{n-1}).$$

We then have the obvious formula

$$d(\mathbf{x}, \mathbf{y}) = w(\mathbf{x} - \mathbf{y}). \tag{3.53}$$

In the binary case, $\mathbf{x} + \mathbf{y} = \mathbf{x} - \mathbf{y}$, and

$$d(\mathbf{x}, \mathbf{y}) = w(\mathbf{x} + \mathbf{y}). \tag{3.54}$$

For two sets $A, B \subseteq \mathbb{Z}_q^n$ or $A, B \subseteq \mathbb{F}_q^n$ we denote

$$A + B = \{\mathbf{a} + \mathbf{b} : \mathbf{a} \in A, \mathbf{b} \in B\}.$$

For $\mathbf{x} \in \mathbb{Z}_q^n$ or $\mathbf{x} \in \mathbb{F}_q^n$ the set

$$\mathbf{x} + A = \{\mathbf{x} + \mathbf{a} : \mathbf{a} \in A\}$$

is called a *translate* of A.

In $\mathbb{F}_q^n$ we can also define a *scalar multiplication*: if $\mathbf{x} = (x_0, x_1, \ldots, x_{n-1}) \in \mathbb{F}_q^n$ and $\alpha \in \mathbb{F}_q$, then

$$\alpha \mathbf{x} = (\alpha x_0, \alpha x_1, \ldots, \alpha x_{n-1}).$$

A code $C \subseteq \mathbb{F}_q^n$ is called *linear* if all the pairwise sums and scalar multiples of code words belong to the code. This means that C is a linear subspace of $\mathbb{F}_q^n$. Thus, we can find a basis consisting of, say, k linearly independent code words $\mathbf{g}_0, \mathbf{g}_1, \ldots, \mathbf{g}_{k-1}$. The $k \times n$ matrix

$$\mathbf{G} = \mathbf{G}(C) = \begin{pmatrix} \mathbf{g}_0 \\ \mathbf{g}_1 \\ \vdots \\ \mathbf{g}_{k-1} \end{pmatrix}$$

is called a *generator matrix* of C. The code words of C are exactly the q^k linear combinations of the rows of $\mathbf{G}$, and the code is said to have *dimension k*. A linear code $C \subseteq \mathbb{F}_q^n$ with dimension k, minimum distance d, and covering radius R is called an $[n, k, d]_q R$ code.

Clearly, permuting the coordinates of all the code words does not change the parameters of the code. Apart from the order of the coordinates, we can always put the generator matrix in the form $(\mathbf{I}_k, \mathbf{P})$ using the Gaussian elimination method. Here $\mathbf{I}_k$ stands for the $k \times k$ identity matrix.

If $\mathbf{x}, \mathbf{y} \in \mathbb{F}_q^n$, their *scalar product* $\langle \mathbf{x}, \mathbf{y} \rangle$ is defined by

$$\langle \mathbf{x}, \mathbf{y} \rangle = x_0 y_0 + x_1 y_1 + \ldots + x_{n-1} y_{n-1}.$$

The vectors $\mathbf{x}$ and $\mathbf{y}$ are called *orthogonal* if $\langle \mathbf{x}, \mathbf{y} \rangle = 0$.

When $q = 2$, we have

$$w(\mathbf{x} + \mathbf{y}) = w(\mathbf{x}) + w(\mathbf{y}) - 2w(\mathbf{x} * \mathbf{y}). \tag{3.55}$$

Therefore in a binary linear code either all or exactly half of the code words are even. Indeed, if the code C has at least one odd code word $\mathbf{c}$, then $C_o = \mathbf{c} + C_e$ by the previous formula and C_e is an $[n, k-1]$ code.

The *dual* $C^\perp$ of a linear code $C \subseteq \mathbb{F}_q^n$ consists of all the vectors $\mathbf{x} \in \mathbb{F}_q^n$ such that $\langle \mathbf{x}, \mathbf{c} \rangle = 0$ for all $\mathbf{c} \in C$. Assume that $\mathbf{G} = \mathbf{G}(C) = (\mathbf{I}_k, \mathbf{P})$. Clearly, for each

$\mathbf{x}_2 \in \mathbb{F}_q^{n-k}$ there is a unique $\mathbf{x}_1 \in \mathbb{F}_q^k$, such that the vector $(\mathbf{x}_1, \mathbf{x}_2)$ is orthogonal to all the rows of $\mathbf{G}$. Hence the dimension of the dual code is $n - k$. The minimum distance of $C^\perp$ is called the *dual distance* $d^\perp$ of the code C. Let $\mathbf{H} = (-\mathbf{P}^T, \mathbf{I}_{n-k})$. A routine check shows that $\mathbf{GH}^T$ is the all-zero matrix and therefore $C^\perp$ has generator matrix $\mathbf{H}$. Any matrix $\mathbf{H} = \mathbf{H}(C)$ which is a generator matrix of $C^\perp$ is called a *parity check matrix* of C. Clearly $(C^\perp)^\perp = C$. Therefore if $\mathbf{H}$ is any parity check matrix of C, the code C can also be defined as

$$C = \{\mathbf{x} \in \mathbb{F}^n : \mathbf{Hx}^T = \mathbf{0}\}.$$

For every $\mathbf{x} \in \mathbb{F}^n$, we call the vector $\mathbf{Hx}^T \in \mathbb{F}^{n-k}$ the *syndrome* of $\mathbf{x}$. Hence the code consists of the vectors with syndrome equal to $\mathbf{0}$.

From (3.53), the minimum distance of a linear code is the smallest nonzero weight of a code word, i.e., the *minimum weight* of a nonzero vector $\mathbf{x} \in \mathbb{F}_q^n$ such that $\mathbf{Hx}^T = \mathbf{0}$. In particular, in the binary case we have the following theorem.

Theorem 3.48 *Let C be a binary $[n, k]$ code with parity check matrix $\mathbf{H}$. The minimum distance of C is the smallest positive integer d such that the sum of some d columns of $\mathbf{H}$ is $\mathbf{0}$.* □

3.4.3 The MacWilliams identities

In this section I introduce the MacWilliams transform. It is an important tool in the analysis of possible distance distributions of codes. Here the proofs are given only for the binary case, which makes the presentation simpler.

We start with characters. Let $\mathbf{x}$ and $\mathbf{y}$ be two vectors in $\mathbb{F}^n$. We define the *additive character* ψ on $\mathbb{F}^n$ as

$$\psi_{\mathbf{x}}(\mathbf{y}) = (-1)^{\langle \mathbf{x}, \mathbf{y} \rangle}.$$

The *projection* $\mathbf{p}_{\mathbf{x}}(\mathbf{y})$ of $\mathbf{y}$ on $\mathbf{x}$ is the vector of length $w(\mathbf{x})$ obtained from $\mathbf{y}$ by deleting all the positions of $\mathbf{y}$ not belonging to $\text{supp}(\mathbf{x})$. The projection of a set $G \subseteq \mathbb{F}^n$ on $\mathbf{x} \in \mathbb{F}^n$ is

$$P_{\mathbf{x}}(G) = \{\mathbf{p}_{\mathbf{x}}(\mathbf{g}) : \mathbf{g} \in G\},$$

i.e., the set of projections of the vectors in G. Note that the character equals 1 if $\mathbf{p}_{\mathbf{x}}(\mathbf{y})$ is even, and is -1 otherwise. Evidently, for all $\mathbf{x}, \mathbf{y}, \mathbf{z} \in \mathbb{F}^n$ we have

$$\psi_{\mathbf{x}}(\mathbf{y} + \mathbf{z}) = \psi_{\mathbf{x}}(\mathbf{y}) \, \psi_{\mathbf{x}}(\mathbf{z}). \tag{3.56}$$

For arbitrary $G \subseteq \mathbb{F}^n$ and $\mathbf{x} \in \mathbb{F}^n$ define

$$\psi_{\mathbf{x}}(G) = \sum_{\mathbf{g} \in G} \psi_{\mathbf{x}}(\mathbf{g}). \tag{3.57}$$

To make this definition more intuitive, notice that this expression just calculates the difference between the number of even and odd vectors in the projection of G on $\mathbf{x}$.

Let $C \subseteq \mathbb{F}^n$ be a binary linear code. Its *weight distribution* $\mathbf{W}(C) = \mathbf{W} = (W_0, W_1, \ldots, W_n)$ is a vector of dimension $n + 1$, where

$$W_j = |\{\mathbf{c} \in C : w(\mathbf{c}) = j\}|,$$

i.e., the jth component of $\mathbf{W}(C)$ is the number of code words of weight j in C.

The *distance distribution* $\mathbf{B}(C) = \mathbf{B} = (\mathrm{B}_0, \mathrm{B}_1, \ldots, \mathrm{B}_n)$ is defined by

$$\mathrm{B}_j = \frac{1}{|C|} |\{\mathbf{c}_1, \mathbf{c}_2 \in C : d(\mathbf{c}_1, \mathbf{c}_2) = j\}|,$$

i.e., the jth component of $\mathrm{B}(C)$ is the average number of code words being at distance i from a code word of C. In the case of linear codes the vectors $\mathbf{W}$ and $\mathbf{B}$ coincide, a fact that does not hold in general for nonlinear codes.

If C is a linear code, the distance distribution $\mathbf{B}(C^\perp)$ of the dual code $C^\perp$ is denoted by $\mathbf{B}^\perp$ and is called the *dual spectrum* of C. The following theorem shows that $\mathbf{B}^\perp$ is uniquely determined by $\mathbf{B}$.

Theorem 3.49 *(MacWilliams identities)*   *For a linear code C of length n*

$$\mathrm{B}_j^\perp = \frac{1}{|C|} \sum_{k=0}^{n} \mathrm{B}_k K_j^n(k), \tag{3.58}$$

where

$$K_j^n(k) = K_j(k) = \sum_{\ell=0}^{j} (-1)^\ell \binom{k}{\ell}\binom{n-k}{j-\ell}$$

is the Krawtchouk polynomial of degree j.

Proof   By linearity, the projection of C on any vector $\mathbf{x} \in \mathbb{F}^m$, $m \leq n$, contains only even vectors, or the same number of even and odd vectors. Moreover, $P_\mathbf{x}(C)$ consists of only even vectors if and only if $\mathbf{x} \in C^\perp$. Thus,

$$\psi_\mathbf{x}(C) = \begin{cases} |C| & \text{if } \mathbf{x} \in C^\perp, \\ 0 & \text{if } \mathbf{x} \notin C^\perp. \end{cases}$$

Furthermore, if S_j stands for the set of n-tuples of weight j, we have

$$\sum_{\mathbf{x} \in S_j} \psi_\mathbf{x}(C) = |C|\,\mathrm{B}_j^\perp. \tag{3.59}$$

On the other hand, the same sum can be estimated in a different way, namely,

$$\sum_{\mathbf{x} \in S_j} \psi_\mathbf{x}(C) = \sum_{\mathbf{x} \in S_j} \sum_{\mathbf{c} \in C} \psi_\mathbf{x}(\mathbf{c}) = \sum_{\mathbf{c} \in C} \sum_{\mathbf{x} \in S_j} \psi_\mathbf{x}(\mathbf{c}). \tag{3.60}$$

The inner sum in the last expression does not depend on the particular choice of **c**, but only on the weight $w(\mathbf{c})$. If $w(\mathbf{c}) = k$, then each of the $\binom{k}{\ell}\binom{n-k}{j-\ell}$ vectors of weight j with exactly ℓ 1s in common with **c** contributes $(-1)^\ell$ to the inner sum. Continuing (3.60) we get

$$\sum_{\mathbf{x} \in S_j} \psi_{\mathbf{x}}(C) = \sum_{i=0}^{n} \sum_{\mathbf{c} \in C, w(\mathbf{c})=k} \sum_{\mathbf{x} \in S_j} \psi_{\mathbf{x}}(\mathbf{c})$$

$$= \sum_{k=0}^{n} B_k \sum_{\ell=0}^{j} (-1)^\ell \binom{k}{\ell}\binom{n-k}{j-\ell} = \sum_{k=0}^{n} B_k K_j(k).$$

Comparing the result with (3.59) we get the claim. $\qquad\square$

If C is a nonlinear (n, K) code, we may formally define the MacWilliams transform of the $n + 1$-tuple $\mathbf{B}(C)$, using expression (3.58). In general, the result of the transform, $\mathbf{B}^\perp$, does not correspond to the distance distribution of any code. Nevertheless, an interpretation can be given to some values appearing in the transform. Notice that $K_0(k) = 1$, and

$$B_0^\perp = \frac{1}{|C|} \sum_{k=0}^{n} B_k K_0(k) = 1.$$

However, several first components of $\mathbf{B}^\perp$ with positive index could be zero. Define the *dual distance* $d^\perp(C) = d^\perp$ of the code C as the minimum nonzero index of a nonzero component of $\mathbf{B}^\perp$. For linear codes, this is just the minimum distance of the dual code, $d(C^\perp)$. Let $\mathbf{C}$ be the $K \times n$ array having as its rows all the code words of C.

Theorem 3.50 *(i)* $B_j^\perp \geq 0$ *for* $j \in [0, n]$; *(ii) any set of* $r \leq d^\perp - 1$ *columns of* $\mathbf{C}$ *contains each* r-*tuple exactly* $\frac{K}{2^r}$ *times, and* $d^\perp$ *is the largest integer with this property.*

Proof As in the previous proof, we see that

$$B_j^\perp = \frac{1}{|C|} \sum_{k=0}^{n} B_i K_j(k) = \frac{1}{|C|^2} \sum_{i=0}^{n} \sum_{\mathbf{a},\mathbf{b} \in C, w(\mathbf{a}+\mathbf{b})=k} K_j(k)$$

$$= \frac{1}{|C|^2} \sum_{i=0}^{n} \sum_{\mathbf{a},\mathbf{b} \in C, w(\mathbf{a}+\mathbf{b})=i} \sum_{\mathbf{y} \in S_j} \psi_{\mathbf{y}}(\mathbf{a}+\mathbf{b})$$

$$= \frac{1}{|C|^2} \sum_{\mathbf{y} \in S_j} \sum_{\mathbf{a} \in C} \sum_{\mathbf{b} \in C} \psi_{\mathbf{y}}(\mathbf{a}+\mathbf{b})$$

$$= \frac{1}{|C|^2} \sum_{\mathbf{y} \in S_j} |\psi_{\mathbf{y}}(C)|^2 \geq 0,$$

and (i) is proved. Moreover, $B_j^\perp = 0$ means that $\psi_\mathbf{y}(C) = 0$ for every $\mathbf{y} \in S_j$. For $j = 1$ this gives that every column in $\mathbf{C}$ has $\frac{K}{2}$ zeros and $\frac{K}{2}$ ones.

If $B_2^\perp = B_1^\perp = 0$, in any two columns all four 2-tuples appear exactly $\frac{K}{4}$ times each. Indeed, consider two arbitrary columns of $\mathbf{C}$. Let $a_{00}, a_{01}, a_{10}, a_{11}$ stand for the number of occurrences of $00, 01, 10$ and 11, respectively, and let $\mathbf{y} \in S_2$ be a vector with its ones in the corresponding coordinates. Then, solving the system

$$a_{00} - a_{01} - a_{10} + a_{11} = 0,$$

$$a_{00} + a_{01} = a_{10} + a_{11} = a_{00} + a_{10} = a_{01} + a_{11} = \frac{K}{2},$$

where the first equality follows from $\psi_\mathbf{y}(C) = 0$, concludes the case $j = 2$.

Clearly, the same can be done up to $d^\perp - 1$. On the other hand, when $j = d^\perp$ there exists a vector $\mathbf{y} \in S_{d^\perp}$ such that $\psi_\mathbf{y}(C) \neq 0$. $\square$

Arrays with the property that all $r \leq s$ columns contain all possible r-tuples an equal number of times are called *orthogonal arrays of strength s* or just *codes of strength s*. Thus, a code with dual distance $d^\perp$ is an orthogonal array of strength $d^\perp - 1$.

3.4.4 Krawtchouk polynomials

Krawtchouk polynomials play a special role in the MacWilliams transform. Actually, Krawtchouk polynomials can be defined for nonbinary alphabets as well. In what follows, we will use an extended definition of the binomial coefficient: for real x and m an integer,

$$\binom{x}{m} = \begin{cases} \frac{x(x-1)\ldots(x-m+1)}{m!} & \text{if } m > 0, \\ 1 & \text{if } m = 0, \\ 0 & \text{otherwise}, \end{cases}$$

where $m! = m(m-1)(m-2)\ldots 1$ and $0! = 1$.

The q-ary Krawtchouk polynomials $K_k^n(x)$ (of degree k) are defined by the following generating function:

$$\sum_{k=0}^{\infty} K_k^n(x) z^k = (1-z)^x (1 + (q-1)z)^{n-x}. \tag{3.61}$$

Usually n is fixed, and is omitted if no confusion arises. An explicit expression for Krawtchouk polynomials is given by

$$K_k^n(x) = \sum_{j=0}^{k} (-1)^j (q-1)^{k-j} \binom{x}{j} \binom{n-x}{k-j}. \tag{3.62}$$

From now on, we consider only binary Krawtchouk polynomials ($q = 2$). Features of nonbinary Krawtchouk polynomials are very similar to those of their binary counterparts.

The properties listed below can be derived by straightforward calculations using (3.61). There are several explicit expressions for Krawtchouk polynomials:

$$K_k^n(x) = \sum_{j=0}^{k} (-1)^j \binom{x}{j}\binom{n-x}{k-j} = \sum_{j=0}^{k} (-2)^j \binom{x}{j}\binom{n-j}{k-j}$$

$$= \sum_{j=0}^{k} (-1)^j 2^{k-j} \binom{n-x}{k-j}\binom{n-k+j}{j}. \tag{3.63}$$

From Cauchy's integral formula we get from (3.61) for nonnegative integer x:

$$K_k(x) = \frac{1}{2\pi x} \oint \frac{(1+z)^{n-x}(1-z)^x}{z^{k+1}} dz$$

$$= \frac{(-\iota)^x}{2\pi} \int_0^{2\pi} e^{\iota(\frac{n}{2}-k)\theta} \left(\cos\frac{\theta}{2}\right)^{n-x} \left(\sin\frac{\theta}{2}\right)^x d\theta.$$

A remarkable property of Krawtchouk polynomials is that they satisfy a linear inductive relation in every variable:

$$(k+1)K_{k+1}^n(x) = (n-2x)K_k^n(x) - (n-k+1)K_{k-1}^n(x), \tag{3.64}$$

$$(n-x)K_k^n(x+1) = (n-2k)K_k^n(x) - xK_k^n(x-1), \tag{3.65}$$

$$(n-k+1)K_k^{n+1}(x) = (3n-2k-2x+1)K_k^n(x) - 2(n-x)K_k^{n-1}(x). \tag{3.66}$$

The first Krawtchouk polynomials are

$$K_0(x) = 1, \quad K_1(x) = n - 2x, \quad K_2(x) = \frac{(n-2x)^2 - n}{2}, \tag{3.67}$$

$$K_3(x) = \frac{(n-2x)\left((n-2x)^2 - 3n + 2\right)}{6}.$$

Here are several values of Krawtchouk polynomials:

$$K_k(0) = \binom{n}{k}, \quad K_k(1) = \left(1 - \frac{2k}{n}\right)\binom{n}{k} \tag{3.68}$$

$$K_k(n/2) = 0, \text{ for } k \text{ odd}; \quad K_k(n/2) = (-1)^{k/2}\binom{n/2}{k/2}, \text{ for } k \text{ even}. \tag{3.69}$$

If $K_k(x) = \sum_{j=0}^{k} c_j x^j$, then

$$c_k = \frac{(-2)^k}{k!}, \quad c_{k-1} = \frac{(-2)^{k-1}n}{(k-1)!}. \tag{3.70}$$

The following relations, derived from the definition of Krawtchouk polynomials by rearranging binomial coefficients, reflect some symmetry with respect to their parameters:

$$\binom{n}{x} K_k(x) = \binom{n}{k} K_x(k), \quad \text{(for a nonnegative integer } x), \tag{3.71}$$

$$K_k(x) = (-1)^k K_k(n - x), \tag{3.72}$$

$$K_k(x) = (-1)^x K_{n-k}(x), \quad \text{(for an integer } x, \ 0 \le x \le n). \tag{3.73}$$

Theorem 3.51 *The following orthogonality relations hold:*

$$\sum_{j=0}^{n} \binom{n}{j} K_k(j) K_\ell(j) = \delta_{k\ell} \binom{n}{k} 2^n, \tag{3.74}$$

$$\sum_{j=0}^{n} K_k(j) K_j(\ell) = \delta_{k\ell} 2^n. \tag{3.75}$$

Proof By (3.61) the left-hand side of (3.74) is the coefficient of $z^k y^\ell$ in

$$\sum_{j=0}^{n} \binom{n}{j} (1 - z)^j (1 + z)^{n-j} (1 - y)^j (1 + y)^{n-j}$$

$$= \sum_{j=0}^{n} \binom{n}{j} \big((1 - z)(1 - y)\big)^j \big((1 + z)(1 + y)\big)^{n-j}$$

$$= \big((1 - z)(1 - y) + (1 + z)(1 + y)\big)^n$$

$$= 2^n (1 + zy)^n,$$

and (3.75) then follows from (3.71).   □

Theorem 3.52 *Every polynomial $\alpha(x)$ of degree s possesses a unique Krawtchouk expansion:*

$$\alpha(x) = \sum_{k=0}^{s} \alpha_k K_k(x), \tag{3.76}$$

where the coefficients are

$$\alpha_k = 2^{-n} \sum_{j=0}^{n} \alpha(j) K_j(k). \tag{3.77}$$

Proof The uniqueness follows from the fact that each Krawtchouk polynomial $K_k(x)$ is of degree k. By (3.75),

$$2^{-n} \sum_{\ell=0}^{n} \alpha(\ell) K_i(k) = 2^{-n} \sum_{\ell=0}^{n} \sum_{j=0}^{s} \alpha_j K_j(i) K_i(k)$$

$$= \sum_{j=0}^{s} \alpha_j \left(2^{-n} \sum_{\ell=0}^{n} K_j(\ell) K_\ell(k) \right) = \alpha_k,$$

proving (3.77). □

In particular, (3.68) yields

$$\alpha_0 = 2^{-n} \sum_{i=0}^{n} \alpha(i) \binom{n}{i}. \tag{3.78}$$

3.4.5 Families of error-correcting codes

The essential parameters of codes are alphabet, length, cardinality (or dimension for linear codes) and minimum distance. There are several methods to construct new codes from given ones. I present a few of them, in their binary version. Assume that C is an (n, K, d) or $[n, k, d]$ code.

Shortening: For a linear code, choose one coordinate and take the subcode of C consisting of the code words having 0 in this position. Deleting the chosen coordinate in every code word of the subcode gives a linear code C° with parameters $[n - 1, \geq k - 1, \geq d]$. For a nonlinear code, pick the largest of the two subcodes of C with either 0 or 1 in this coordinate. The resulting code C° is an $(n - 1, \geq \lceil K/2 \rceil, \geq d)$ code.

Puncturing: If $d \geq 2$, deleting one coordinate gives an $(n - 1, K, \geq d - 1)$ or $[n - 1, k, \geq d - 1]$ code C^*.

Extending: For d odd, adding the overall parity check (sum modulo 2 of all the symbols) to every code word results in the code $\widehat{C}$ having parameters $(n + 1, K, d + 1)$ or $[n + 1, k, d + 1]$.

Now we consider several essential classes of error-correcting codes. In what follows, $q = 2^m$, except for the Reed–Solomon codes where $q = p^m$, p prime.

Cyclic codes: A binary linear code C of length n is *cyclic* if for every $\mathbf{c} = (c_0, c_1, \ldots, c_{n-1}) \in C$, the vector $(c_{n-1}, c_0, \ldots, c_{n-2})$ is also a code word. In what follows, we identify the vector $\mathbf{a} = (a_0, a_1, \ldots, a_{n-1})$ and the polynomial in $\mathbb{F}[x]$ $a(x) = a_0 + a_1 x + \ldots + a_{n-1} x^{n-1}$.

The primitive case: Assume that $n = q - 1$ and α is a primitive element of the field $\mathbb{F}_q$. Remember that two elements α and β in $\mathbb{F}_q$ are *conjugates* if for some integer ℓ, $\alpha = \beta^{2^\ell}$. Pick a subset $\aleph = \{\alpha^{i_1}, \alpha^{i_2}, \ldots, \alpha^{i_s}\}$ of $\mathbb{F}_q$ not containing two conjugate elements.

Then all the binary polynomials of degree at most $q - 2$ whose set of roots contains $\aleph$ (and the conjugates of elements in $\aleph$) constitute a *primitive cyclic code*. To see that the code is cyclic, notice that the polynomial $x^n - 1$ has all nonzero elements of $\mathbb{F}_q$ as its roots, and so $xc(x) \bmod x^n - 1$ has the same set of nonzero roots as $c(x)$. But $xc(x) \bmod x^n - 1$ is the cyclic shift of $c(x)$.

By definition, the cyclic code C consists of all binary words $\mathbf{c}$ of length $n = q - 1$ satisfying $\mathbf{Hc}^T = \mathbf{0}$, where

$$
\mathbf{H} = \begin{pmatrix}
1 & \alpha^{i_1} & \alpha^{2i_1} & \ldots & \alpha^{(n-1)i_1} \\
1 & \alpha^{i_2} & \alpha^{2i_2} & \ldots & \alpha^{(n-1)i_2} \\
\ldots & \ldots & \ldots & \ldots & \ldots \\
1 & \alpha^{i_s} & \alpha^{2i_s} & \ldots & \alpha^{(n-1)i_s}
\end{pmatrix}. \tag{3.79}
$$

The matrix $\mathbf{H}$ is called a parity-check matrix *over* $\mathbb{F}_q$. To obtain the binary parity check matrix, replace every element of $\mathbb{F}_q$ with the corresponding binary column vector of size m. The length of the code is thus $n = 2^m - 1$, and its dimension is at least $n - ms$ (the rows of $\mathbf{H}$ may be dependent!). Encoding of cyclic codes can be easily implemented using shift registers with feedback.

The problem of estimating the minimum distance is difficult in general, but, as we shall see later, easy for some particular choices of $\aleph$.

The nonprimitive case: Let β' be a nonprimitive element of $\mathbb{F}_q$. Then β' is conjugate to $\beta = \alpha^h$ for some factor h of $q - 1$, $h > 1$, where α is a suitable primitive element and

$$
\beta^{(q-1)/h} = 1.
$$

Let all the degrees of α in $\aleph$ be multiples of h,

$$
\aleph = \{\alpha^{hi_1}, \alpha^{hi_2}, \ldots, \alpha^{hi_s}\}
$$
$$
= \{\beta^{i_1}, \beta^{i_2}, \ldots, \beta^{i_s}\}.
$$

Then the matrix

$$
\mathbf{H}(C) = \begin{pmatrix}
1 & \beta^{i_1} & \beta^{2i_1} & \ldots & \beta^{(n-1)i_1} \\
1 & \beta^{i_2} & \beta^{2i_2} & \ldots & \beta^{(n-1)i_2} \\
\ldots & \ldots & \ldots & \ldots & \ldots \\
1 & \beta^{i_s} & \beta^{2i_s} & \ldots & \beta^{(n-1)i_s}
\end{pmatrix} \tag{3.80}
$$

where $n = (2^m - 1)/h$, defines a *nonprimitive cyclic code*. It has length n and dimension $k \geq n - ms$.

Hamming codes: The parity check matrix of the *Hamming code*, $\mathcal{H}_m$, consists of all nonzero columns of size m in some order. The parameters of $\mathcal{H}_m$ are

$$[n = 2^m - 1, k = 2^m - m - 1, d = 3].$$

The codes $\mathcal{H}_m$ are perfect. If we define $\aleph = \{\alpha\}$, for some primitive $\alpha \in \mathbb{F}_q$, then we get the Hamming codes in a cyclic form.

The *extended Hamming codes* have parameters

$$[n = 2^m, k = 2^m - m - 1, d = 4],$$

the *shortened Hamming codes* have parameters

$$[n, k = n - \lfloor \log_2 n \rfloor - 1, d = 3],$$

and the *extended shortened Hamming codes* are

$$[n + 1, k = n - \lfloor \log_2 n \rfloor - 1, d = 4]$$

codes. All these codes have the largest minimum distance among linear codes with the same length and dimension.

Reed–Muller codes: The *Reed–Muller code* $\mathcal{RM}(r, m)$ of order r, $r = 0, 1, \ldots, m$, has parameters

$$\left[n = 2^m, k = \sum_{i=0}^{r} \binom{m}{i}, d = 2^{m-r} \right].$$

The binary Boolean function in m variables $x_1, \ldots, x_m$, of degree r is the sum of products of variables such that in any product there are at most r variables, and there is a product with exactly r variables. For example, $x_1 x_3 x_4 + x_2 x_5 + x_1 x_6$ is a Boolean function in six variables of degree 3. Evaluation of a Boolean function $f(x_1, \ldots, x_m)$ is a vector of length 2^m having at the jth position, $j = 0, 1, \ldots, 2^m - 1$, the value of $f(j_0, \ldots, j_{m-1})$ where $j_0, \ldots, j_{m-1}$ are the bits of the binary expansion of j. The code $\mathcal{RM}(r, m)$ is defined as the set of vectors being evaluations of Boolean functions in m variables of degree at most r.

The dual code of $\mathcal{RM}(r, m)$ is $\mathcal{RM}(m - r - 1, m)$. The Reed–Muller codes constitute a family of nested codes, namely,

$$\mathcal{RM}(0, m) \subset \mathcal{RM}(1, m) \subset \ldots \subset \mathcal{RM}(m, m).$$

Particular cases of Reed–Muller codes coincide with some of the aforementioned codes:

$\mathcal{RM}(0, m)$ is the repetition $[2^m, 1, 2^m]$ code;
$\mathcal{RM}(m - 2, m)$ is the extended Hamming code;
$\mathcal{RM}(m - 1, m)$ is the even-weight $[2^m, 2^m - 1, 2]$ code;
$\mathcal{RM}(m, m)$ is the $[2^m, 2^m, 1]$ code consisting of all possible words of length 2^m.

First-order Reed–Muller codes: The codes $\mathcal{RM}(1, m)$ form a class of codes having parameters

$$[n = 2^m, k = m + 1, d = 2^{m-1}].$$

These codes contain **0**, **1** and $2^{m+1} - 2$ code words of weight 2^{m-1}. The first-order Reed–Muller codes can be efficiently decoded using a fast Walsh–Hadamard transform.

Punctured Reed–Muller codes: The code $\mathcal{RM}^*(r, m)$ can be presented in a cyclic form. The set of zeros $\aleph$ consists of all α^i, $1 \le i \le 2^m - 2$, such that $w_2(i)$, the number of ones in the binary expansion of i, satisfies the inequality

$$1 \le w_2(i) \le m - r - 1.$$

The parameters of $\mathcal{RM}^*(r, m)$ are

$$\left[n = 2^m - 1, k = \sum_{i=0}^{r} \binom{m}{i}, d = 2^{m-r} - 1 \right].$$

Simplex codes = M-sequences: The *simplex codes* $\mathcal{SIM}_m$ can be obtained by shortening the code $\mathcal{RM}(1, m)$, and have parameters

$$[n = 2^m - 1, k = m, d = 2^{m-1}].$$

All nonzero code words in $\mathcal{SIM}_m$ have the same weight, 2^{m-1}. These are cyclic codes with

$$\aleph = \mathbb{F}_q^* \setminus \{\alpha^1, \alpha^2, \alpha^4, \ldots\}.$$

The codes $\mathcal{SIM}_m$ are the duals of the Hamming codes $\mathcal{H}_m$ and have generator matrices consisting of all the nonzero columns of size m. The simplex codes are also called M-sequences.

Primitive BCH codes: The (narrow-sense) *primitive BCH codes* $\mathcal{BCH}(e, m)$ are cyclic codes with

$$\aleph = \{\alpha^1, \alpha^3, \ldots, \alpha^{2e-1}\}.$$

Their parameters are

$$[n = 2^m - 1, k \ge n - me, d \ge 2e + 1].$$

The BCH codes constitute a nested family of codes:

$$\mathcal{BCH}(e+1,m) \subseteq \mathcal{BCH}(e,m).$$

The BCH codes contain punctured Reed–Muller codes as subcodes, namely, for every $e \leq 2^i - 1$

$$\mathcal{RM}^*(m-i-1,m) \subseteq \mathcal{BCH}(e,m).$$

If $2e - 2 < 2^{\lceil m/2 \rceil}$, then the dimension of $\mathcal{BCH}(e,m)$ is exactly $n - me$.

Nonprimitive BCH codes: If $n = (2^m - 1)/h$ and β is an nth root of unity in $\mathbb{F}_q$, then a (narrow-sense) *nonprimitive BCH code* $\mathcal{BCH}_h(e,m)$ is defined as a nonprimitive cyclic code with $\aleph = \{\beta^1, \beta^3, \ldots, \beta^{2e-1}\}$. Its parameters are

$$[n = (2^m - 1)/h, k \geq n - me, d \geq 2e + 1].$$

Duals of BCH codes: Let $\alpha \in \mathbb{F}_q$ be a primitive element. Then $\mathcal{BCH}^\perp(e,m)$ consists of all vectors

$$\left(Tr(f(0)), Tr(f(1)), Tr(f(\alpha)), \ldots, Tr(f(\alpha^{q-2}))\right),$$

where $f(x) \in \mathbb{F}_q[x]$ is of the form

$$f(x) = a_1 x + a_2 x^3 + \ldots + a_e x^{2e-1}.$$

The minimum distance of $\mathcal{BCH}^\perp(e,m)$ can be estimated by using the Carlitz–Uchiyama bound, namely,

$$d\left(\mathcal{BCH}^\perp(e,m)\right) \geq 2^{m-1} - (e-1)2^{m/2}.$$

Encoding of dual BCH codes can be easily implemented using linear shift registers with feedback.

Reed–Solomon codes: These are q-ary codes, with $q = p^m$, p prime. One way of defining the *Reed–Solomon code* $\mathcal{RS}(k,q)$ is by its polynomial representation: for every q, $\mathcal{RS}(k,q)$ consists of vectors

$$\left(f(1), f(\alpha), f(\alpha^2), \ldots, f(\alpha^{q-2})\right),$$

where $f(x) \in \mathbb{F}_q[x]$ runs through all polynomials of, at most, degree $k - 1$. This code has parameters

$$[n = q - 1, k, d = n - k + 1]_q.$$

If we add an extra coordinate containing $f(0)$, we get the *extended Reed–Solomon code* with parameters

$$[n = q, k, d = n - k + 1]_q.$$

It is possible to extend the Reed–Solomon code further, thus getting the *doubly extended Reed–Solomon code* with parameters

$$[n = q + 1, k, d = n - k + 1]_q.$$

If $q = 2^m$, there exist *triply extended Reed–Solomon codes* with parameters

$$[n = q + 2, k = 3, d = q]_q,$$

and

$$[n = q + 2, k = q - 1, d = 4]_q.$$

For every $n < q - 1$, there exist *shortened Reed–Solomon codes* with parameters

$$[n, k, d = n - k + 1]_q.$$

All these codes achieve the maximum possible minimum distance for a given length and size by the Singleton bound:

Theorem 3.53 (Singleton bound)   *For every $(n, K, d)_q$ code,*

$$d \leq n - \log_q K + 1.$$

Proof   Puncturing an $(n, K, d)_q$ code $d - 1$ times yields a code of length $n - d + 1$, minimum distance at least one, and cardinality, K.   □

Codes achieving the Singleton bound are called *maximum distance separable* (MDS), so the Reed–Solomon codes are MDS.

The Reed–Solomon and extended Reed–Solomon codes constitute a nested family of codes, namely,

$$\mathcal{RS}(k, q) \subset \mathcal{RS}(k + 1, q).$$

Hadamard codes: A ± 1-matrix $\mathbf{H}_n^{\pm}$ is a *Hadamard $n \times n$ matrix* if

$$\mathbf{H}_n^{\pm}(\mathbf{H}_n^{\pm})^T = n\,\mathbf{I}_n.$$

Hadamard matrices conjecturally exist for every n being a multiple of 4.

A *Hadamard code* $\mathcal{HAD}_n$ consists of the rows of $\mathbf{H}_n^{\pm}$ and their complements after substitutions $1 \to 0$ and $-1 \to 1$. The parameters of $\mathcal{HAD}_n$ are

$$(n, K = 2n, d = n/2).$$

The code $\mathcal{RM}(1, m)$ is a Hadamard code for $n = 2^m$. I define here $\mathbf{A} \otimes \mathbf{B}$, the *Kronecker product* of two square matrices $\mathbf{A} = (a_{i,j})$ and $\mathbf{B}$ of dimension n_A and n_B, respectively: $\mathbf{A} \otimes \mathbf{B}$ is the square matrix of dimension $n_A n_B$ obtained from $\mathbf{A}$ by replacing every entry $a_{i,j}$ by $a_{i,j}\mathbf{B}$. The matrix $\mathbf{H}_{n_1} \otimes \mathbf{H}_{n_2}$ is also a Hadamard matrix of dimension $n_1 n_2$. Hadamard matrices of size 2^m can be obtained as an m-times Kronecker product $\mathbf{H}_2 \otimes \mathbf{H}_2 \otimes \ldots \otimes \mathbf{H}_2$ of the matrices

$$\mathbf{H}_2 = \begin{pmatrix} 1 & 1 \\ 1 & -1 \end{pmatrix}.$$

Convolutional codes: In contrast to block codes, where the information is partitioned to information vectors of length k and its $n - k$ parity bits depend only on these k bits, in the convolutional codes the redundant bits depend on some prehistory. Namely, let G_j, $j = 0, 1, \ldots, m$, be $k_0 \times n_0$ matrices. Then the corresponding convolutional code is defined by a semi-infinite matrix

$$G = \begin{pmatrix} G_0 & G_1 & G_2 & \ldots & G_m & 0 & 0 & 0 & \ldots \\ 0 & G_0 & G_1 & G_2 & \ldots & G_m & 0 & 0 & \ldots \\ 0 & 0 & G_0 & G_1 & G_2 & \ldots & G_m & 0 & \ldots \\ \ldots & \ldots & \ldots & \ldots & \ldots & \ldots & \ldots & \ldots & \ldots \\ \ldots & \ldots & \ldots & \ldots & \ldots & \ldots & \ldots & \ldots & \ldots \end{pmatrix}.$$

Convolutional codes can easily be encoded using shift registers. Efficient decoding can be done by application of the Viterbi algorithm.

Product codes: The information vector is placed in a rectangular table, and the redundancy is added by encoding the rows and columns of the table. Iterative decoding in rows and columns can be applied.

Low-density parity-check (LDPC) codes: Regular LDPC codes are defined by parity-check matrices with fixed small numbers of ones in each row and column. The concept can be generalized to the irregular case when the spectrum of the values of row and column sums for the parity-check matrix is prescribed. These codes have efficient iterative decoding algorithms.

Turbo codes: In these codes two redundancy sequences are generated, one directly from the information vector, and one from an interleaved information vector. In decoding, the two redundant parts are used in turn to update the likelihoods of the components of the information vector.

Now we consider algebraic codes over Galois rings. Let the size of the alphabet be 2^e, $e \geq 1$.

Reed–Muller codes over $\mathbb{Z}_{2^e}$: A generalized Boolean function in m binary variables is a function having values in $\mathbb{Z}_{2^e}$. Any such function can be uniquely expressed as a linear combination over $\mathbb{Z}_{2^e}$ of the monomials

$$1, x_1, \ldots, x_m, x_1 x_2, x_1 x_3, \ldots, x_{m-1} x_{m-2}, \ldots, x_1 x_2 \ldots x_m,$$

where the coefficient of each monomial belongs to $\mathbb{Z}_{2^e}$. The degree of a function is the maximum number of variables present in a monomial in the linear combination. Then the rth order Reed–Muller code $\mathcal{RM}_{2^e}$ is the set of vectors that are evaluations of the generalized Boolean functions of degree at most r. Analogously, for $e > 1$, the code $\mathcal{ZRM}_{2^e}(r, m)$ is the set of evaluations of generalized Boolean functions composed of monomials of degree at most $r - 1$ and two-times monomials of degree r. The code $\mathcal{RM}(r, m)$ contains $2^{e \sum_{j=0}^{r} \binom{m}{j}}$ code words, while $\mathcal{ZRM}_{2^e}(r, m)$ contains $2^{e \sum_{j=0}^{r-1} \binom{m}{j}} \cdot 2^{(e-1)\binom{m}{r}}$ vectors. The minimum Hamming distance of both codes is 2^{m-r}, while the Lee distance is 2^{m-r} for $\mathcal{RM}(r, m)$ and 2^{m-r+1} for $\mathcal{ZRM}_{2^e}(r, m)$.

Let $f \in R_{e,m}[x]$ be a polynomial. Let β be a generator for the cyclic subgroup of $\mathcal{T}_{e,m}^*$. With f we associate a length $n = 2^m - 1$ vector $\mathbf{c}_f$ whose components are

$$(\mathbf{c}_f)_k = e^{2\pi \iota \frac{\text{Tr}(f(\beta^k))}{2^e}}.$$

Setting $e = 2$, we consider the following codes.

Kerdock codes over $\mathbb{Z}_4$:

$$\mathcal{K} = \{\mathbf{c}_f : f(x) = b_0 x, \quad b_0 \in R_{2,m}\},$$

Delsarte–Goethals codes over $\mathbb{Z}_4$:

$$\mathcal{DG}_t = \left\{ \mathbf{c}_f : f(x) = b_0 x + 2 \sum_{j=1}^{t} b_j x^{1+2^j}, \quad b_0 \in R_{2,m}, b_j \in \mathcal{T}_{2,m} \right\}.$$

Clearly, the quaternary Kerdock codes correspond to the case $t = 0$ in the Delsarte–Goethals codes. The length and the number of words in the Delsarte–Goethals codes are $2^m - 1$ and $2^{(2+t)m}$, respectively.

The minimum Lee distance of the quaternary codes $\mathcal{K}$ and $\mathcal{DG}_t$ is $2^m - 2^{\frac{m}{2}}$ and $2^m - 2^{t+\frac{m}{2}}$, respectively. To prove this, we use Theorem 3.38 and the fact that nonzero code words of the codes are obtained from nondegenerate polynomials having weighted degrees 2 and $2^t + 1$. These bounds on minimum distances can be improved for odd m [156] to obtain the quaternary codes' true minimum Lee distances of $2^m - 2^{\frac{m-1}{2}}$ and $2^m - 2^{t+\frac{m-1}{2}}$ respectively. The quaternary codes can be lengthened by adding a coordinate corresponding to $f(0)$ (an overall parity check) and then adding modulo 4 to every code word multiple of the all-one code word. When m is off, the images under the Gray map of the lengthened quaternary codes are the binary Kerdock and Delsarte–Goethals codes [149].

Weighted degree trace codes: For $t \geq 1$ satisfying $2t - 1 < 2^{\lceil \frac{m}{2} \rceil} + 1$, the code is defined by

$$\mathcal{WD}_t = \left\{ \mathbf{c}_f : f \in R_{e,m}[x], \quad f = \sum_{j=0}^{t-1} f_{2j+1} x^{2j+1}, \quad D_f \leq 2t - 1 \right\}.$$

The code $\mathcal{WD}_t$ has length $n = 2^m - 1$, and for $e = 1$ coincides with $\mathcal{BCH}_t^{\perp}$. It can be shown, using 2-adic expansions and simple counting, that when $e = 2$,

$$|\mathcal{WD}_t| = 2^{(2t-1-\lceil \frac{2t-1}{4} \rceil)m}.$$

Theorem 3.39 can be applied to show that the minimum Lee distance of the quaternary code ($e = 2$) is at least $2^m - (2t - 2)2^{\frac{m}{2}}$.

3.5 Fast computation of the maximum of DFT

One of the most important computational tasks in the peak power control is fast estimation of the maximum of the absolute value of the signal. As it will be shown in the next chapter, this can be done by calculating the maximum of a relevant DFT. The formal statement of the problem is as follows: given an n-vector of complex numbers, $\mathbf{a} = (a_0, a_1, \ldots, a_{n-1})$, find

$$M(\mathbf{a}) = \max_{j=0,1,\ldots,n} \left| \sum_{k=0}^{n-1} a_k e^{2\pi \imath \frac{j}{n}} \right|.$$

The standard approach is first to compute the result of the transform (Fourier spectrum), and then to find the maximum among the absolute values of the spectral components. However, notice that to compute the maximum we do not need to know all the spectral components. This can be used for further simplification of the algorithm.

We start with a description of the standard algorithm. It is based on the Cooley–Tukey FFT procedure, see, e.g., [29]. Let $n = n_1 \cdot n_2$.

FFT algorithm *Input*: $\mathbf{a} = (a_0, a_1, \ldots, a_{n-1})$; *Output*: $M(\mathbf{a})$.

1. Arrange the entries of $\mathbf{a}$ in matrix $G = [g_{k,j}]$, $k = 0, 1, \ldots, n_1 - 1$; $j = 0, 1, \ldots, n_2 - 1$; as follows:

$$G = \begin{pmatrix} a_0 & a_{n_1} & \cdots & a_{n_2 n_1 - n_1} \\ a_1 & a_{n_1+1} & \cdots & a_{n_2 n_1 - n_1 + 1} \\ \cdots & \cdots & \cdots & \cdots \\ a_{n_1-1} & a_{2n_1-1} & \cdots & a_{n_2 n_1 - 1} \end{pmatrix}.$$

2. Implement DFT of size n_1 on the columns of G. The resulting matrix $T = [t_{k,j}]$, $k = 0, 1, \ldots, n_1 - 1$; $j = 0, 1, \ldots, n_2 - 1$, is computed as follows:

$$t_{k,j} = \sum_{\ell=0}^{n_1-1} g_{\ell,j} e^{2\pi i \ell \frac{k}{n_1}}.$$

3. Compute matrix $R = [r_{k,j}]$, $k = 0, 1, \ldots, n_1 - 1$; $j = 0, 1, \ldots, n_2 - 1$, as follows:

$$r_{k,j} = t_{k,j} e^{2\pi i \frac{kj}{n}}.$$

4. Implement DFT of size n_2 on the rows of R. The resulting matrix $F = [f_{\ell,j}]$, $\ell = 0, 1, \ldots, n_1 - 1$; $j = 0, 1, \ldots, n_2 - 1$, is computed as follows:

$$f_{\ell,j} = \sum_{j=0}^{n_2-1} r_{\ell,j} e^{2\pi i j \frac{\ell}{n_2}}.$$

5. Output

$$M = \max_{\ell=0,1,\ldots,n_1; j=0,1,\ldots,n_2} |f_{\ell,j}|.$$

6. Stop.   □

The algorithm requires $n(n_1 + n_2 + 1)$ complex additions, $n(n_1 + n_2 - 1)$ complex multiplications, and $n - 1$ comparisons. Recurrent use of the factors of n_1 and n_2 leads to further simplification. Especially simple is the case of n being a power of 2. Then, choosing $n_1 = 2$ at each recursive step, we arrive at an algorithm of complexity of order $n \log n$.

Further simplification can be achieved by truncating the results of intermediate steps of the FFT. Indeed, each of the spectrum elements depends on all the entries of the vector **a**. However, if one looks at the results of an intermediate step of the FFT, each of the spectrum elements depends only on a part of them. Taking the sum of the squares of absolute values of the intermediate results influencing a particular spectrum element as an (upper) estimate of the square of its absolute value, we may discard intermediate results having a small impact.

A variant of such an algorithm [15] is presented in what follows. Let $n = n_1 n_2 \ldots n_m$, and, for $k = 0, 1, \ldots, m - 1$, denote $q_k = n_{k+1} \ldots n_m$. We have $n = q_0$, and assume $q_m = 1$.

Truncated FFT algorithm *Input*: $\mathbf{a} = (a_0, a_1, \ldots, a_{n-1})$; *Output*: M.

1. For $\ell = 1$ up to m, perform steps 2–6.

2. Arrange the entries of **a** in matrix $G = [g_{k,j}]$, $k = 0, 1, \ldots, n_\ell - 1$, $j = 0, 1, \ldots, q_\ell - 1$,

$$G = \begin{pmatrix} a_0 & a_1 & \cdots & a_{q_\ell-1} \\ a_{q_\ell} & a_{q_\ell+1} & \cdots & a_{2q_\ell-1} \\ \cdots & \cdots & \cdots & \cdots \\ a_{(n_\ell-1)q_\ell} & a_{(n_\ell-1)q_\ell+1} & \cdots & a_{n_\ell q_\ell-1} \end{pmatrix}.$$

3. Implement the DFT of dimension n_ℓ on the columns of G. The resulting matrix $T = [t_{k,j}]$, $k = 0, 1, \ldots, n_\ell$, $j = 0, 1, \ldots, q_\ell - 1$, is computed as follows:

$$t_{k,j} = \sum_{r=0}^{n_\ell-1} g_{r,j} e^{2\pi i r \frac{k}{n_\ell}}.$$

4. Compute the $\mathbf{p} = (p_0, p_1, \ldots, p_{n_\ell-1})$,

$$p_k = \sum_{j=0}^{q_\ell-1} t_{k,j} t_{k,j}^*.$$

5. Find $b_\ell \in \{0, 1, \ldots, q_\ell - 1\}$, the index of the maximal entry in **p**. If $\ell = m$ go to step 7.
6. Compute the new vector $\mathbf{a} = (a_0, a_1, \ldots, a_{q_\ell-1})$,

$$a_j = t_{b_\ell,j} e^{2\pi i b_\ell \frac{j}{n}}.$$

7. Output $M = p_{b_m}$.
8. Stop. □

The algorithm requires

$$\sum_{j=1}^{m} (2n_j + 1) q_i$$

complex multiplications,

$$\sum_{j=1}^{m} ((n_j - 1)q_j + (q_j - 1)n_j)$$

complex additions, and

$$\sum_{j=1}^{m} (n_j - 1)$$

comparisons. In particular, for $n = 2^m$ there are $5n - 5$ complex multiplications, $3n - 2m - 3$ complex additions, and m comparisons, i.e., the complexity is linear in n. Recall that the FFT's complexity is of order $n \log n$. In contrast with the FFT algorithm the complexity of the truncated algorithm depends on the ordering of the factors of n. Direct verification shows that to minimize the complexity it is necessary to have increasing orders of factors.

Actually this algorithm does not always give the right value of the maximum. However, there is experimental evidence that it is correct for most of the sequences. Further modifications of the algorithm are possible. For example, one could discard only those intermediate results that have the value of p_j (see step 4 of the truncated algorithm) below some threshold and continuing with the rest of the intermediate results. Since p_i is an upper bound on the square of the absolute value of the corresponding spectrum elements, such an algorithm guarantees detection of the maximum exceeding the threshold. Another option is to use some other function in step 4 of the algorithm, e.g. instead of $t_{k,j}t_{k,j}^*$ one could use $(t_{k,j}t_{k,j}^*)^s$ for some positive, not necessarily integer, s. The analysis of performance of the mentioned algorithms (complexity as a function of the error probability) is an open problem.

3.6 Notes

Section 3.1 Introductory textbooks on harmonic analysis are Katznelson [198] and Deitmar [89]. For signal processing aspects of harmonic analysis see Oppenheim *et al.* [307]. Generalizations of the Parseval identity are discussed in Montgomery [275]. For inequalities see Beckenbach and Bellman [22]. The Bernstein inequality and other polynomial inequalities are treated in Pólya and Szegö [331], Borwein and Erdélyi [38], and references therein. The theory of Chebyshev polynomials can be found in Szegö [392] and Mason [258].

Section 3.2 Standard texts on probability are Feller [110] and Papoulis and Pillai [313]. Although the theory of stochastic processes will be used in what follows, I have omitted introduction to it since it can be found in standard engineering textbooks, e.g. Proakis [335], Proakis and Salehi [336], and Wong and Hajek [432]. A mathematical theory of stochastic processes is presented, e.g., in Papoulis and Pillai [313] and Ross [345].

Section 3.3 A general introduction to algebraic structures can be found in Dummit and Foote [98] and Anderson [4]. For the theory of finite fields, see McEliece [263] and Lidl and Niederreiter [242]. Exponential sums over finite fields are considered in Schmidt [358]. Galois rings are treated in Wan [424]. Exponential sums over rings were considered by Kumar *et al.* [224], Helleseth *et al.* [156], and Ling and Ozbudak [245].

Section 3.4 For the theory of error correcting codes see MacWilliams and Sloane [257], van Lint [247], and Huffman and Pless [169]. A more practically oriented source is Lin and Costello [244]. The theory of convolutional codes can be found in Johannesson and Zigangirov [185]. Codes over rings are considered by Hammons *et al.* [149] and Wan and Wan [425]. A survey of properties of Krawtchouk polynomials can be found in Krasikov and Litsyn [217].

Section 3.5 On the general theory of fast orthogonal transforms see, e.g., Blahut [29] and Bracewell [42]. The truncated FFT algorithm was proposed by Ashikhmin and Litsyn [15].

4

Discrete and continuous maxima in MC signals

In many situations it is beneficial to deal with a discrete-time "sampled" version of multicarrier signals. This reduction allows passing from the continuous setting to an easier-to-handle discrete one. However, we have to estimate the inaccuracies stemming from the approach. In this chapter, I analyze the ratio between the maximum of the absolute value of a continuous MC signal and the maximum over a set of the signal's samples. We start with considering the ratio when the signal is sampled at the Nyquist frequency, i.e. the number of sampling points equals the number of tones. In this case I show that the maximum of the ratio over all MC signals grows with the number of subcarriers (Theorem 4.2). However, if one computes a weighted sum of the maximum of the signal's samples and the maximum of the signal derivative's samples the ratio already is, at most, a constant (Theorem 4.5). I further show that actually the ratio depends on the maximum of the signal; the larger the maximum is the smaller is the ratio (Theorem 4.6). An even better strategy is to use oversampling. Then the ratio becomes constant tending to 1 when the oversampling rate grows (Theorems 4.8, 4.9, 4.10, and 4.11). Furthermore, I tackle the case when we have to use the maximum estimation, projections on specially chosen measuring axes instead of the absolute values of the signal (Theorem 4.14). Finally, I address the problem of relation between the PAPR and the PMEPR and show that the PMEPR estimates the PAPR quite accurately for large values of the carrier frequency (Theorem 4.19).

4.1 Nyquist sampling

For an MC signal,

$$F_{\mathbf{a}}(t) = \sum_{k=0}^{n-1} a_k e^{2\pi \iota k t},$$ (4.1)

68

where $a_k \in \mathbb{C}$, for $k = 0, \ldots, n-1$, let

$$\mathcal{M}_d(F_{\mathbf{a}}) = \max_{j=0,\ldots,n-1} \left| F_{\mathbf{a}}\left(\frac{j}{n}\right) \right|, \tag{4.2}$$

$$\mathcal{M}_c(F_{\mathbf{a}}) = \max_{t \in [0,1)} |F_{\mathbf{a}}(t)|, \tag{4.3}$$

be the discrete and continuous maxima of $F_{\mathbf{a}}(t)$, respectively. Clearly,

$$\mathcal{M}_c(F_{\mathbf{a}}) \geq \mathcal{M}_d(F_{\mathbf{a}}),$$

i.e. the continuous maximum is not less than the discrete one. We address the problem of estimating the maximum ratio between the two maxima.

For the sake of simplicity, we consider signals of constant energy, i.e. satisfying

$$\sum_{k=0}^{n-1} |a_k|^2 = n. \tag{4.4}$$

We will need the following simple lower bound on $\mathcal{M}_d(F_{\mathbf{a}})$.

Lemma 4.1

$$\mathcal{M}_d(F_{\mathbf{a}}) \geq \sqrt{n}. \tag{4.5}$$

Proof By the Parseval identity (3.3),

$$n = \sum_{k=0}^{n-1} |a_k|^2 = \frac{1}{n} \sum_{j=0}^{n-1} \left| F_{\mathbf{a}}\left(\frac{j}{n}\right) \right|^2 \leq \frac{1}{n} \cdot n(\mathcal{M}_d(F_{\mathbf{a}}))^2$$

and the claim follows. □

We will extensively apply an interpolation of $F_{\mathbf{a}}(t)$ using its discrete samples. Let us first transform the expression for $F_{\mathbf{a}}(t)$,

$$F_{\mathbf{a}}(t) = \sum_{k=0}^{n-1} a_k e^{2\pi \imath k t} = \sum_{k_1=0}^{n-1} \sum_{k_2=0}^{n-1} a_{k_1} e^{2\pi \imath k_2 t} \cdot \frac{1}{n} \sum_{j=0}^{n-1} e^{2\pi \imath j \frac{(k_1 - k_2)}{n}}.$$

Here we used the orthogonality relation (see Theorem 3.1). This yields

$$F_{\mathbf{a}}(t) = \frac{1}{n} \sum_{j=0}^{n-1} \left(\sum_{k_1=0}^{n-1} a_{k_1} e^{2\pi \imath k_1 \frac{j}{n}} \right) \cdot \left(\sum_{k_2=0}^{n-1} e^{2\pi \imath k_2 \left(t - \frac{j}{n}\right)} \right)$$

$$= \frac{1}{n} \sum_{j=0}^{n-1} F_{\mathbf{a}}\left(\frac{j}{n}\right) D_n\left(t - \frac{j}{n}\right), \tag{4.6}$$

and here

$$D_n(t) = \sum_{k=0}^{n-1} e^{2\pi \imath k t} \tag{4.7}$$

is called the *Dirichlet kernel*. Figure 4.1 shows a typical behavior of $|D_n(t)|$.

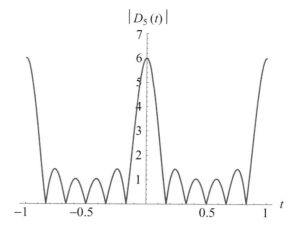

Figure 4.1 Typical behavior of the absolute value of the Dirichlet kernel

Clearly,

$$|D_n(t)| = \left| \sum_{k=0}^{n-1} e^{2\pi \imath kt} \right| = \left| \frac{e^{2\pi \imath nt} - 1}{e^{2\pi \imath t} - 1} \right| = \left| \frac{2\imath e^{\pi \imath nt}(e^{\pi \imath nt} - e^{-\pi \imath nt})}{2\imath e^{\pi \imath t}(e^{\pi \imath t} - e^{-\pi \imath t})} \right|$$

$$= \left| \frac{(e^{\pi \imath nt} - e^{-\pi \imath nt})}{2\imath} \cdot \frac{2\imath}{(e^{\pi \imath t} - e^{-\pi \imath t})} \right| = \left| \frac{\sin \pi nt}{\sin \pi t} \right|. \qquad (4.8)$$

The following result shows that there are MC signals having the ratio between the continuous and discrete maxima growing with n.

Theorem 4.2 *For $n > 3$,*

$$\frac{2}{\pi} \ln n + 0.603 - \frac{1}{6n} < \max_{F_a} \left\{ \frac{\mathcal{M}_c(F_a)}{\mathcal{M}_d(F_a)} \right\} < \frac{2}{\pi} \ln n + 1.132 + \frac{4}{n}. \qquad (4.9)$$

□

The proof consists of two parts in which we obtain the upper and the lower bounds from (4.9).

4.1.1 Upper bound

We start with the upper bound. In the next section it will be shown that this upper bound cannot be essentially improved.

Lemma 4.3   *For every MC signal $F_a(t)$, $n > 3$,*

$$\mathcal{M}_c(F_a) < \left(\frac{2}{\pi}\ln n + 1.132 + \frac{4}{n}\right)\mathcal{M}_d(F_a). \tag{4.10}$$

Proof   By (4.6),

$$|F_a(t)| \leq \max_{j=0,\ldots,n-1}\left|F_a\left(\frac{j}{n}\right)\right| \cdot \frac{1}{n}\sum_{j=0}^{n-1}\left|D_n\left(t-\frac{j}{n}\right)\right|, \tag{4.11}$$

and thus

$$\mathcal{M}_c(F_a) \leq \mathcal{M}_d(F_a)\max_{t=[0,1]}\frac{1}{n}\sum_{j=0}^{n-1}\left|D_n\left(t-\frac{j}{n}\right)\right|. \tag{4.12}$$

So our problem reduces to estimating

$$\phi(n) = \max_t \frac{1}{n}\sum_{j=0}^{n-1}\left|D_n\left(t-\frac{j}{n}\right)\right|. \tag{4.13}$$

In what follows we will derive an asymptotically exact bound on $\phi(n)$. Though an upper bound on $\phi(n)$ is our goal, we start with a derivation of a lower bound. This will allow us to demonstrate the accuracy of our approach, as well as an argument used in its derivation that will be employed later.

Lower bound on $\phi(n)$. From the definition (4.7) of D_n, evaluating $\phi(n)$ at $t = \frac{1}{2n}$ we obtain using (4.8),

$$\phi(n) \geq \frac{1}{n}\cdot\sum_{j=0}^{n-1}\left|D_n\left(\frac{1}{2n}-\frac{j}{n}\right)\right| = \frac{1}{n}\cdot\sum_{j=0}^{n-1}\left|\frac{\sin \pi n\left(\frac{1}{2n}-\frac{j}{n}\right)}{\sin \pi\left(\frac{1}{2n}-\frac{j}{n}\right)}\right|$$

$$= \frac{1}{n}\cdot\sum_{j=0}^{n-1}\frac{1}{\left|\sin \pi\left(\frac{1}{2n}-\frac{j}{n}\right)\right|} = \frac{1}{n}\cdot\sum_{j=0}^{n-1}\frac{1}{\sin \pi\left(\frac{1}{2n}+\frac{j}{n}\right)}$$

$$= \begin{cases} \frac{2}{n}\cdot\sum_{j=0}^{(n-2)/2}\frac{1}{\sin \pi\left(\frac{1}{2n}+\frac{j}{n}\right)}, & \text{if } n \text{ even,} \\[2mm] \frac{1}{n}\cdot\left(1 + 2\cdot\sum_{j=0}^{(n-3)/2}\frac{1}{\sin \pi\left(\frac{1}{2n}+\frac{j}{n}\right)}\right), & \text{if } n \text{ odd.} \end{cases} \tag{4.14}$$

Let us find the error if we replace the sines in (4.14) with their arguments. We will use the following two inequalities (see Theorem 3.17), which are valid for $|x| \leq \frac{1}{2}$,

$$|\sin \pi x - \pi x| \leq \frac{\pi^3 x^3}{6}$$

and

$$|\sin \pi x| \geq 2|x|.$$

Thus we obtain (for $a = 2$ if n is even, or 3 if n is odd)

$$\left| \sum_{j=0}^{(n-a)/2} \left(\frac{1}{\sin \pi \left(\frac{1}{2n} + \frac{j}{n} \right)} - \frac{1}{\pi \left(\frac{1}{2n} + \frac{j}{n} \right)} \right) \right|$$

$$\leq \sum_{j=0}^{(n-a)/2} \left| \frac{\pi \left(\frac{1}{2n} + \frac{j}{n} \right) - \sin \pi \left(\frac{1}{2n} + \frac{j}{n} \right)}{\pi \left(\frac{1}{2n} + \frac{j}{n} \right) \cdot \sin \pi \left(\frac{1}{2n} + \frac{j}{n} \right)} \right|$$

$$\leq \sum_{j=0}^{(n-a)/2} \frac{\pi^3 \left(\frac{1}{2n} + \frac{j}{n} \right)^3}{6 \cdot 2 \cdot \left(\frac{1}{2n} + \frac{j}{n} \right) \cdot \pi \left(\frac{1}{2n} + \frac{j}{n} \right)}$$

$$= \frac{\pi^2}{24n} \cdot \sum_{j=0}^{(n-a)/2} (1 + 2j) = \begin{cases} \frac{n\pi^2}{96}, & \text{if } n \text{ even,} \\ \frac{(n-1)^2\pi^2}{96n}, & \text{if } n \text{ odd.} \end{cases} \qquad (4.15)$$

Thus,

$$\sum_{j=0}^{(n-a)/2} \frac{1}{\sin \pi \left(\frac{1}{2n} + \frac{j}{n} \right)} \geq -\frac{n\pi^2}{96} + \frac{2n}{\pi} \sum_{j=0}^{(n-a)/2} \frac{1}{1 + 2j}. \qquad (4.16)$$

We use the following known inequality,

$$\sum_{j=0}^{m} \frac{1}{2j+1} \geq \ln(2m+1) - \frac{1}{2} \ln m + \frac{\gamma}{2} - \frac{1}{16m},$$

where $\gamma = 0.577216\ldots$ is the Euler constant. From (4.14) and (4.16), we obtain, for even n,

$$\phi(n) \geq \frac{2}{\pi} \left(2 \ln(n-1) - \ln(n-2) + \ln 2 + \gamma - \frac{1}{4(n-2)} \right) - \frac{\pi^2}{48},$$

and, for odd n,

$$\phi(n) \geq \frac{2}{\pi} \left(2 \ln(n-2) - \ln(n-3) + \ln 2 + \gamma - \frac{1}{4(n-3)} \right) + \frac{1}{n} - \frac{\pi^2}{48}.$$

Now we apply the following easy-to-check inequalities

$$2 \ln(n-1) - \ln(n-2) > \ln n,$$

$$2 \ln(n-2) - \ln(n-3) > \ln n - \frac{1}{n-1},$$

$$\frac{1}{2\pi(n-2)} < \frac{1}{6n},$$

and arrive at the bound valid for even and odd n,

$$\phi(n) \geq \frac{2}{\pi} \ln n + 0.6031 - \frac{1}{6n}. \tag{4.17}$$

Upper bound on $\phi(n)$. Indeed, since $|D_n(t)|$ is even in t,

$$
\begin{aligned}
\phi(n) &= \max_t \frac{1}{n} \cdot \sum_{j=0}^{n-1} \left| \frac{\sin \pi n \left(t + \frac{j}{n} \right)}{\sin \pi \left(t + \frac{j}{n} \right)} \right| \\
&= \max_{0 \leq t \leq \frac{1}{2n}} \frac{1}{n} \cdot \sum_{j=0}^{n-1} \left| \frac{\sin \pi n \left(t + \frac{j}{n} \right)}{\sin \pi \left(t + \frac{j}{n} \right)} \right| \\
&\leq \max_{0 \leq t \leq \frac{1}{2n}} \frac{1}{n} \cdot \left(\frac{\sin \pi n t}{\sin \pi t} + \sum_{j=1}^{n-1} \frac{1}{\sin \pi \left(t + \frac{j}{n} \right)} \right) \\
&\leq 1 + \max_{0 \leq t \leq \frac{1}{2n}} \frac{2}{n} \cdot \left(\frac{1}{\cos \pi t} + \sum_{j=1}^{\lfloor (n-1)/2 \rfloor} \frac{1}{\sin \pi \left(t + \frac{j}{n} \right)} \right) \\
&\leq 1 + \frac{3}{n} + \max_{0 \leq t \leq \frac{1}{2n}} \frac{2}{n} \cdot \left(\sum_{j=1}^{\lfloor (n-1)/2 \rfloor} \frac{1}{\sin \pi \left(t + \frac{j}{n} \right)} \right).
\end{aligned}
$$

In the inequality before the last we used (see Theorem 3.18)

$$\frac{\sin \pi n t}{\sin \pi t} \leq n,$$

and (holding trivially for $n \geq 2$)

$$\frac{1}{\cos \frac{\pi}{2n}} \leq \frac{3}{2}.$$

In the range of summation $t + \frac{j}{n} \leq \frac{1}{2}$. However, $\sin x$ is monotonically increasing on $[0, \pi/2]$ and thus

$$\phi(n) \leq 1 + \frac{3}{n} + \frac{2}{n} \sum_{j=1}^{\lfloor (n-1)/2 \rfloor} \frac{1}{\sin \pi \frac{j}{n}}.$$

Replacing

$$\frac{2}{n} \sum_{j=1}^{\lfloor (n-1)/2 \rfloor} \left(\sin \pi \frac{j}{n} \right)^{-1} \quad \text{with} \quad \frac{2}{n} \sum_{j=1}^{\lfloor (n-1)/2 \rfloor} \left(\pi \frac{j}{n} \right)^{-1},$$

with an error of absolute magnitude (using arguments analogous to (4.15)) at most $\pi^2/48$, we get

$$\phi(n) \leq 1 + \frac{3}{n} + \frac{2}{n} \sum_{j=1}^{\lfloor (n-1)/2 \rfloor} \frac{n}{\pi j} + \frac{\pi^2}{48}$$

$$= 1 + \frac{3}{n} + \frac{2}{\pi} \sum_{j=1}^{\lfloor (n-1)/2 \rfloor} \frac{1}{j} + \frac{\pi^2}{48}$$

$$\leq 1 + \frac{2}{n} + \frac{2}{\pi} \left(\ln \frac{n-1}{2} + \gamma + \frac{1}{n-1} \right) + \frac{\pi^2}{48}$$

$$< \frac{2}{\pi} \ln n + 1.132 + \frac{4}{n}.$$

In the last inequality we used

$$\ln(n-1) \leq \ln n - \frac{1}{n}.$$

$\square$

4.1.2 Lower bound

Lemma 4.4   *For every $n > 2$ there exists a signal $F_\mathbf{a}(t)$ such that*

$$\mathcal{M}_c(F_\mathbf{a}) \geq \left(\frac{2}{\pi} \ln n + 0.603 - \frac{1}{6n} \right) \cdot \mathcal{M}_d(F_\mathbf{a}). \tag{4.18}$$

Proof   We again use (4.6) and get

$$F_\mathbf{a}(t) = \frac{1}{n} \sum_{j=0}^{n-1} F_\mathbf{a}\left(\frac{j}{n} \right) D_n\left(t - \frac{j}{n} \right),$$

where $D_n(t)$ is the Dirichlet kernel (4.7).

   We construct $F_\mathbf{a}(t)$ in such a way that, for $j = 0, \ldots, n-1$,

$$F_\mathbf{a}\left(\frac{j}{n} \right) = \sqrt{n} \cdot \mathrm{e}^{-2\pi \iota \varphi_j}, \tag{4.19}$$

where

$$\mathrm{e}^{2\pi \iota \varphi_j} = \mathrm{e}^{\iota \arg D_n(t-j/n)}.$$

By (4.19) for such a polynomial,

$$\mathcal{M}_d(F_\mathbf{a}) = \sqrt{n}. \tag{4.20}$$

Provided we managed to construct such a polynomial, we obtain from (4.6)

$$F_{\mathbf{a}}(t) = \frac{1}{\sqrt{n}} \sum_{j=0}^{n-1} \left| D_n \left(t - \frac{j}{n} \right) \right|. \tag{4.21}$$

Now we will show that the system of equations

$$F_{\mathbf{a}} \left(\frac{j}{n} \right) = \sqrt{n} e^{-2\pi i \varphi_j}, \quad j = 0, \ldots, n-1, \tag{4.22}$$

is soluble. Indeed,

$$F_{\mathbf{a}}(t) = \sum_{k=0}^{n-1} a_k e^{2\pi i k t},$$

and thus the system has the following form

$$
\begin{aligned}
a_0 + a_1 + \cdots + a_{n-1} &= \sqrt{n} e^{-2\pi i \varphi_0} \\
a_0 + a_1 e^{2\pi i \frac{1}{n}} + \cdots + a_{n-1} e^{2\pi i \frac{n-1}{n}} &= \sqrt{n} e^{-2\pi i \varphi_1} \\
&\cdots \qquad\qquad \cdots \\
a_0 + a_1 e^{2\pi i \frac{j}{n}} + \cdots + a_{n-1} e^{2\pi i \frac{j(n-1)}{n}} &= \sqrt{n} e^{-2\pi i \varphi_j} \\
&\cdots \qquad\qquad \cdots \\
a_0 + a_1 e^{2\pi i \frac{n-1}{n}} + \cdots + a_{n-1} e^{2\pi i \frac{(n-1)(n-1)}{n}} &= \sqrt{n} e^{-2\pi i \varphi_{n-1}}.
\end{aligned} \tag{4.23}
$$

Or, in the matrix form, it is

$$E \mathbf{a}^T = \varphi^T,$$

where $\mathbf{a}$ and φ are row vectors of size n with components a_j and $\sqrt{n} e^{-2\pi i \varphi_j}$ for $j = 0, \ldots, n-1$, and E is an $n \times n$ matrix with entries

$$e_{j,k} = e^{2\pi i \frac{jk}{n}}, \quad j = 0, \ldots, n-1; \; k = 0, \ldots, n-1.$$

Denote the columns of matrix E by $\mathbf{e}_0, \ldots, \mathbf{e}_{n-1}$. Since matrix E is the matrix of the DFT, the system (4.23) has the unique solution described by

$$a_j = \frac{1}{n} (\varphi, \mathbf{e}_j). \tag{4.24}$$

Moreover, by Parseval's equality,

$$\sum_{j=0}^{n-1} |a_j|^2 = \frac{1}{n^2} \sum_{j=0}^{n-1} |(\varphi, \mathbf{e}_j)|^2 = n, \tag{4.25}$$

and we proved that the energy of $F_\mathbf{a}(t)$ is n. Thus we have constructed a polynomial possessing $\mathcal{M}_d(F_\mathbf{a}) = \sqrt{n}$, and

$$\mathcal{M}_c(F_\mathbf{a}) = \frac{1}{n} \sum_{j=0}^{n-1} \max_t \left| D_n \left(t - \frac{j}{n} \right) \right|. \tag{4.26}$$

A lower bound on the right-hand side is given in (4.17), which accomplishes the proof. □

4.2 Estimating the continuous maximum from the discrete one and its derivative

In this section, I consider the problem of estimating the continuous maximum when the sampled values of a signal are given. As we know, to compute the sample values of the signal we have to apply DFT to the vector $\mathbf{a}$ of the coefficients. However, it is almost as simple to obtain the sample values of the derivative of the signal as the DFT of $(0 \cdot a_0, 1 \cdot a_1, \ldots, (n-1) \cdot a_{n-1})$. Though we have seen in the previous section that the maximum over the samples at the Nyquist frequency provides a poor estimate of the continuous maximum, a weighted combination of the maxima of the sampled signal and its derivative yields a much tighter bound, guaranteeing a constant ratio between the lower and upper bounds. In this section we consider only signals satisfying $|a_k| = 1$, $k = 0, 1, \ldots, n-1$.

Theorem 4.5   *Let $F_\mathbf{a}(t) = \sum_{k=0}^{n-1} a_k e^{2\pi i k \frac{t}{n}}$, and $|a_k| = 1$, $k = 0, 1, \ldots, n-1$.
Define*

$$\mathcal{M}_d(F_\mathbf{a}) = \max_{j=0,\ldots,n-1} \left| \sum_{k=0}^{n-1} a_k e^{2\pi i k \frac{j}{n}} \right|,$$

$$\mathcal{M}'_d(F_\mathbf{a}) = \max_{j=0,\ldots,n-1} \frac{1}{2\pi} \left(\frac{d}{dx} \sum_{k=0}^{n-1} a_k e^{2\pi i k x} \right) \bigg|_{x=j/n} = \max_{j=0,\ldots,n-1} \left| \sum_{k=0}^{n-1} k a_k e^{2\pi i k \frac{j}{n}} \right|.$$

Then

$$|F_\mathbf{a}(t)| \leq \max_{\theta \in (0,1/2]} \left\{ c_n(\theta) \mathcal{M}_d(F_\mathbf{a}) + \frac{\left| e^{2\pi i \theta} - 1 \right|}{n} \mathcal{M}'_d(F_\mathbf{a}) \right\}$$

$$+ \frac{1}{\pi} (\ln n + 0.5773) + 1.181 \tag{4.27}$$

$$\leq c_n \mathcal{M}_d(F_\mathbf{a}) + \frac{2}{n} \mathcal{M}'_d(F_\mathbf{a}) + \frac{1}{\pi} (\ln n + 0.5773) + 1.181, \tag{4.28}$$

where

$$c_n(\theta) \leq \frac{|e^{2n\pi i\theta} - 1|}{2}\left(\ln\frac{\pi}{2} + \frac{8}{n}\right)$$

$$+ \frac{|e^{2\pi i\theta/n} - 1| \cdot |e^{2\pi i\theta} - 1|}{8} \cdot \frac{\sin\pi\theta - \pi\theta\cos\pi\theta}{2\theta^2\sin\pi\theta}$$

$$+ \frac{e^{2\pi i\theta} - 1}{2} + \left|\frac{1}{n} \cdot \frac{e^{2\pi i\theta} - 1}{e^{2\pi i\theta/n} - 1} - \frac{e^{2\pi i\theta} - 1}{2}\right| + \frac{0.5773}{n}, \tag{4.29}$$

and

$$c_n = \max_{\theta\in(0,1/2]} c_n(\theta), \tag{4.30}$$

where c_n is decreasing when n grows, and

$$c_{10} < 3.659, \quad c_{100} < 2.749, \quad c_{1000} < 2.658, \quad c_n < 2.648,$$

for n big enough.

Proof Note that from (4.5), $\mathcal{M}_d(F_\mathbf{a}) \geq \sqrt{n}$, thus the ln n term does not affect behavior of the estimate of $|F_\mathbf{a}(t)|$ for big lengths. Another remark is that (4.27) gives a more accurate estimate when we optimize the sum for given $\mathcal{M}_d(F)$ and $\mathcal{M}'_d(F)$. In the second estimate (4.28) we maximize the coefficients in the sum independently, but derive a result that is valid for any values of $\mathcal{M}_d(F)$ and $\mathcal{M}'_d(F)$.

We invoke (4.6) again,

$$F_\mathbf{a}(t) = \frac{1}{n}\sum_{j=0}^{n-1}\left(\sum_{k_1=0}^{n-1} a_{k_1}e^{2\pi i\frac{k_1 j}{n}}\right) \cdot \left(\sum_{k_2=0}^{n-1} e^{2\pi i k_2\left(t - \frac{j}{n}\right)}\right).$$

In the first bracket we have the values of $F_\mathbf{a}(t)$ on a discrete set of points uniformly distributed over the circle. Let us analyze the influence of the expression in the second bracket.

Set

$$t = \frac{j_0}{n} + \frac{\theta}{n}$$

where j_0 is an integer in the interval $[0, n-1]$, and $|\theta| \leq \frac{1}{2}$. Then, by (4.6), we have, using notation $s = j - j_0$,

$$F_\mathbf{a}(t) = \frac{1}{n}\sum_{j=0}^{n-1} F_\mathbf{a}\left(\frac{j}{n}\right) \cdot \left(\sum_{k=0}^{n-1} e^{2\pi i k\left(-\frac{j_0}{n} + \frac{\theta}{n} - \frac{j}{n}\right)}\right)$$

$$= \frac{1}{n}\sum_{-n/2\leq s\leq n/2} F_\mathbf{a}\left(\frac{j_0 + s}{n}\right) \cdot \left(\sum_{k=0}^{n-1} e^{2\pi i k\left(-\frac{s}{n} + \frac{\theta}{n}\right)}\right). \tag{4.31}$$

Evidently, we can assume that $\theta \neq 0$, otherwise we obtain a value of the function at a sample point. Now,

$$\sum_{k=0}^{n-1} e^{2\pi \imath k(\frac{\theta}{n} - \frac{s}{n})} = \frac{e^{2\pi \imath (\frac{\theta}{n} - \frac{s}{n})n} - 1}{e^{2\pi \imath (\frac{\theta}{n} - \frac{s}{n})} - 1} = \frac{e^{2\pi \imath \theta} - 1}{e^{2\pi \imath (\frac{\theta}{n} - \frac{s}{n})} - 1}.$$

Inserting this into (4.31), we have

$$F_{\mathbf{a}}(t) = \frac{e^{2\pi \imath \theta} - 1}{n} \cdot \sum_{-n/2 < s \leq n/2} F_{\mathbf{a}}\left(\frac{j_0 + s}{n}\right) \cdot \frac{1}{e^{2\pi \imath (\frac{\theta}{n} - \frac{s}{n})} - 1}. \tag{4.32}$$

We now partition the sum into two sums where the summation will be undertaken over two nonintersecting intervals:

$$\begin{aligned}
F_{\mathbf{a}}(t) &= \frac{e^{2\pi \imath \theta} - 1}{n} \sum_{|s| \leq H} F_{\mathbf{a}}\left(\frac{j_0 + s}{n}\right) \cdot \frac{1}{e^{2\pi \imath (\frac{\theta}{n} - \frac{s}{n})} - 1} \\
&+ \frac{e^{2\pi \imath \theta} - 1}{n} \sum_{|s| > H} F_{\mathbf{a}}\left(\frac{j_0 + s}{n}\right) \cdot \frac{1}{e^{2\pi \imath (\frac{\theta}{n} - \frac{s}{n})} - 1} \\
&= \Sigma_1 + \Sigma_2, \tag{4.33}
\end{aligned}$$

where $H = \varepsilon n$, and ε is a positive constant, which will be chosen later.

An estimate for Σ_2. Indeed,

$$|\Sigma_2| \leq \frac{\left|e^{2\pi \imath \theta} - 1\right|}{n} \cdot \mathcal{M}_d(F_{\mathbf{a}}) \cdot \sum_{H < |s| \leq n/2} \frac{1}{\left|e^{2\pi \imath (\frac{\theta}{n} - \frac{s}{n})} - 1\right|}.$$

Furthermore,

$$e^{2\pi \imath \beta} - 1 = e^{\pi \imath \beta} \frac{2\imath(e^{\pi \imath \beta} - e^{-\pi \imath \beta})}{2\imath} = 2\imath e^{\pi \imath \beta} \sin \pi \beta,$$

and, for every β, $|\sin \pi \beta| \geq 2\|\beta\|$, where $\|\beta\|$ is the distance from β to the closest integer. Thus,

$$\begin{aligned}
|\Sigma_2| &\leq \frac{\left|e^{2\pi \imath \theta} - 1\right|}{n} \mathcal{M}_d(F_{\mathbf{a}}) \cdot \sum_{H < |s| \leq n/2} \frac{1}{4\left\|\frac{\theta - s}{n}\right\|} \\
&\leq \mathcal{M}_d(F_{\mathbf{a}}) \cdot \frac{\left|e^{2\pi \imath \theta} - 1\right|}{2} \cdot \sum_{s=H-1}^{n/2+1} \frac{1}{s}.
\end{aligned}$$

Since $\sum_{s=1}^{L} \frac{1}{n} = \ln L + \gamma + \frac{\xi}{2L}$, where γ is the Euler constant and $0 < \xi \leq 1$, we have

$$\begin{aligned}
|\Sigma_2| &\leq \mathcal{M}_d(F_{\mathbf{a}}) \cdot \frac{\left|e^{2\pi \imath \theta} - 1\right|}{2} \cdot \left(\ln \frac{(n+2)}{2(H-1)} + \frac{1}{n+2}\right) \\
&= \mathcal{M}_d(F_{\mathbf{a}}) \cdot \frac{\left|e^{2\pi \imath \theta} - 1\right|}{2} \cdot \left(\ln \frac{1}{2\varepsilon} + \ln \left(1 + \frac{2 + 1/\varepsilon}{n - 1/\varepsilon}\right) + \frac{1}{n+2}\right). \tag{4.34}
\end{aligned}$$

An estimate for Σ_1. We have

$$
\Sigma_1 = \frac{1}{n} \cdot \frac{e^{2\pi \imath \theta} - 1}{e^{2\pi \imath \frac{\theta}{n}} - 1} F_{\mathbf{a}}\left(\frac{j_0}{n}\right)
$$
$$
+ \frac{e^{2\pi \imath \theta} - 1}{n} \sum_{1 \le |s| \le H} F_{\mathbf{a}}\left(\frac{j_0 + s}{n}\right) \cdot \frac{1}{e^{2\pi \imath \left(\frac{\theta}{n} - \frac{s}{n}\right)} - 1} \qquad (4.35)
$$
$$
= \frac{1}{n} \cdot \frac{e^{2\pi \imath \theta} - 1}{e^{2\pi \imath \frac{\theta}{n}} - 1} F_{\mathbf{a}}\left(\frac{j_0}{n}\right) + \Sigma_1^1 .
$$

To estimate Σ_1^1, let us compute the error E if we replace $\frac{1}{e^{2\pi \imath \left(\frac{\theta}{n} - \frac{s}{n}\right)} - 1}$ in (4.35) with $\frac{1}{e^{2\pi \imath \frac{s}{n}} - 1}$. Indeed,

$$
E \le \frac{\left|e^{2\pi \imath \theta} - 1\right|}{n} \sum_{1 \le |s| \le H} \left|F_{\mathbf{a}}\left(\frac{j_0 + s}{n}\right)\right| \cdot \left|\frac{1}{e^{2\pi \imath \left(\frac{\theta}{n} - \frac{s}{n}\right)} - 1} - \frac{1}{e^{-2\pi \imath \frac{s}{n}} - 1}\right|
$$
$$
\le \left|e^{2\pi \imath \theta} - 1\right| \mathcal{M}_d(F_{\mathbf{a}}) \cdot \frac{1}{n} \cdot \sum_{1 \le |s| \le H} \frac{\left|1 - e^{2\pi \imath \frac{\theta}{n}}\right|}{\left|e^{2\pi \imath \left(\frac{\theta}{n} - \frac{s}{n}\right)} - 1\right| \cdot \left|e^{-2\pi \imath \frac{s}{n}} - 1\right|}.
$$

Since

$$
\left|e^{2\pi \imath \left(\frac{\theta}{n} - \frac{s}{n}\right)} - 1\right| \ge 4 \left\|\frac{\theta - s}{n}\right\|, \text{ and } \left|e^{-2\pi \imath \frac{s}{n}} - 1\right| \ge 4 \left\|\frac{s}{n}\right\|,
$$

$$
E \le \left|e^{2\pi \imath \theta} - 1\right| \mathcal{M}_d(F_{\mathbf{a}}) \cdot \frac{1}{n} \cdot \sum_{1 \le |s| \le H} \frac{\left|1 - e^{2\pi \imath \frac{\theta}{n}}\right|}{4 \left\|\frac{\theta - s}{n}\right\| \cdot 4 \left\|\frac{s}{n}\right\|}
$$
$$
\le \frac{\left|1 - e^{2\pi \imath \frac{\theta}{n}}\right| \cdot \left|e^{2\pi \imath \theta} - 1\right|}{16} \cdot \mathcal{M}_d(F_{\mathbf{a}}) \cdot \frac{1}{n} \cdot \sum_{1 \le |s| \le H} \frac{1}{\left\|\frac{\theta - s}{n}\right\| \cdot \left\|\frac{s}{n}\right\|}
$$
$$
= \frac{\left|1 - e^{2\pi \imath \frac{\theta}{n}}\right| \cdot \left|e^{2\pi \imath \theta} - 1\right|}{16} \cdot \mathcal{M}_d(F_{\mathbf{a}}) \cdot \frac{1}{n^2} \left(\sum_{s=1}^{H} \frac{1}{\frac{s - \theta}{n} \cdot \frac{s}{n}} + \sum_{s=-1}^{-H} \frac{1}{\frac{-s + \theta}{n} \cdot \frac{-s}{n}}\right)
$$
$$
= \frac{\left|1 - e^{2\pi \imath \frac{\theta}{n}}\right| \left|e^{2\pi \imath \theta} - 1\right|}{16} \cdot \mathcal{M}_d(F_{\mathbf{a}}) \cdot \sum_{s=1}^{H} \frac{1}{s} \left(\frac{1}{s + \theta} + \frac{1}{s - \theta}\right)
$$
$$
= \frac{\left|1 - e^{2\pi \imath \frac{\theta}{n}}\right| \left|e^{2\pi \imath \theta} - 1\right|}{8} \cdot \mathcal{M}_d(F_{\mathbf{a}}) \cdot \sum_{s=1}^{H} \frac{1}{s^2 - \theta^2}
$$
$$
\le \frac{\left|1 - e^{2\pi \imath \frac{\theta}{n}}\right| \cdot \left|e^{2\pi \imath \theta} - 1\right|}{8} \cdot \mathcal{M}_d(F_{\mathbf{a}}) \cdot \sum_{s=1}^{\infty} \frac{1}{s^2 - \theta^2}
$$
$$
= \frac{\left|1 - e^{2\pi \imath \frac{\theta}{n}}\right| \cdot \left|e^{2\pi \imath \theta} - 1\right|}{8} \cdot \frac{\sin \pi \theta - \pi \theta \cos \pi \theta}{2\theta^2 \sin \pi \theta} \cdot \mathcal{M}_d(F_{\mathbf{a}}). \qquad (4.36)
$$

Thus from (4.35) we have obtained

$$
\begin{aligned}
\Sigma_1 &= \frac{1}{n} \cdot \frac{e^{2\pi i\theta} - 1}{e^{2\pi i \frac{\theta}{n}} - 1} \cdot F_{\mathbf{a}}\left(\frac{j_0}{n}\right) \\
&\quad + \frac{e^{2\pi i\theta} - 1}{n} \cdot \sum_{1 \le |s| \le H} F_{\mathbf{a}}\left(\frac{j_0 + s}{n}\right) \cdot \frac{1}{e^{-2\pi i \frac{s}{n}} - 1} + E \\
&= A_0 + A_1 + E,
\end{aligned}
\tag{4.37}
$$

and $|E|$ is estimated in (4.36).

Let us estimate the error E^* if in A_1 we replace $\frac{1}{e^{-2\pi i \frac{s}{n}} - 1}$ with $\frac{1}{-2\pi i \frac{s}{n}}$. From the inequality (see Theorem 3.19)

$$
\left| e^{it} - 1 - it \right| \le \frac{t^2}{2},
$$

we have

$$
\begin{aligned}
|E^*| &\le \left| \frac{e^{2\pi i\theta} - 1}{n} \right| \cdot \sum_{1 \le |s| \le H} \left| F_{\mathbf{a}}\left(\frac{j + s}{n}\right) \right| \cdot \left| \frac{1}{e^{-2\pi i \frac{s}{n}} - 1} - \frac{1}{-2\pi i \frac{s}{n}} \right| \\
&\le \frac{\left| e^{2\pi i\theta} - 1 \right|}{n} \mathcal{M}_d(F_{\mathbf{a}}) \cdot \sum_{1 \le |s| \le H} \frac{\left| -2\pi i \frac{s}{n} + 1 - e^{-2\pi i \frac{s}{n}} \right|}{\left| e^{-2\pi i \frac{s}{n}} - 1 \right| \cdot 2\pi \frac{|s|}{n}} \\
&\le \left| e^{2\pi i\theta} - 1 \right| \cdot \mathcal{M}_d(F_{\mathbf{a}}) \cdot \frac{1}{2\pi n} \cdot \sum_{1 \le |s| \le H} \frac{\frac{4\pi^2 s^2}{2n^2}}{4 \left\| \frac{s}{n} \right\| \cdot \frac{|s|}{n}} \\
&\le \frac{\pi \left| e^{2\pi i\theta} - 1 \right|}{2} \cdot \mathcal{M}_d(F_{\mathbf{a}}) \cdot \frac{H}{n}.
\end{aligned}
\tag{4.38}
$$

Thus we may substitute A_1 by the expression

$$
A_1^* = \frac{e^{2\pi i\theta} - 1}{-2\pi i} \sum_{1 \le |s| \le H} \frac{1}{s} \cdot F_{\mathbf{a}}\left(\frac{j_0 + s}{n}\right)
\tag{4.39}
$$

with error E^* of absolute value estimated in (4.38). The sum in the last expression can be estimated as follows:

$$
\begin{aligned}
\sum_{1 \le |s| \le H} \frac{1}{s} F_{\mathbf{a}}\left(\frac{j_0 + s}{n}\right) &= \sum_{1 \le |s| \le H} \frac{1}{s} \left(\sum_{k=0}^{n-1} a_k e^{2\pi i \frac{k(j_0+s)}{n}} \right) \\
&= \sum_{k=0}^{n-1} a_k e^{2\pi i \frac{kj_0}{n}} \cdot \sum_{1 \le |s| \le H} \frac{1}{s} e^{2\pi i \frac{ks}{n}}
\end{aligned}
$$

$$= \sum_{k=0}^{n-1} a_k e^{2\pi i \frac{kj_0}{n}} \cdot \left(\sum_{s=1}^{H} \frac{1}{s} e^{2\pi i \frac{ks}{n}} + \sum_{s=1}^{H} -\frac{1}{s} e^{-2\pi i \frac{ks}{n}} \right)$$

$$= 2i \sum_{k=0}^{n-1} a_k e^{2\pi i \frac{kj_0}{n}} \cdot \sum_{s=1}^{H} \frac{1}{s} \cdot \sin 2\pi \frac{ks}{n}.$$

Let us denote $x = k/n$, and we can then estimate $\sum_{s=1}^{H} \frac{1}{s} \sin 2\pi s x$. Then

$$\sum_{s=1}^{H} \frac{\sin 2\pi s x}{s} = 2\pi \int_0^x \sum_{s=1}^{H} \cos 2\pi s t \, dt = 2\pi \int_0^x \left(\frac{\sin(2H+1)\pi t}{2 \sin \pi t} - \frac{1}{2} \right) dt$$

$$= -\pi x + 2\pi \int_0^x \left(\frac{1}{2 \sin \pi t} - \frac{1}{2\pi t} \right) \cdot \sin(2H+1)\pi t \, dt$$

$$+ 2\pi \int_0^x \frac{\sin(2H+1)\pi t}{2\pi t} \, dt$$

$$= -\pi x + I_1 + I_2.$$

Furthermore,

$$I_2 = \int_0^x \frac{\sin \pi(2H+1)t}{t} \, dt = \int_0^{(2H+1)\pi x} \frac{\sin u}{u} \, du = \frac{\pi}{2} + \frac{\xi_1}{\pi(2H+1)x},$$

and $|\xi_1| \leq 1$. Moreover, for $x = 0$, $I_2 = 0$.

Now,

$$I_1 = 2\pi \int_0^x \left(\frac{1}{2 \sin \pi t} - \frac{1}{2\pi t} \right) \sin(2H+1)\pi t \, dt$$

$$= \pi \int_0^x \left(\frac{1}{\sin \pi t} - \frac{1}{\pi t} \right) \sin \pi(2H+1)t \, dt.$$

Integrating by parts we get

$$I_1 = \pi \left(-\frac{1}{\pi(2H+1)} \cos \pi(2H+1)t \cdot \left(\frac{1}{\sin \pi t} - \frac{1}{\pi t} \right) \Big|_0^x \right.$$

$$\left. + \frac{1}{\pi(2H+1)} \int_0^x \cos \pi(2H+1)t \cdot \left(\frac{1}{\sin \pi t} - \frac{1}{\pi t} \right)' dt \right).$$

We assume $x \in (0, \frac{1}{2})$, and thus

$$I_1 = \frac{\xi_2}{2H+1} \left(\max_{0 < x \leq 1/2} \left| \frac{1}{\sin \pi x} - \frac{1}{\pi x} \right| \right)$$

$$+ \frac{\xi_2}{2\pi(2H+1)} \cdot \max_{0 < x \leq 1/2} \left| \left(\frac{1}{\sin \pi x} - \frac{1}{\pi x} \right)' \right|$$

$$= \frac{\xi_2}{2H+1} \left(1 - \frac{2}{\pi} + \frac{2}{\pi^2} \right),$$

where $|\xi_2| \le 1$. Thus, for $x \ne 0$,

$$\sum_{s=1}^{H} \frac{\sin 2\pi s x}{s} = -\pi x + \frac{\pi}{2} + \frac{\xi_1}{\pi(2H+1)x} + \frac{\xi_2}{2H+1}\left(1 - \frac{2}{\pi} + \frac{2}{\pi^2}\right).$$

For $x = 0$, this sum is 0.

Returning to (4.39) we have

$$A_1^* = \frac{e^{2\pi i\theta} - 1}{-\pi} \cdot \sum_{k=1}^{n-1} a_k e^{2\pi i \frac{kj_0}{n}}$$

$$\cdot \left(-\frac{\pi k}{n} + \frac{\pi}{2} + \frac{\xi_1 n}{\pi(2H+1)k} + \frac{\xi_2}{2H+1}\left(1 - \frac{2}{\pi} + \frac{2}{\pi^2}\right)\right). \quad (4.40)$$

Collecting the intermediate results, we arrive at the following:

$$A_0 + A_1^* = \left(\frac{1}{n}\frac{e^{2\pi i\theta} - 1}{e^{2\pi i\frac{\theta}{n}} - 1} - \frac{e^{2\pi i\theta} - 1}{2}\right) \cdot \left(\sum_{k=0}^{n-1} a_k e^{2\pi i k \frac{j_0}{n}}\right) + \frac{e^{2\pi i\theta} - 1}{2} \cdot a_0$$

$$+ \frac{e^{2\pi i\theta} - 1}{n} \sum_{k=1}^{n-1} k a_k e^{2\pi i k \frac{j_0}{n}} - \frac{\xi_1(e^{2\pi i\theta} - 1)}{2\pi^2 \varepsilon} \cdot \left(\sum_{k=1}^{n-1} \frac{1}{k} a_k e^{2\pi i k \frac{j_0}{n}}\right)$$

$$- \frac{\xi_2\left(1 - \frac{2}{\pi} + \frac{2}{\pi^2}\right)}{2\varepsilon n + 1} \cdot \frac{e^{2\pi i\theta} - 1}{\pi} \cdot \left(-a_0 + \sum_{k=0}^{n-1} a_k e^{2\pi i k \frac{j_0}{n}}\right).$$

Next, since for all k, $|a_k| = 1$,

$$\left|\sum_{k=1}^{n-1} \frac{1}{k} a_k e^{2\pi i k \frac{j_0}{n}}\right| \le \sum_{k=1}^{n-1} \frac{1}{k} \le \ln n + \gamma.$$

Thus,

$$|A_0 + A_1^*| \le \left(\left|\frac{1}{n}\frac{e^{2\pi i\theta} - 1}{e^{2\pi i\frac{\theta}{n}} - 1} - \frac{e^{2\pi i\theta} - 1}{2}\right| + \frac{0.181}{\varepsilon n}\right) \cdot \mathcal{M}_d(F_\mathbf{a})$$

$$+ \frac{|e^{2\pi i\theta} - 1|}{n} \mathcal{M}_d'(F_\mathbf{a}) + \frac{|e^{2\pi i\theta} - 1|}{2\pi^2 \varepsilon}(\ln n + 0.5773) + 1.181.$$

Finally, maximizing in θ, we have

$$|F_\mathbf{a}(t)| \le \max_{\theta \in (0, 1/2]} \left\{|\Sigma_2| + |E| + |E^*| + |A_0 + A_1^*|\right\}$$

$$\le \max_{\theta \in (0, 1/2]} \left\{c_n(\theta)\mathcal{M}_d(F_\mathbf{a}) + \frac{|e^{2\pi i\theta} - 1|}{n}\mathcal{M}_d'(F_\mathbf{a})\right\}$$

$$+ \max_{\theta \in (0, 1/2]} \left\{\frac{|e^{2\pi i\theta} - 1|}{2\pi^2 \varepsilon}\right\}(\ln n + 0.5773) + 1.181,$$

where $c_n(\theta)$ satisfies

$$c_n(\theta) \le \frac{\left|e^{2\pi\iota\theta} - 1\right|}{2} \cdot \left(\ln\frac{1}{2\varepsilon} + \ln\left(1 + \frac{2 + 1/\varepsilon}{n - 1/\varepsilon}\right) + \frac{1}{n + 2}\right)$$

$$+ \frac{\left|e^{2\pi\iota\frac{\theta}{n}} - 1\right| \cdot \left|e^{2\pi\iota\theta} - 1\right|}{8} \cdot \frac{|\sin\pi\theta - \pi\theta\cos\pi\theta|}{2\theta^2\sin\pi\theta}$$

$$+ \frac{\pi\varepsilon\left|e^{2\pi\iota\theta} - 1\right|}{2} + \left|\frac{1}{n}\frac{e^{2\pi\iota\theta} - 1}{e^{2\pi\iota\frac{\theta}{n}} - 1} - \frac{e^{2\pi\iota\theta} - 1}{2}\right| + \frac{0.181}{\varepsilon n}.$$

Choosing $\varepsilon = \frac{1}{\pi}$, we arrive at (4.27) and (4.28) as claimed. Direct checking shows that for every θ and n increasing, $c_n(\theta)$ monotonically decreases, thus $\max_\theta c_n(\theta)$ also decreases with n. □

4.3 Dependence of the ratio on the maximum

The upper bound of Theorem 4.2 becomes trivial when $\mathcal{M}_d(F_\mathbf{a})$ is greater than $\frac{\pi}{2}\frac{n}{\ln n}$. This hints at a possibility of having bounds on the ratio between the continuous and discrete maxima depending on $\mathcal{M}_d(F_\mathbf{a})$.

Theorem 4.6 *For $F_\mathbf{a}(t) = \sum_{k=0}^{n-1} a_k e^{2\pi\iota t}$ with $a_k \in \mathbb{C}$, $\sum_{k=0}^{n-1}|a_k|^2 = n$, let*

$$\mathcal{M}_c(F_\mathbf{a}) = \max_t |F_\mathbf{a}(t)|, \quad \mathcal{M}_d(F_\mathbf{a}) = \max_{j=0,\dots,n-1}\left|F_\mathbf{a}\left(\frac{j}{n}\right)\right|.$$

Then

$$\mathcal{M}_c(F_\mathbf{a}) \le \mathcal{M}_d(F_\mathbf{a})\left(\frac{4}{\pi}\ln\frac{n}{\mathcal{M}_d(F_\mathbf{a})} + 4.92\right). \tag{4.41}$$

Remark 4.1 Noticing that the derivative in $\mathcal{M}_d$ of the right-hand side of (4.41) is positive, we conclude that $\mathcal{M}_d(F_\mathbf{a})$ in the bound can be substituted by any upper estimate.

Proof We start with noticing that every vector satisfying the Parseval identity corresponds to a legal vector of coefficients. Another ingredient of the proof is the following inequality. Let $(a_0, a_1, \dots, a_{n-1})$ and $(b_0, b_1, \dots, b_{n-1})$ be two nonnegative vectors such that $a_0 \ge a_1 \ge \dots \ge a_{n-1}$. Then $\sum_{k=0}^{n-1} a_k b_{\sigma(k)}$, where σ is a permutation on $\{0, 1, \dots, n-1\}$, attains a maximum if $b_{\sigma(0)} \ge b_{\sigma(1)} \ge \dots \ge b_{\sigma(n-1)}$. In other words, to maximize the scalar product of two nonnegative vectors defined up to permutation, one should line up their entries in decreasing order. This fact is proved in Theorem 3.13.

Let $\psi_j = \arg D_n\left(t - \frac{j}{n}\right)$. Then any MC signal $\hat{F}_n(t)$ can be converted to another legal MC signal by setting

$$F_{\mathbf{a}}\left(\frac{j}{n}\right) = e^{-\imath \psi_j}\left|\hat{F}_n\left(\frac{j}{n}\right)\right|,$$

which, by the interpolation formula (4.6), satisfies

$$|F_{\mathbf{a}}(t)| = \frac{1}{n}\sum_{j=0}^{n-1}\left|F_{\mathbf{a}}\left(\frac{j}{n}\right)\right| \cdot \left|D_n\left(t - \frac{j}{n}\right)\right| \leq \frac{1}{n}\sum_{j=0}^{n-1}|F_j| \cdot |D_j|. \qquad (4.42)$$

Here $|F_0| \geq |F_1| \geq \ldots \geq |F_{n-1}|$ and $|D_0| \geq |D_1| \geq \ldots \geq |D_{n-1}|$ are the correspondingly ordered values of $F_{\mathbf{a}}\left(\frac{j}{n}\right)$ and $D_n\left(t - \frac{j}{n}\right)$.

For every t, we have

$$\frac{1}{n}\sum_{j=0}^{n-1}\left|D_n\left(t - \frac{j}{n}\right)\right|^2 = n. \qquad (4.43)$$

Indeed,

$$D_n\left(t - \frac{j}{n}\right) = \sum_{k=0}^{n-1} e^{2\pi \imath k t} \cdot e^{-2\pi \imath k \frac{j}{n}},$$

and thus $D_n\left(t - \frac{j}{n}\right)$ are the components of DFT of the vector

$$\left(1, e^{2\pi \imath t}, e^{2\pi \imath 2 t}, \ldots, e^{2\pi \imath (n-1) t}\right).$$

Now (4.43) follows from the Parseval identity (Theorem 3.3).

Therefore, by the Cauchy–Schwartz inequality (3.4), we have

$$F_{\mathbf{a}}(t) \leq \frac{1}{n} \cdot \sqrt{\left(\sum_{j=0}^{n-1}|F_j|^2\right) \cdot \left(\sum_{j=0}^{n-1}|D_j|^2\right)} = n.$$

Though this inequality is trivial we can derive from it a condition on the attainability of the upper bound. Indeed, the Cauchy–Schwartz inequality is tight if, and only if, for all j and some nonzero λ, $|F_j| = \lambda|D_j|$. Since $\sum_{j=0}^{n-1}|F_j|^2 = \sum_{j=0}^{n-1}|D_j|^2$, $\lambda = 1$, and we conclude that in this case $|F_j| = |D_j|$ for all j. Therefore, for some t_0,

$$\max_{j=0,1,\ldots,n-1}\left|D_n\left(t_0 - \frac{j}{n}\right)\right| = \mathcal{M}_d(F_{\mathbf{a}}).$$

Recall that the closest to 0 zeros of the function $|D_n(t)|$ are $-\frac{1}{n}$ and $\frac{1}{n}$. Let ψ_1 and ψ_2 be the two solutions in t of $\mathcal{M}_d(F_{\mathbf{a}}) = |D_n(t)|$ belonging to $\mathcal{I} = \left(-\frac{1}{n}, \frac{1}{n}\right)$. Since $|D_n(t)|$ is an even function in t, $\psi_1 = -\psi_2$. Moreover, since the size of $\mathcal{I}$ is

$\frac{2}{n}$, it contains exactly two points of the form, say, $\alpha_1 = t_0 - \frac{i}{n}$ and $\alpha_2 = t_0 - \frac{j-1}{n}$. Therefore, the values of $|D_n(\alpha_1)|$ and $|D_n(\alpha_2)|$ have to be at most $\mathcal{M}_d(F_{\mathbf{a}})$. However, if $\psi_2 - \psi_1 > \frac{1}{n}$, there is such j^* that $\left|D_n\left(t_0 - \frac{j^*}{n}\right)\right| > \mathcal{M}_d(F_{\mathbf{a}})$, a contradiction. Therefore, to achieve the maximum of n one should have $\psi_2 - \psi_1 \leq \frac{1}{n}$ which is equivalent to $\psi_2 \leq \frac{1}{2n}$. This is equivalent to

$$\mathcal{M}_d(F_{\mathbf{a}}) \geq \left|D_n\left(\frac{1}{2n}\right)\right| = \frac{1}{\sin\frac{\pi}{2n}}.$$

Notice that the right-hand side converges to $\frac{2n}{\pi}$ when n increases. The maxima of $|D_n(t)|$ are decreasing when $|t|$ increases, and the second maximum is less than $\frac{1}{\sin\frac{\pi}{2n}}$. Thus, we have proved that if, and only if, $\mathcal{M}_d(F_{\mathbf{a}}) \geq \frac{1}{\sin\frac{\pi}{2n}}$, there exists a polynomial with $\mathcal{M}_c(F_{\mathbf{a}}) = n$.

Now assume that $\max_{j=0,1,\dots,n-1}\left|\left|F_{\mathbf{a}}\left(\frac{j}{n}\right)\right|\right| < \frac{1}{\sin\frac{\pi}{2n}}$. Let us continue our study of the right-hand side of (4.42). We have concluded that there is no λ such that for all $j = 0, 1, \dots, n-1$, $\lambda|F_j| = |D_j|$. Consider the second and third terms of the right-hand side of (4.42),

$$I = |F_1| \cdot |D_1| + |F_2| \cdot |D_2|,$$

under the constraint $|F_1|^2 + |F_2|^2 = R^2$. Then the maximum of I is achieved when

$$\frac{|F_1|}{|D_1|} = \frac{|F_2|}{|D_2|}. \tag{4.44}$$

Then

$$|F_1|^2 + |F_2|^2 = |F_1|^2\left(1 + \left(\frac{|D_2|}{|D_1|}\right)^2\right).$$

If

$$\frac{R^2}{1 + \left(\frac{|D_2|}{|D_1|}\right)^2} < \mathcal{M}_d(F_{\mathbf{a}}),$$

the maximum is achieved for $|F_1|$ and $|F_2|$ satisfying (4.44), otherwise, $|F_1| = \mathcal{M}_d$. The same argument works for any pair of indices k and j, $k > j$, from $\{1, 2, \dots, n-1\}$, and

$$|F_k| \cdot |D_k| + |F_j| \cdot |D_j|. \tag{4.45}$$

Therefore, there exists J such that

$$\begin{aligned}
\text{for } j &= 0, \dots, J-1, & |F_j| &= \mathcal{M}_d(F_{\mathbf{a}}); \\
\text{for } j &= J, \dots, n-1, & \frac{|F_j|}{|D_j|} &= \frac{|F_{j+1}|}{|D_{j+1}|} = \dots = \frac{|F_{n-1}|}{|D_{n-1}|}.
\end{aligned}$$

Then

$$|F_{\mathbf{a}}(t)| \leq \mathcal{M}_d(F_{\mathbf{a}}) \cdot \frac{1}{n} \sum_{j=0}^{J-1} |D_j| + |F_J| \cdot \frac{1}{n|D_J|} \sum_{j=J}^{n-1} |D_j|^2. \qquad (4.46)$$

Now we would like to determine the possibilities for ordering the values of

$$\left| D_n \left(t - \frac{j}{n} \right) \right| = \left| \frac{\sin \pi n \left(t - \frac{j}{n} \right)}{\sin \pi \left(t - \frac{j}{n} \right)} \right| = \left| \frac{\sin \pi n t}{\sin \pi \left(t - \frac{j}{n} \right)} \right|.$$

Since $|\sin \pi t|$ is periodic with period 1 and even, the ordering of

$$\frac{|\sin \pi n t|}{|\sin \pi t|}, \quad \frac{|\sin \pi n t|}{\left| \sin \pi \left(t - \frac{1}{n} \right) \right|}, \quad \dots \quad \frac{|\sin \pi n t|}{\left| \sin \pi \left(t - \frac{n-1}{n} \right) \right|}$$

can be considered for the interval $t \in \left[0, \frac{1}{2} \right]$. Indeed, for $t = 0$ we have ordering $n, 0, \dots, 0$, and the trivial estimate $|F_{\mathbf{a}}(t)| \leq \mathcal{M}_d(F_{\mathbf{a}})$. Assume $t \neq 0$. Evidently the ordering then is

$$\frac{|\sin \pi n t|}{|\sin \pi t|}, \quad \frac{|\sin \pi n t|}{\left| \sin \pi \left(t - \frac{1}{n} \right) \right|}, \quad \frac{|\sin \pi n t|}{\left| \sin \pi \left(t + \frac{1}{n} \right) \right|}, \quad \frac{|\sin \pi n t|}{\left| \sin \pi \left(t - \frac{2}{n} \right) \right|}, \quad \frac{|\sin \pi n t|}{\left| \sin \pi \left(t + \frac{2}{n} \right) \right|}, \dots$$

Therefore, from (4.46),

$$\begin{aligned}
|F_{\mathbf{a}}(t)| \leq &\ \mathcal{M}_d(F_{\mathbf{a}}) \cdot \frac{1}{n} \sum_{|j| \leq \frac{J-1}{2}} \frac{|\sin \pi n t|}{\left| \sin \pi \left(t + \frac{j}{n} \right) \right|} \\
&+ |F_J| \cdot \frac{1}{n} \cdot \frac{\left| \sin \pi \left(t + \frac{J+1}{2n} \right) \right|}{|\sin \pi n t|} \cdot \sum_{\frac{J+1}{2} \leq |j| \leq \frac{n-1}{2}} \frac{|\sin \pi n t|^2}{\left| \sin \pi \left(t + \frac{j}{n} \right) \right|^2} \\
\leq &\ \mathcal{M}_d(F_{\mathbf{a}}) \cdot \frac{1}{n} \left(n + 2 \sum_{j=1}^{\lceil \frac{J-1}{2} \rceil} \frac{1}{\left| \sin \pi \left(t + \frac{j}{n} \right) \right|} \right) \\
&+ |F_J| \cdot \frac{1}{n} \cdot \left| \sin \pi \left(t + \frac{J+1}{2n} \right) \right| \cdot \sum_{\lfloor \frac{J+1}{2} \rfloor \leq |j| \leq \frac{n-1}{2}} \frac{1}{\left| \sin \pi \left(t + \frac{j}{n} \right) \right|^2}.
\end{aligned}$$
$$(4.47)$$

We estimate the first summand exactly as in the derivation of the upper bound on $\phi(n)$, see p. 74, and obtain

$$\frac{1}{n} \left(n + 2 \sum_{j=1}^{\lceil \frac{J-1}{2} \rceil} \frac{1}{\left| \sin \pi \left(t + \frac{j}{n} \right) \right|} \right) \leq \frac{2}{\pi} \ln J + 1.132 + \frac{2}{\pi(J-1)}.$$

As for the second summand, we have

$$\max_{0 \le t \le \frac{1}{2n}} \left| \sin \pi \left(t + \frac{J+1}{2n} \right) \right| \le \frac{\pi (J+2)}{2n},$$

and

$$\max_{0 \le t \le \frac{1}{2n}} \sum_{\lfloor \frac{J+1}{2} \rfloor \le |j| \le \frac{n-1}{2}} \frac{1}{\left| \sin \pi \left(t + \frac{j}{n} \right) \right|^2} \le 2 \cdot \sum_{j = \lfloor \frac{J+1}{2} \rfloor}^{\frac{n-1}{2}} \frac{1}{\left| \sin \pi \frac{j}{n} \right|^2}$$

$$\le 2 \cdot \sum_{j = \lfloor \frac{J+1}{2} \rfloor}^{\frac{n-1}{2}} \frac{n^2}{4 j^2}$$

$$\le \frac{n^2}{2} \cdot \sum_{j = \frac{J}{2}}^{\infty} \frac{1}{j^2} \le \frac{n^2}{J}.$$

Plugging these estimates into (4.47), and substituting $|F_J|$ by $\mathcal{M}_d$, we obtain

$$|F_{\mathbf{a}}(t)| \le \mathcal{M}_d(F_{\mathbf{a}}) \left(\frac{2}{\pi} \ln J + 4.92 \right),$$

which is valid for $J \ge 2$. Let us estimate J from above. Indeed, by the Parseval identity, $\mathcal{M}_d^2(F_{\mathbf{a}}) J \le n^2$, and

$$J \le \left(\frac{n}{\mathcal{M}_d(F_{\mathbf{a}})} \right)^2,$$

which gives

$$|F_{\mathbf{a}}(t)| \le \mathcal{M}_d(F_{\mathbf{a}}) \left(\frac{4}{\pi} \ln \frac{n}{\mathcal{M}_d} + 4.92 \right).$$

$\square$

Actually, we see that the main term of (4.41) coincides with the main term of (4.9), $\frac{2}{\pi} \ln n$, only for $\mathcal{M}_d(F_{\mathbf{a}}) = \sqrt{n}$ (and it cannot be less than $\sqrt{n}$, see (4.5)). For larger $\mathcal{M}_d(F_{\mathbf{a}})$, the derived bound improves on (4.41). However, we may guarantee a constant ratio between the discrete and continuous maxima only if $\mathcal{M}_d(F_{\mathbf{a}})$ is proportional to n.

4.4 Oversampling

In this section, I consider the problem of approximating the maximum of a function when the number of sampling points is essentially greater than the degree of the polynomial corresponding to the signal. Denote, for an MC signal $F_\mathbf{a}(t)$,

$$\mathcal{M}_d(F_\mathbf{a}, M) = \max_{j=0,\ldots,M-1} \left| F_\mathbf{a}\left(\frac{j}{M}\right) \right|. \tag{4.48}$$

Clearly, $\mathcal{M}_d(F_\mathbf{a}, n) = \mathcal{M}_d(F_\mathbf{a})$. Note that computation of $\mathcal{M}_d(F_\mathbf{a}, M)$ can be implemented using DFT of size M over the vector $\mathbf{a} = (a_0, a_1, \ldots, a_{n-1}, 0, \ldots, 0)$ of length M. Indeed, for $j = 0, 1, \ldots, M-1$,

$$\sum_{k=0}^{M} a_k e^{2\pi \iota k \frac{j}{M}} = \sum_{k=0}^{n-1} a_k e^{2\pi \iota k \frac{j}{M}} = F_\mathbf{a}\left(\frac{j}{M}\right).$$

We start with a simple bound derived using Bernstein's inequality; see Theorem 3.26. Though this bound will be improved upon later, essential conclusions can already be derived from this one.

Theorem 4.7 *For $F_\mathbf{a}(t) = \sum_{k=0}^{n-1} a_k e^{2\pi \iota k t}$, and $M > n$,*

$$\mathcal{M}_c(F_\mathbf{a}) \leq \mathcal{M}_d(F_\mathbf{a}, M) \cdot \sqrt{\frac{M}{M - \pi(n-1)}}. \tag{4.49}$$

Proof By Lagrange's theorem for any real continuous differentiable function $f(x)$, and $x_2 > x_1$,

$$|f(x_1) - f(x_2)| \leq (x_2 - x_1) \max_{x_1 \leq x \leq x_2} |f'(x)|. \tag{4.50}$$

Let $f(\theta) = \left| F_\mathbf{a}\left(\frac{\theta}{2\pi}\right) \right|^2$. By Lemma 3.24, $f(\theta)$ is then a real trigonometric polynomial of order $(n-1)$. Thus, by Bernstein's inequality, (3.40), (4.50) can be rewritten as

$$|f(\theta_1) - f(\theta_2)| \leq (n-1)(\theta_2 - \theta_1) \max_{\theta \in [0, 2\pi)} f(\theta),$$

which is valid for any $0 \leq \theta_1 < \theta_2 \leq 2\pi$. Let $\max_{\theta \in [0, 2\pi]} f(\theta) = \mathcal{M}_c^2(F_\mathbf{a})$ be achieved at θ^*. The distance from the closest point of the form $\frac{2\pi j}{M}$ to θ^*, say $\hat\theta$, where $j = 0, 1, \ldots, M$, is at most $\frac{\pi}{M}$. Therefore,

$$\mathcal{M}_c^2(F_\mathbf{a}) - \mathcal{M}_d^2(F_\mathbf{a}, M) \leq f(\theta^*) - f(\hat\theta) \leq \frac{\pi(n-1)}{M} \mathcal{M}_c^2(F_\mathbf{a}),$$

and we obtain (4.49).

$\square$

An easy consequence of the theorem is that if one samples the signal in the number of points essentially greater than n, a constant ratio between the discrete and continuous maxima is guaranteed. Let us pass to a derivation of tighter bounds.

4.4.1 Large oversampling rates

Theorem 4.8 *Let $F_\mathbf{a}(t) = \sum_{k=0}^{n-1} e^{2\pi\imath kt}$ be a MC signal, and $r = \frac{M}{n}$ be the oversampling factor. Then*

$$\mathcal{M}_c(F_\mathbf{a}) \leq \frac{1}{\cos\frac{\pi}{2r}}\mathcal{M}_d(F_\mathbf{a}, M). \tag{4.51}$$

The equality in (4.51) is attained if, and only if, n is even and $r \geq 2$ is integer.

Proof We start with considering real trigonometric polynomials of order m, for simplicity scaled to

$$\lambda_0 + \lambda_1 \cos\frac{\pi}{m}t + \mu_1 \sin\frac{\pi}{m}t + \lambda_2 \cos\frac{2\pi}{m}t$$
$$+ \mu_2 \sin\frac{2\pi}{m}t + \ldots + \lambda_m \cos\pi t + \mu_m \sin\pi t.$$

The critical sampling of such a polynomial requires $2m + 1$ equidistant samples. Indeed, it cannot be less than that since otherwise there exists a nonzero real trigonometric polynomial of order m vanishing in all points. We denote by M, $M \geq 2m + 1$, the number of equidistant sample points, and by $C_r = \frac{\mathcal{M}_c}{\mathcal{M}_d}$ the sought maximum ratio between the discrete and continuous maxima of such polynomials. The considered polynomials have period $2m$. Thus the intervals between the sampling points are of length $2\frac{m}{M}$.

Let $f(t)$ be a polynomial on which the maximum ratio is achieved, and $f(t^*) = \mathcal{M}_c(f)$, i.e., the continuous maximum attained at t^*. Clearly, if $f(t)$ is a real trigonometric polynomial then $f(t - t_0)$, for any t_0, is such as well. Choose $t_0 = 2j \cdot \frac{m}{M}$ for some integer j, so that $f(t^* - t_0)$ belongs to the interval $\left(-\frac{m}{M}, \frac{m}{M}\right]$. Clearly for $f(t - t_0)$ the ratio between the discrete and continuous maximum is the same as for $f(t)$. Therefore, we may assume in what follows that t^* belongs to $\left(-\frac{m}{M}, \frac{m}{M}\right] \in (-1, 1)$.

Define

$$\Delta(t) = f(t + t^*) - \mathcal{M}_c(f)\cos\pi t.$$

Notice that $\Delta(0) = 0$ by the definition, and $\Delta'(0) = 0$ since f has a local extremum at t^*. The samples of Δ in integer points are

$$\Delta(k) = f(k + t^*) - \mathcal{M}_c(f)(-1)^k,$$

and since for all t, $|f(t)| \leq \mathcal{M}_c(f)$, we have

$$\Delta(k) \leq 0 \text{ for odd } k,$$
$$\Delta(k) \geq 0 \text{ for even } k.$$

Thus, we may rewrite

$$\Delta(k) = \xi_k(-1)^{k+1}, \quad |k| = 1, 2, \ldots,$$

where $\xi_k \geq 0$. For band-limited signals $g(t)$ with the Fourier spectrum supported on $[-\omega, \omega]$, to provide convergence on the supremum norm the following sampling series given by Schönhage [360] should be used,

$$g(t) = g'(0)\frac{\sin \omega t}{\omega} + g(0)\frac{\sin \omega t}{\omega t} + t \cdot \sum_{k=-\infty, k \neq 0}^{\infty} \frac{g\left(\frac{\pi}{\omega}\right)}{k} \cdot \frac{\sin \omega \left(t - \frac{k\pi}{\omega}\right)}{\omega \left(t - \frac{k\pi}{\omega}\right)}.$$

Applying it to $\Delta(t)$, and setting $\omega = \pi$, we get

$$\Delta(t) = \Delta'(0)\frac{\sin \pi t}{\pi} + \Delta(0)\frac{\sin \pi t}{\pi t} + t \cdot \sum_{k=-\infty, k \neq 0}^{\infty} \frac{\Delta(k)}{k} \cdot \frac{\sin \pi(t - k)}{\pi(t - k)}$$

$$= t \cdot \sum_{k=-\infty, k \neq 0}^{\infty} \frac{\Delta(k)}{k} \cdot \frac{\sin \pi(t - k)}{\pi(t - k)}$$

$$= t \frac{\sin \pi t}{\pi} \sum_{k=-\infty, k \neq 0}^{\infty} \frac{\Delta(k)}{k} \frac{(-1)^k}{(t - k)}$$

$$= t \frac{\sin \pi t}{\pi} \sum_{k=1}^{\infty} \frac{\Delta(k)}{k} \frac{(-1)^k}{(t - k)} + t \frac{\sin \pi t}{\pi} \sum_{k=-\infty}^{-1} \frac{\Delta(k)}{k} \frac{(-1)^k}{(t - k)}. \qquad (4.52)$$

We consider $\Delta(t)$ for $t \in (-1, 1)$. In this interval, $t \cdot \frac{\sin \pi t}{\pi} \geq 0$. Furthermore, the first sum in (4.52) is

$$\sum_{k=1}^{\infty} \frac{\Delta(k)}{k} \frac{(-1)^k}{(t - k)} = \sum_{k=1}^{\infty} \frac{\xi_k(-1)^{k+1}}{k} \frac{(-1)^k}{(t - k)} = \sum_{k=1}^{\infty} \frac{\xi_k}{k(k - t)} \geq 0.$$

As for the second sum in (4.52), we have

$$\sum_{k=-\infty}^{-1} \frac{\Delta(k)}{k} \frac{(-1)^k}{(t - k)} = - \sum_{k=-\infty}^{-1} \frac{\xi_k}{k(t - k)} = \sum_{k=1}^{\infty} \frac{\xi_k}{k(k + t)} \geq 0.$$

Thus, we have proved that $\Delta(t) \geq 0$ for $t \in (-1, 1)$. Equality $\Delta(t) = 0$ is achieved for some $t \in (-1, 1)/\{0\}$ if, and only if, $\xi_k = 0$ for all nonzero k, i.e., $f(t + t^*) = \mathcal{M}_c(f) \cos \pi t$. This means that $\mathcal{M}_c(f) \cos \pi(t - t^*)$ always has a discrete maximum not exceeding that of any other real trigonometric polynomial of order m. Indeed, the normalized value of any other real trigonometric at $m/M < 1$ is greater than the corresponding value of $\cos \pi(t - t^*)$, for which it is the discrete maximum.

We choose t^* so as to minimize $\mathcal{M}_d(f, M)$. This happens if the maximum of $\cos \pi(t - t^*)$ is in the middle of an interval between two sampling points, e.g.,

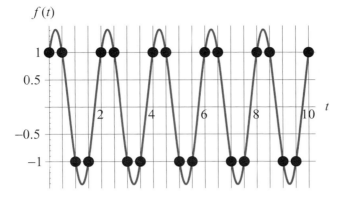

Figure 4.2 Extremal function

$t^* = \frac{m}{M}$ (see Fig. 4.2). Finally, we have

$$\mathcal{M}_d(f, M) \geq \mathcal{M}_c(f)\cos\frac{\pi m}{M}. \tag{4.53}$$

This inequality is exact if M is even, and $\frac{M}{2m}$ is integer. Only then do all the discrete samples of the properly shifted $\cos\pi t$ have the same value.

Now we pass to MC signals, $F_\mathbf{a}(t) = \sum_{k=0}^{n-1} a_k e^{2\pi\imath kt}$. Let n be even. Introduce the function $\Phi_n(t)$,

$$\Phi_n(t) = e^{-2\pi\imath\frac{n-2}{2}t} F_\mathbf{a}(t).$$

Clearly, $|F_\mathbf{a}(t)| = |\Phi_n(t)|$. Furthermore,

$$\Phi_n(t) = \sum_{k=-\frac{n-2}{2}}^{\frac{n}{2}} a_{k+\frac{n-2}{2}} e^{2\pi\imath kt}$$

$$= \sum_{k=-\frac{n-2}{2}}^{\frac{n}{2}} a_{k+\frac{n-2}{2}}(\cos 2\pi kt + \imath\sin 2\pi kt)$$

$$= \sum_{k=0}^{\frac{n}{2}} a_{k+\frac{n-2}{2}}(\cos 2\pi kt + \imath\sin 2\pi kt)$$

$$+ \sum_{k=1}^{\frac{n-2}{2}} a_{-k+\frac{n-2}{2}}(\cos 2\pi kt - \imath\sin 2\pi kt)$$

$$= a_{\frac{n-2}{2}} + \sum_{k=1}^{\frac{n-2}{2}}\left(\left(a_{-k+\frac{n-2}{2}} + a_{k+\frac{n-2}{2}}\right)\cos 2\pi kt\right.$$

$$+ \imath\left(-a_{-k+\frac{n-2}{2}} + a_{k+\frac{n-2}{2}}\right)\sin 2\pi kt\right)$$

$$+ a_{n-1}\cos 2\pi\frac{n}{2}t + \imath a_{n-1}\sin 2\pi\frac{n}{2}t.$$

Denoting

$$b_0 = a_{\frac{n-2}{2}}, b_{\frac{n}{2}} = a_{n-1}, b_k = a_{-k+\frac{n-2}{2}} + a_{k+\frac{n-2}{2}}, k = 1, 2, \ldots, \frac{n-2}{2},$$

and $\theta = 2\pi t$, we obtain

$$\Phi_n(\theta) = b_0 + \sum_{k=1}^{\frac{n-2}{2}} (b_k \cos k\theta + \imath b_k \sin kt) + b_{\frac{n}{2}} \cos \frac{n}{2}t + \imath b_{\frac{n}{2}} \sin \frac{n}{2}t.$$

Clearly $\Phi_n(\theta)$ is a complex trigonometric polynomial. Let $\max_{\theta \in [0,2\pi)} |\Phi_n(\theta)|$ be attained at $\theta = \theta^*$. Then the polynomial

$$\Phi^*(\theta) = e^{-\arg \Phi_n(\theta^*)} \Phi(\theta)$$

has real value at t^*. Moreover,

$$\max_{t \in [0,1)} |F_{\mathbf{a}}(t)| = \max_{\theta \in [0,2\pi)} |\Phi_n(\theta)| = \max_{\theta \in [0,2\pi)} |\Phi_n^*(\theta)| = \Phi_n^*(\theta^*).$$

Now, similarly to Lemma 3.23, $\Re(\Phi_n^*(\theta))$ is a real trigonometric polynomial of order $\frac{n}{2}$ with $\mathcal{M}_c(\Re(\Phi_n^*(\theta))) = \mathcal{M}_c(F_{\mathbf{a}})$. Since, for all θ,

$$\Re(\Phi_n^*(\theta)) \le |\Phi_n^*(\theta)| = |\Phi_n(\theta)| = \left| F_{\mathbf{a}}\left(\frac{\theta}{2\pi}\right) \right|,$$

we conclude that

$$C_r = \frac{\mathcal{M}_c(F_{\mathbf{a}})}{\mathcal{M}_d(F_{\mathbf{a}}, M)} \le \frac{\mathcal{M}_c(\Re(\Phi_n^*(\theta)))}{\mathcal{M}_d(\Re(\Phi_n^*(\theta)), M)} \le \frac{1}{\cos \frac{\pi n}{2M}},$$

where the last inequality follows from (4.53). The treatment for odd n is analogous.

The derived inequality is exact if n is even and $r \ge 2$ is an integer. Indeed, it is attained for the signal $F_{\mathbf{a}}(t) = e^{2\pi \imath (n-1)\left(t - \frac{1}{2M}\right)}$. □

4.4.2 Bounds from de la Vallée–Poussin kernels

As an improvement on the previous section, bounds can be obtained for the range of oversampling rates between 1 and 2.

Theorem 4.9   *For every MC signal, $F_{\mathbf{a}}(t) = \sum_{k=0}^{n-1} a_k e^{2\pi \imath kt}$, and $r = \frac{M}{n}$,*

$$\mathcal{M}_c(F_{\mathbf{a}}) \le \sqrt{\frac{r-1}{r}} \cdot \mathcal{M}_d(F_{\mathbf{a}}, M).$$

Proof   Let

$$T_{M,n}(t) = \frac{1}{(M-n)} \cdot D_M(t) \cdot D_{M-n}(-t),$$

where $D_j(t)$ is the Dirichlet kernel, defined in (4.7),

$$D_j(t) = \sum_{k=0}^{j-1} e^{2\pi \imath kt}.$$

Then

$$T_{M,n}(t) = \frac{1}{(M-n)} \cdot \left(\sum_{k=0}^{M-1} e^{2\pi \imath kt} \right) \left(\sum_{j=0}^{M-n-1} e^{-2\pi \imath jt} \right)$$

$$= \frac{1}{(M-n)} \cdot \sum_{k=0}^{M-1} \sum_{j=0}^{M-n-1} e^{2\pi \imath (k-j)t} = \sum_{k=-M+n+1}^{M-1} c_k e^{2\pi \imath kt},$$

where

$$c_k = \begin{cases} \frac{k}{M-n} + 1, & \text{if } k = n-M+1, n-M+2, \ldots, -1; \\ 1, & \text{if } k = 0, 1, \ldots, n-1; \\ -\frac{k}{M-n+1} + \frac{M}{M-n+1}, & \text{if } k = n, n+1, \ldots, M-1. \end{cases} \tag{4.54}$$

The kernel $T_{M,n}$ belongs to the class of *de la Vallée–Poussin kernels*.

Now

$$\frac{1}{M} \sum_{j=0}^{M-1} F_{\mathbf{a}} \left(\frac{j}{M} \right) \cdot T_{M,n} \left(t - \frac{j}{M} \right)$$

$$= \frac{1}{M} \sum_{k_1=0}^{n-1} \sum_{k_2=n-M+1}^{M-1} a_{k_1} c_{k_2} \sum_{j=0}^{M-1} e^{2\pi \imath k_1 \frac{j}{M}} \cdot e^{2\pi \imath k_2 \left(t - \frac{j}{M} \right)}$$

$$= \sum_{k_1=0}^{n-1} \sum_{k_2=n-M+1}^{M-1} a_{k_1} c_{k_2} \cdot e^{2\pi \imath k_2 t} \cdot \frac{1}{M} \cdot \sum_{j=0}^{M-1} e^{2\pi \imath (k_1-k_2) \frac{j}{M}}.$$

The last inner sum equals M if, and only if, $k_1 - k_2 \equiv 0 \mod m$, and is 0 otherwise. However, in the interval under consideration this congruence is satisfied only if $k_1 = k_2$. Therefore,

$$\frac{1}{M} \sum_{j=0}^{M-1} F_{\mathbf{a}} \left(\frac{j}{M} \right) \cdot T_{M,n} \left(t - \frac{j}{M} \right) = \sum_{k=0}^{n-1} a_k b_k e^{2\pi \imath kt} = F_{\mathbf{a}}(t).$$

Here, in the last step, we used the fact that $b_k = 1$ in the range of summation, see (4.54). Thus, we have proved that

$$F_{\mathbf{a}}(t) = \frac{1}{M} \sum_{j=0}^{M-1} F_{\mathbf{a}} \left(\frac{j}{M} \right) \cdot \frac{1}{(M-n)} \cdot D_M \left(t - \frac{j}{M} \right) \cdot D_{M-n} \left(t - \frac{j}{M} \right).$$

This yields

$$\max_{t\in[0,1)}|F_{\mathbf{a}}(t)| \le \frac{1}{M(M-n)} \cdot \max_{j=0,\dots,M-1}\left|F_{\mathbf{a}}\left(\frac{j}{M}\right)\right|$$

$$\cdot \max_{t\in[0,1)} \sum_{j=0}^{M-1}\left|D_M\left(t-\frac{j}{M}\right)\right| \cdot \left|D_{M-n}\left(t-\frac{j}{M}\right)\right|$$

$$\le \frac{1}{M-n} \cdot \max_{j=0,\dots,M-1}\left|F_{\mathbf{a}}\left(\frac{j}{M}\right)\right| \cdot \max_{t\in[0,1)}\left(\frac{1}{M}\sum_{j=0}^{M-1}\left|D_M\left(t-\frac{j}{M}\right)\right|^2\right)^{\frac{1}{2}}$$

$$\cdot \left(\frac{1}{M}\sum_{j=0}^{M-1}\left|D_{M-n}\left(t-\frac{j}{M}\right)\right|^2\right)^{\frac{1}{2}}.$$

On the last expression, the Cauchy–Schwartz inequality is used.

For every t and $s \le M$, we have

$$\frac{1}{M}\sum_{j=0}^{M-1}\left|D_s\left(t-\frac{j}{M}\right)\right|^2 = s. \tag{4.55}$$

Indeed,

$$D_s\left(t-\frac{j}{M}\right) = \sum_{k=0}^{s-1}\mathrm{e}^{2\pi\imath kt}\cdot\mathrm{e}^{-2\pi\imath k\frac{j}{M}} + \sum_{k=s}^{M-1}0\cdot\mathrm{e}^{-2\pi\imath k\frac{j}{M}},$$

and thus $D_s\left(t-\frac{j}{M}\right)$ are the components of DFT of the size M vector

$$\left(1, \mathrm{e}^{2\pi\imath t}, \mathrm{e}^{2\pi\imath 2t}, \dots, \mathrm{e}^{2\pi\imath(s-1)t}, 0, \dots, 0\right).$$

Now, (4.55) follows from the Parseval identity (Theorem 3.3).

Finally, we have

$$\mathcal{M}_c(F_{\mathbf{a}}) \le \mathcal{M}_d(F_{\mathbf{a}}, M) \cdot \frac{1}{M-n} \cdot \sqrt{M} \cdot \sqrt{M-n}$$

$$= \mathcal{M}_d(F_{\mathbf{a}}, M) \cdot \sqrt{\frac{M}{M-n}},$$

and we are done.   □

A better estimate can be obtained for specific values $r \in (1, 2)$. We start with $r = 3/2$.

Theorem 4.10 *For every MC signal, $F_{\mathbf{a}}(t) = \sum_{k=0}^{n-1}a_k\mathrm{e}^{2\pi\imath kt}$,*

$$\mathcal{M}_c(F_{\mathbf{a}}) \le \frac{5}{3}\cdot\mathcal{M}_d\left(F_{\mathbf{a}}, 3\left\lceil\frac{n-1}{2}\right\rceil\right). \tag{4.56}$$

Moreover, when n is even, the estimate is tight.

Proof   Let n be even, $n = 2m$. Denote $M = 3m$. Define a new function

$$\Phi_n(t) = F_{\mathbf{a}}(t)e^{-\pi\imath nt}. \tag{4.57}$$

Notice that

$$
\begin{aligned}
\mathcal{M}_d(F_{2m}, 3m) &= \max_{0 \le j \le 3m-1} \left| F_{\mathbf{a}}\left(\frac{j}{3m}\right) \right| \\
&= \max_{0 \le j \le 3m-1} \left| F_{\mathbf{a}}\left(\frac{j}{3m}\right) \cdot e^{-\pi\imath 2m\frac{j}{3m}} \right| \\
&= \max_{0 \le j \le 3m-1} \left| \Phi_n\left(\frac{j}{3m}\right) \right| = \mathcal{M}_d(\Phi_{2m}, 3m).
\end{aligned}
\tag{4.58}
$$

Thus, we may consider the function $\Phi_{2m}(t)$ instead of $F_{2m}(t)$.

Let

$$T_n(t) = \sum_{|k| \le n} c_k e^{2\pi\imath kt}, \tag{4.59}$$

where

$$c_k = \begin{cases} 1, & \text{if } |k| \le m, \\ 2 - \frac{|k|}{m}, & \text{if } m < |k| \le 2m. \end{cases} \tag{4.60}$$

Now,

$$
\begin{aligned}
&\frac{1}{M}\sum_{j=0}^{M-1}\Phi_n\left(\frac{j}{M}\right) \cdot T_n\left(t - \frac{j}{M}\right) \\
&= \frac{1}{M}\sum_{j=0}^{M-1}\left(\sum_{k_1=0}^{n-1} a_{k_1} e^{2\pi\imath(k_1-m)\cdot\frac{j}{M}}\right) \cdot \sum_{|k_2|\le n} c_k e^{2\pi\imath k_2\left(t-\frac{j}{M}\right)} \\
&= \sum_{k_1=0}^{n-1}\sum_{|k_2|\le n} a_{k_1} \cdot c_{k_2} \cdot e^{2\pi\imath k_2 t} \cdot \frac{1}{M}\sum_{j=0}^{M-1} e^{2\pi\imath(k_1-m-k_2)\frac{j}{M}}.
\end{aligned}
$$

The inner sum gives 0 in all cases except when

$$k_1 - m - k_2 \equiv 0 \,(\mathrm{mod}\ M),$$

and thus we have proved

$$\Phi_n(t) = \frac{1}{3m}\sum_{j=0}^{3m-1}\Phi_n\left(\frac{j}{3m}\right) \cdot T_{2m}\left(t - \frac{j}{3m}\right). \tag{4.61}$$

Therefore, using (4.58), we get

$$
|\Phi_n(t)| \leq \left(\max_{0 \leq j \leq 3m-1} \left| \Phi_n\left(\frac{j}{3m}\right) \right| \right) \cdot \frac{1}{3m} \sum_{j=0}^{3m-1} \left| T_{2m}\left(t - \frac{j}{3m}\right) \right|
$$

$$
= \mathcal{M}_d(F_{2m}, 3m) \cdot \frac{1}{3m} \sum_{j=0}^{3m-1} \left| T_{2m}\left(t - \frac{j}{3m}\right) \right|. \tag{4.62}
$$

Now notice that

$$
T_{2m}\left(t - \frac{j}{3m}\right) = \sum_{|k| \leq m} \left(1 - \frac{|k|}{m}\right) \cdot e^{2\pi \imath k\left(t - \frac{j}{3m}\right)} \cdot e^{-2\pi \imath m\left(t - \frac{j}{3m}\right)}
$$

$$
+ \sum_{|k| \leq m} \left(1 - \frac{|k|}{m}\right) \cdot e^{2\pi \imath k\left(t - \frac{j}{3m}\right)}
$$

$$
+ \sum_{|k| \leq m} \left(1 - \frac{|k|}{m}\right) \cdot e^{2\pi \imath k\left(t - \frac{j}{3m}\right)} \cdot e^{2\pi \imath m\left(t - \frac{j}{3m}\right)}
$$

$$
= \left(\sum_{|k| \leq m} \left(1 - \frac{|k|}{m}\right) \cdot e^{2\pi \imath k\left(t - \frac{j}{3m}\right)} \right)
$$

$$
\cdot \left(e^{2\pi \imath m\left(t - \frac{j}{3m}\right)} + 1 + e^{-2\pi \imath m\left(t - \frac{j}{3m}\right)} \right).
$$

In the last product, we denote the first term

$$
K_m\left(t - \frac{j}{3m}\right) = \sum_{|k| \leq m} \left(1 - \frac{|k|}{m}\right) \cdot e^{2\pi \imath k\left(t - \frac{j}{3m}\right)}, \tag{4.63}
$$

which is called the *Fejér kernel*. For all t, the Fejér kernel is real and

$$
K_m\left(t - \frac{j}{3m}\right) \geq 0. \tag{4.64}
$$

This is true, since

$$
K_m\left(t - \frac{j}{3m}\right) = \sum_{k_1=-\frac{m-1}{2}}^{\frac{m-1}{2}} \sum_{k_2=-\frac{m-1}{2}}^{\frac{m-1}{2}} e^{2\pi \imath (k_1-k_2)\left(t - \frac{j}{3m}\right)}
$$

$$
= \left(\sum_{k_1=-\frac{m-1}{2}}^{\frac{m-1}{2}} e^{2\pi \imath k_1\left(t - \frac{j}{3m}\right)} \right) \left(\sum_{k_2=-\frac{m-1}{2}}^{\frac{m-1}{2}} e^{-2\pi \imath k_2\left(t - \frac{j}{3m}\right)} \right)
$$

$$
= \left| \sum_{k_1=-\frac{m-1}{2}}^{\frac{m-1}{2}} e^{2\pi \imath k_1\left(t - \frac{j}{3m}\right)} \right|^2.
$$

Thus, by (4.62) and (4.64), we have

$$|\Phi_n(t)| \leq \mathcal{M}_d(F_{2m}, 3m) \cdot \frac{1}{3m} \cdot \sum_{j=0}^{3m-1} K_m\left(t - \frac{j}{3m}\right) \cdot \left|1 + 2\cos 2\pi m\left(t - \frac{j}{3m}\right)\right|.$$
$$(4.65)$$

Let

$$I = \frac{1}{3m} \sum_{j=0}^{3m-1} K_m\left(t - \frac{j}{3m}\right)\left|1 + 2\cos 2\pi\left(mt - \frac{j}{3}\right)\right|.$$

Then

$$\left|1 + 2\cos 2\pi\left(mt - \frac{j}{3}\right)\right|$$

$$= \begin{cases} |1 + 2\cos 2\pi mt|, & \text{for } j \equiv 0 \bmod 3, \\ |1 - \cos 2\pi mt + \sqrt{3}\sin 2\pi mt|, & \text{for } j \equiv 0 \bmod 3, \\ |1 - \cos 2\pi mt - \sqrt{3}\sin 2\pi mt|, & \text{for } j \equiv 0 \bmod 3. \end{cases}$$

Therefore,

$$I = \frac{1}{3m}|1 + 2\cos 2\pi mt| \cdot \sum_{j \equiv 0 \bmod 3} K_m\left(t - \frac{j}{3m}\right)$$

$$+ \frac{1}{3m}|1 - \cos 2\pi mt + \sqrt{3}\sin 2\pi mt| \cdot \sum_{j \equiv 1 \bmod 3} K_m\left(t - \frac{j}{3m}\right)$$

$$+ \frac{1}{3m}|1 - \cos 2\pi mt - \sqrt{3}\sin 2\pi mt| \cdot \sum_{j \equiv 2 \bmod 3} K_m\left(t - \frac{j}{3m}\right).$$

Now,

$$\frac{1}{3m} \sum_{j \equiv h \bmod 3} K_m\left(t - \frac{j}{3m}\right)$$

$$= \frac{1}{3m} \sum_{j \equiv h \bmod 3} \sum_{|k| \leq m}\left(1 - \frac{|k|}{m}\right) \cdot e^{2\pi \imath k\left(t - \frac{j}{3m}\right)}$$

$$= \sum_{|k| \leq m}\left(1 - \frac{|k|}{m}\right) \cdot e^{2\pi \imath k\left(t - \frac{j}{3m}\right)} \cdot \frac{1}{3m} \sum_{j \equiv h \bmod 3} e^{2\pi \imath \frac{kj}{3m}}$$

$$= \frac{1}{3},$$

since

$$\frac{1}{3m} \sum_{j \equiv h \bmod 3} e^{2\pi \imath \frac{kj}{3m}} = \begin{cases} \frac{1}{3} & \text{if } k = 0, \\ 0 & \text{otherwise.} \end{cases}$$

Thus,

$$I = \frac{1}{3}\Big(|1 + 2\cos 2\pi mt|$$

$$+ |1 - \cos 2\pi mt + \sqrt{3}\sin 2\pi mt| + |1 - \cos 2\pi mt - \sqrt{3}\sin 2\pi mt|\Big)$$

$$\le \frac{1}{3}\max_{t}\Big\{|1 + 2\cos 2\pi mt| + |1 - \cos 2\pi mt + \sqrt{3}\sin 2\pi mt|$$

$$+ |1 - \cos 2\pi mt - \sqrt{3}\sin 2\pi mt|\Big\}$$

$$= \frac{5}{3}.$$

Combining this result with (4.65), and noticing that $|\Phi_n(t)| = |F_a(t)|$, we obtain the claimed result. For odd n, the same arguments work.

For the lower bound when n is even it is enough to consider the polynomials $1 - \frac{1}{2}e^{\pi i(n-1)t} + e^{2\pi i(n-1)t}$, and check that they attain the claimed upper bound. $\quad\square$

This approach can be further developed for r having the form $\frac{s}{s+1}$.

Theorem 4.11 *For every MC signal, $F_a(t) = \sum_{k=0}^{n-1} a_k e^{2\pi i k t}$, and integer s, $s \ge 1$,*

$$\mathcal{M}_c(F_a) \le \frac{\gamma_s}{s+1}\cdot\mathcal{M}_d\left(F_a,\,(s+1)\left\lceil\frac{n-1}{s}\right\rceil\right), \qquad (4.66)$$

where

$$\gamma_s = \max_{t\in[0,1)}\sum_{j=0}^{s}\left|\sum_{k=0}^{s}e^{2\pi i k\left(t - \frac{j}{s+1}\right)}\right|.$$

Proof This is similar to the proof of the previous theorem. Again we construct the kernel analogous to (4.59) and being the sum of $s + 1$ Fejér kernels (4.63) with shifted arguments. We omit the details. $\quad\square$

For example, if $s = 1$, we have

$$\gamma_1 = \left|1 + e^{2\pi i t}\right| + \left|1 - e^{2\pi i t}\right| = 2\sqrt{2}.$$

4.5 Projections on measuring axes

In the previous sections I assumed that we can measure the absolute value of an MC signal in the sampling points. This is not always possible, and we have to use in our estimates only values of the real or imaginary part of the signal, or, more generally, projections of the signal on specially chosen axes. Let $F_n^R(t)$ and $F_n^I(t)$ be the real and imaginary parts of $F_a(t) = \sum_{k=0}^{n-1} a_k e^{2\pi i k t}$, respectively. Then, using notation r for the oversampling rate and C_r for the maximum ratio between the discrete and continuous maxima of MC signals, we have

Lemma 4.12 *Let $rn \in \mathbb{N}$. Then*

$$\max_{t \in [0,1)} |F_{\mathbf{a}}(t)| \leq C_r \sqrt{2} \max_{j=0,1,\ldots,rn-1} \left\{ F_{\mathbf{a}}^R \left(\frac{j}{rn} \right), F_{\mathbf{a}}^I \left(\frac{j}{rn} \right) \right\}.$$

Proof This is trivial. □

In fact, this approach can be improved. Indeed, instead of projecting on two axes (real and imaginary) we may pick a greater number, say h, of evenly distributed lines passing through the origin, $r_\ell(\varphi) = re^{(\frac{2\pi\ell}{h} + \varphi)i}$, $r, \varphi \in \mathbb{R}$, $\ell = 0, 1, \ldots, h - 1$. For a complex number c, let $c^{(\ell)}(\varphi)$ be its orthogonal projection on $r_\ell(\varphi)$, $|c^{(\ell)}(\varphi)| = |\langle c, r_\ell(\varphi) \rangle|$. We can also write $c^{(\ell)}(\varphi) = \Re\left(c \cdot e^{-(\frac{2\pi\ell}{h} + \varphi)i}\right)$.

Straightforward analysis similar to the proof of the previous lemma then gives the following statement.

Lemma 4.13 *For any complex number c,*

$$|c| \leq \max_{\ell=0,1,\ldots,h-1} \frac{\left|c^{(\ell)}(\varphi)\right|}{\cos\left(\frac{\pi}{2h}\right)}. \tag{4.67}$$

If c belongs to the set $R = \{r_h e^{\frac{2\pi i}{s} h}\}$, $r_h \in \mathbb{R}$, $h \in \mathbb{Z}$, and h divides s, choosing $\varphi = \frac{\pi}{2h} - \frac{\pi}{s}$, we get

$$|c| \leq \max_{\ell=0,1,\ldots,h-1} \frac{\left|c^{(\ell)}(\varphi)\right|}{\cos\left(\frac{\pi}{2}\left(\frac{1}{h} - \frac{2}{s}\right)\right)}. \tag{4.68}$$

□

This yields the following useful result.

Theorem 4.14 *Let $rn \in \mathbb{N}$ and $\varphi \in \mathbb{R}$. Then*

$$\max_{t \in [0,1)} |F_{\mathbf{a}}(t)| \leq C_r \max_{j=0,1,\ldots,rn-1} \max_{\ell=0,1,\ldots,h-1} \frac{\left|\left(F_{\mathbf{a}}\left(\frac{j}{rn}\right)\right)^{(\ell)}(\varphi)\right|}{\cos\left(\frac{\pi}{2h}\right)}.$$

□

Notice that Lemma 4.12 is a special case of Theorem 4.14 when $h = 2$ and $\varphi = 0$.

So far, I have discussed measuring the maximum of the absolute value. However, there are situations when we have to take into account the sign of the projection value. Analogously to the previous theorem we deduce the following result.

Theorem 4.15 *Let $rn \in \mathbb{N}$ and $\varphi \in \mathbb{R}$. Then*

$$\max_{t \in [0,1)} |F_{\mathbf{a}}(t)| \leq C_r \hat{C}_h \max_{j=0,1,\ldots,rn-1} \max_{\ell=0,1,\ldots,h-1} \left(F_{\mathbf{a}}\left(\frac{j}{rn}\right)\right)^{(\ell)}(\varphi),$$

where

$$\hat{C}_h = \begin{cases} \frac{1}{\cos(\frac{\pi}{h})} & h > 3, h \text{ even}, \\ \frac{3-\cos(\frac{\pi}{h})}{1+\cos(\frac{\pi}{h})} & h \geq 3, h \text{ odd}. \end{cases}$$

$\square$

4.6 Relation between PAPR and PMEPR

Recalling the definitions of PAPR and PMEPR, for an n-dimensional complex-valued vector $\mathbf{a}$, we deal with two functions,

$$F_\mathbf{a}(t) = \sum_{k=0}^{n-1} a_k e^{2\pi \iota k t},$$

and

$$S_\mathbf{a}(t) = \sum_{k=0}^{n-1} a_k e^{2\pi \iota (\zeta+k)t} = e^{2\pi \iota \zeta t} \cdot F_\mathbf{a}(t).$$

Then

$$\text{PAPR}(\mathbf{a}) = \frac{1}{n} \cdot \max_{t \in [0,1)} |\Re(S_\mathbf{a}(t))|,$$

and

$$\text{PMEPR}(\mathbf{a}) = \frac{1}{n} \cdot \max_{t \in [0,1)} |F_\mathbf{a}(t)|.$$

We know that PAPR always does not exceed PMEPR, see (2.5). We will show that for large values of ζ in comparison with n, PAPR cannot be essentially smaller than PMEPR. We will need the following result, elaborating on arguments used in the proof of (4.49).

We start by proving that the phase of $F_\mathbf{a}(t)$ cannot change too fast.

Lemma 4.16   *Let*

$$F_\mathbf{a}(t) = \sum_{k=0}^{n-1} a_k e^{2\pi \iota k t} = |F_\mathbf{a}| \cdot e^{\iota \arg F_\mathbf{a}(t)}$$

be an MC signal, and $\max |F_\mathbf{a}(t)|$ *be attained at* $t = t_0$. *Then for any* $t \in [t_0 - \Delta, t_0 + \Delta]$,

$$|\arg F_\mathbf{a}(t_0) - \arg F_\mathbf{a}(t^*)| \leq \frac{\pi^2 n^{\frac{3}{2}}}{2} \Delta.$$

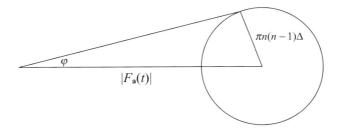

Figure 4.3 The triangle

Proof

$$F_{\mathbf{a}}(t + \Delta) = \sum_{k=0}^{n-1} e^{2\pi \imath k(t+\Delta)} = \sum_{k=0}^{n-1} e^{2\pi \imath kt} \cdot e^{2\pi \imath k\Delta}$$

$$= \sum_{k=0}^{n-1} e^{2\pi \imath kt} \cdot (1 + \gamma_k) = F_{\mathbf{a}}(t) + \sum_{k=0}^{n-1} \gamma_k e^{2\pi \imath kt}.$$

Using (see (3.27))

$$|\gamma_k| = |e^{2\pi \imath k\Delta} - 1| \leq 2\pi k\Delta,$$

we obtain

$$|F_{\mathbf{a}}(t) - F_{\mathbf{a}}(t + \Delta)| \leq 2\pi \Delta \sum_{k=0}^{n-1} k = \pi n(n - 1)\Delta.$$

Consider a triangle in the complex plane with vertices in 0, $F_{\mathbf{a}}(t + \Delta)$ and $F_{\mathbf{a}}(t)$. The three sides of the triangle have lengths $|F_{\mathbf{a}}(t + \Delta)|$, $|F_{\mathbf{a}}(t)|$ and $\leq \pi n(n - 1)\Delta$. An easy trigonometric exercise, see Fig. 4.3, is that the triangle's angle, φ, opposite to the side connecting $F_{\mathbf{a}}(t + \Delta)$ and $F_{\mathbf{a}}(t)$ satisfies

$$\frac{2}{\pi}\varphi \leq \sin \varphi \leq \frac{\pi n(n - 1)\Delta}{|F_{\mathbf{a}}(t)|},$$

were we used (3.23) in the first inequality. Taking into account (see (4.5)) the fact that $|F_{\mathbf{a}}(t_0)| \geq \sqrt{n}$, we accomplish the proof. $\square$

Corollary 4.17 *There exists $t^* \in \left[t_0 - \frac{1}{4\zeta - \pi n^{\frac{3}{2}}}, t_0 + \frac{1}{4\zeta - \pi n^{\frac{3}{2}}}\right]$ such that*

$$\arg S_{\mathbf{a}}(t^*) = 0.$$

Proof Let $\arg S_{\mathbf{a}}(t_0) \in \left[-\frac{\pi}{2}, 0\right)$. For $t_1 > t_0$, $t_1 - t_0 = \Delta$, we have

$$\arg S_{\mathbf{a}}(t_1) - \arg S_{\mathbf{a}}(t_0) \geq 2\pi \zeta \Delta - \frac{\pi^2 n^{\frac{3}{2}}}{2}\Delta.$$

If this difference is greater than $\pi/2$, there is a point $t^* \in [t_0, t_1]$ for which $\arg S_\mathbf{a}(t^*) = 0$. This is satisfied if

$$\Delta \geq \frac{1}{4\zeta - \pi n^{\frac{3}{2}}}.$$

The other cases of $\arg S_\mathbf{a}(t_0)$ are treated analogously.   □

Lemma 4.18   *Let* $F_\mathbf{a}(t) = \sum_{k=0}^{n-1} a_k e^{2\pi \imath k t}$ *be an MC signal, and* t_0 *be such that* $|F_\mathbf{a}(t_0)| = \mathcal{M}_c(F_\mathbf{a})$. *Then for any* $t \in [t_0 - \Delta, t_0 + \Delta]$, *we have*

$$|F_\mathbf{a}(t)| \geq \mathcal{M}_c(F_\mathbf{a}) \cdot \sqrt{1 - 2\pi^2(n-1)^2\Delta^2}.$$

Proof   Define $s(\theta) = \left| F_\mathbf{a}\left(\frac{\theta}{2\pi}\right) \right|^2$. By Lemma 3.24, $s(\theta)$ is a real trigonometric polynomial of order $(n-1)$. Let $\theta_0 = 2\pi t_0$. Since $s'(\theta_0) = 0$, we have the following second-order Taylor expansion at θ_0,

$$s(\theta) = s(\theta_0) + \frac{1}{2}(\theta - \theta_0)^2 |s''(\theta^*)|,$$

where θ^* is a point between θ and θ_0. Applying twice Bernstein's inequality for real trigonometric polynomials we obtain

$$s(\theta) \geq s(\theta_0)\left(1 - \frac{(n-1)^2}{2}(\theta - \theta_0)^2\right),$$

or

$$|F_\mathbf{a}(t)|^2 \geq \mathcal{M}_c^2(F_\mathbf{a})\left(1 - 2\pi^2(n-1)^2(t - t_0)^2\right),$$

and we have proved the claim.   □

Theorem 4.19   *For any MC signal defined by* $\mathbf{a}$,

$$\mathrm{PAPR}(\mathbf{a}) \geq \mathrm{PMEPR}(\mathbf{a}) \cdot \sqrt{1 - \left(\frac{\sqrt{2}\,\pi n}{\left(4\zeta - \pi n^{\frac{3}{2}}\right)}\right)^2}. \tag{4.69}$$

Proof   Combine Corollary 4.17 with Lemma 4.18.   □

For $\zeta \gg n^{\frac{3}{2}}$, (4.69) simplifies to

$$\mathrm{PAPR}(\mathbf{a}) \geq \mathrm{PMEPR}(\mathbf{a}) \cdot \left(1 - \left(\frac{\pi n}{4\zeta}\right)^2\right).$$

4.7 Notes

Section 4.1   Theorem 4.2 in the form presented is from Litsyn and Yudin [248]. For the critical sampling of real trigonometric polynomials, the corresponding ratio (the Lebesgue constant) is estimated in Schönhage [359] and Ehlich and Zeller [103]. An additive constant slightly worse than in the upper bound of (4.9) was derived by Paterson and Tarokh [328]. Other works on the ratio between discrete and continuous maxima are those of Wulich [435], Jedwab [177], Minn *et al.* [271], Tellambura [405, 409], Sharif and Khalaj [377], and Sharif *et al.* [366].

Section 4.2   Theorem 4.5 is by Litsyn and Yudin [248].

Section 4.3   Theorem 4.6 is by Litsyn and Yudin [253].

Section 4.4   Theorem 4.7 is by Jetter *et al.* [182]. In the proof of Theorem 4.8 the ratio for real trigonometric polynomials was first considered by M. Riesz in 1914 [342]. Ehlich and Zeller [102] elaborated on this result. The bound for complex trigonometric polynomials was given by Wunder and Boche [442], see also [31, 32, 33, 438, 440, 441]. Theorem 4.9 was proven by Jetter *et al.* [182]. Theorem 4.10 is by Litsyn and Yudin [248]. Theorem 4.11 is by Litsyn and Yudin [253].

Section 4.5   Theorem 4.14 appears in Litsyn and Shpunt [251]. Wunder and Boche [442] used a similar estimate. However, they allow projections only in the positive direction, and thus have twice as many axes than in Theorem 4.14, for the even number of axes.

Section 4.6   Lemma 4.18 is from Sharif *et al.* [366].

5

Statistical distribution of peak power in MC signals

In this chapter, I consider estimates for the probability distribution of peaks in MC signals. This is done under the assumption that in each subcarrier the choice of the constellation point is made independently of the choices in other subcarriers, and all the points in the constellation are equiprobable. In Section 5.1, I apply the Chernoff bound to construct an upper bound on the probability of a peak exceeding some prescribed value. In Section 5.2, I demonstrate that the distribution of PMEPR in n-subcarrier MC signals is concentrated around $\ln n$. Section 5.3 deals with approximations stemming from modeling MC signals as stochastic processes. In Section 5.4, I show that there are many MC signals with constant PMEPR. Finally, in Section 5.5, I give a simple criterion for a signal to have an essentially high peak, of size linear in n, and estimate the number of MC signals with such peaks.

Throughout, I mean by $f(x) = O(g(x))$ that $\left| \frac{f(x)}{g(x)} \right| \le c$ for all x, where $c > 0$ is a real constant.

5.1 Upper bounds for PMEPR distribution

We consider a multicarrier signal $F_{\mathbf{a}}(t) = \sum_{k=0}^{n-1} a_k e^{2\pi \imath k t}$, defined by the coefficient vector $\mathbf{a} = (a_0, a_1, ..., a_{n-1})$. For convenience, we define $t_{j,r} = \frac{j}{rn}$ where $r > 1$ is the oversampling factor and rn is an integer. Note that the case $r = 1$ corresponds to Nyquist-rate sampling. We need the following results from Chapter 4.

Let the samples of $F_{\mathbf{a}}(t)$ be

$$\{F_{\mathbf{a}}(0), F_{\mathbf{a}}(t_{1,r}), \ldots, F_{\mathbf{a}}(t_{rn-1,r})\}.$$

Then (see Theorem 4.8) the inequality

$$\max_{0 \le t < 1} |F_{\mathbf{a}}(t)| < C_r \cdot \max_{j=0,1,...,rn-1} |F_{\mathbf{a}}(t_{j,r})|,$$

where

$$C_r = \frac{1}{\cos \frac{\pi}{2r}}, \tag{5.1}$$

holds.

Let $\alpha_{s,h} = \frac{2\pi s}{h}$ for a natural $h > 2$ and $s = 0, 1, \ldots, h - 1$. Then (see Theorem 4.15)

$$\max_{\alpha \in [0, 2\pi)} \Re(F_{\mathbf{a}}(t)e^{\iota \alpha}) \leq \hat{C}_h \cdot \max_{s = 0, 1, \ldots, h-1} \Re(F_{\mathbf{a}}(t)e^{\iota \alpha_{s,h}}),$$

where

$$\hat{C}_h = \begin{cases} \frac{1}{\cos(\frac{\pi}{h})} & h > 3, h \text{ even}, \\ \frac{3 - \cos(\frac{\pi}{h})}{1 + \cos(\frac{\pi}{h})} & h \geq 3, h \text{ odd}. \end{cases} \tag{5.2}$$

Let

$$\mathcal{G} = \{(t_{j_1, r}, \alpha_{j_2, h}), j_1 = 0, 1, \ldots, rn - 1; j_2 = 0, 1, \ldots, h - 1\} \tag{5.3}$$

be a lattice in the square $[0, 1) \times [0, 2\pi)$, and suppose that the probabilities

$$\Pr(\Re(F_{\mathbf{a}}(t)e^{\iota \alpha}) > \lambda \sqrt{n}),$$

(t, α) running through $\mathcal{G}$, are given. Using the equality

$$|F_{\mathbf{a}}(t)| = \max_{\alpha \in [0, 2\pi)} \Re(F_{\mathbf{a}}(t)e^{\iota \alpha})$$

we have

$$\mathcal{F}(\lambda) = \Pr\left(\max_{t \in [0,1)} |F_{\mathbf{a}}(t)| > \lambda \sqrt{n}\right)$$

$$= \Pr\left(\max_{t \in [0,1), \alpha \in [0, 2\pi)} \Re(F_{\mathbf{a}}(t)e^{\iota \alpha}) > \lambda \sqrt{n}\right)$$

$$\leq \min_{r > 1, h > 2} \Pr\left(\max_{(t, \alpha) \in \mathcal{G}} \Re(F_{\mathbf{a}}(t_{j_1, r})e^{\iota \alpha_{j_2, h}}) > \frac{\lambda \sqrt{n}}{C_r \hat{C}_h}\right)$$

and by the union bound

$$\mathcal{F}(\lambda) \leq \min_{r > 1, h > 2} \sum_{j_1 = 0}^{rn - 1} \sum_{j_2 = 0}^{h - 1} \Pr\left(\Re(F_{\mathbf{a}}(t_{j_1, r})e^{\iota \alpha_{j_2, h}}) > \frac{\lambda \sqrt{n}}{C_r \hat{C}_h}\right). \tag{5.4}$$

The probability terms can be replaced with the Chernoff bound, i.e.,

$$\Pr(\Re(F_{\mathbf{a}}(t)e^{\iota \alpha}) > \mu) \leq E\left(e^{\varepsilon(\Re(F_{\mathbf{a}}(\theta)e^{\iota \alpha}) - \mu)}\right), \tag{5.5}$$

valid for any $\varepsilon > 0$, and the expectation is taken over all possible choices of **a**. Furthermore,

$$E\left(e^{\varepsilon(\Re(F_a(\theta)e^{\iota\alpha})-\mu)}\right) = e^{-\varepsilon\mu}E\left(e^{\varepsilon\left(\Re\left(\sum_{k=0}^{n-1}a_k e^{\iota(2\pi kt+\alpha)}\right)\right)}\right)$$

$$= e^{-\varepsilon\mu}\prod_{k=0}^{n-1}E\left(e^{\varepsilon\left(\Re\left(a_k e^{\iota(2\pi kt+\alpha)}\right)\right)}\right),$$

where the expectation is already over all equally probable choices of a_k from the constellation. Summarizing we obtain the following result.

Theorem 5.1   *Let $F_a(t)$ belong to the ensemble of MC signals defined by the vectors $\mathbf{a} = (a_0, a_1, \ldots, a_{n-1})$ with the coefficients a_k, $k = 0, 1, \ldots, n-1$, equiprobably chosen from a constellation Q. Then*

$$\Pr\left(\max_{t\in[0,1)}|F_a(t)| > \lambda\sqrt{n}\right)$$

$$\leq \min_{\varepsilon>0}\min_{r>1, rn\in\mathbb{N}}\min_{h>2, h\in\mathbb{N}} e^{-\varepsilon\mu}\sum_{t,\alpha\in\mathcal{G}}\prod_{k=0}^{n-1}E\left(e^{\varepsilon\left(\Re\left(a_k e^{\iota(2\pi kt+\alpha)}\right)\right)}\right),$$

where the expectation is over Q, $\mu = \frac{\lambda\sqrt{n}}{C_r\hat{C}_h}$, C_r and $\hat{C}_h$ are defined in (5.1) and (5.2), and $\mathcal{G}$ is defined in (5.3).   □

Let us now estimate the derived upper bounds for standard constellations. We start with QAM. We assume

$$M\text{-QAM} = \{A\left((2m_1 - 1) + \iota\,(2m_2 - 1)\right), m_1, m_2 \in \{-m/2 + 1, \ldots, m/2\}\}$$

for a natural $m > 1$, and $M = m^2$.

Theorem 5.2   *Let $Q = M\text{-QAM}$. Then the following upper bound holds:*

$$\Pr\left(\max_{t\in[0,1)}|F_a(t)| > \lambda\sqrt{n}\right)$$

$$\leq \min_{\varepsilon>0}\min_{r>1, rn\in\mathbb{N}}\min_{h>2, h\in\mathbb{N}} rhn \cdot e^{-\varepsilon\mu+\frac{\varepsilon^2 n}{4}} \cdot \epsilon_1^{2n}\left(\varepsilon\sqrt{\frac{3}{2(M-1)}}\right),$$

where $\mu = \frac{\lambda\sqrt{n}}{C_r\hat{C}_h}$, C_r and $\hat{C}_h$ are defined in (5.1) and (5.2), and

$$\epsilon_1(x) = 1 - \frac{\sinh(x)\left(e^{\frac{Mx^2}{24}} - e^{\left(\frac{M}{12}-\frac{1}{3}\right)\frac{x^2}{2}}\right) - \frac{x^3}{6}}{\sinh(x)\,e^{\left(\frac{M}{12}-\frac{1}{3}\right)\frac{x^2}{2}}},$$

satisfying $\epsilon_1(x) \to (1 + O(x^4))$ as $x \to 0$.

Proof   We will show that if $a_k \in M$-QAM then the following upper bound holds:

$$E\left(e^{x\Re(a_k e^{i(2\pi kt+\alpha)})}\right) \le e^{\frac{x^2}{4}}\epsilon_1^2\left(x\sqrt{\frac{3}{2(M-1)}}\right).$$

Notice that the right-hand side of the inequality does not depend on either k, t, or α.

For the proof, we set $A = 1$ and scale x appropriately. Since

$$\Re\left(a_k e^{i(2\pi kt+\alpha)}\right) = \Re(a_k)\cos\left(2\pi kt + \alpha\right) - \Im(a_k)\sin\left(2\pi kt + \alpha\right),$$

we have

$$E\left(e^{x\Re(a_k e^{i(2\pi kt+\alpha)})}\right) = E\left(e^{x\Re(a_k)\cos(2\pi kt+\alpha)}\right) \cdot E\left(e^{-x\Im(a_k)\sin(2\pi kt+\alpha)}\right). \qquad (5.6)$$

Now,

$$E\left(e^{x\Re(a_k)\cos(2\pi kt+\alpha)}\right)$$

$$= \frac{1}{m}\left(e^{-x\cos(2\pi kt+\alpha)}\sum_{s=1}^{\frac{m}{2}}e^{2xs\cos(2\pi kt+\alpha)} + e^{x\cos(2\pi kt+\alpha)}\sum_{s=1}^{\frac{m}{2}}e^{-2xs\cos(2\pi kt+\alpha)}\right).$$

Using the formula for the sum of geometric progression, we have

$$\sum_{s=1}^{\frac{m}{2}}e^{2xs\cos(2\pi kt+\alpha)} = e^{x\left(1+\frac{m}{2}\right)\cos(2\pi kt+\alpha)} \cdot \frac{\sinh\left(\frac{xm}{2}\cos\left(2\pi kt + \alpha\right)\right)}{\sinh\left(x\cos\left(2\pi kt + \alpha\right)\right)}$$

and

$$\sum_{s=1}^{\frac{m}{2}}e^{-2xs\cos(2\pi kt+\alpha)} = e^{-x\left(1+\frac{m}{2}\right)\cos(2\pi kt+\alpha)} \cdot \frac{\sinh\left(\frac{xm}{2}\cos\left(2\pi kt + \alpha\right)\right)}{\sinh\left(x\cos\left(2\pi kt + \alpha\right)\right)}.$$

Analogously processing the second term in (5.6) after straightforward manipulations we may conclude that for $a_k \in M$-QAM,

$$E\left(e^{x\Re(a_k e^{i(2\pi kt+\alpha)})}\right) \le \prod_{s=1}^{2}\frac{2\sinh\left(\frac{xmf_s(2\pi kt+\alpha)}{2}\right)\cosh\left(\frac{xmf_s(2\pi kt+\alpha)}{2}\right)}{m\sinh\left(xf_s\left(2\pi kt + \alpha\right)\right)}$$

where $f_1(x) = \cos(x)$ and $f_2(x) = \sin(x)$. Since

$$\cosh\left(\frac{mx}{2}\right) \le e^{\frac{Mx^2}{8}}$$

and $f_1^2(x) + f_2^2(x) = 1$ for all x, I will show that

$$\frac{\sinh\left(\frac{mx}{2}\right)}{\sinh(x)} \le \frac{m}{2} \cdot e^{\left(\frac{M}{12}-\frac{1}{3}\right)\frac{x^2}{2}} \cdot \epsilon_1(x)$$

and thus

$$E\left(e^{x\Re(a_k e^{i(2\pi kt + \alpha)})}\right) \le e^{\frac{(M-1)}{3} \frac{x^2}{2}} \epsilon_1(x).$$

Then, after rescaling, the result will follow.

First, we want to find $\gamma > 0$ such that

$$\frac{\sinh\left(\frac{mx}{2}\right)}{\sinh(x)} \le \frac{m}{2} e^{\frac{\gamma x^2}{2}}. \tag{5.7}$$

Since $\frac{\sinh\left(\frac{mx}{2}\right)}{\sinh(x)}$ is symmetric around $x = 0$ it suffices to prove inequality (5.7) for $x \ge 0$. Using $\sinh(x) \ge x$ we have

$$\frac{\sinh\left(\frac{mx}{2}\right)}{\sinh(x)} \le \frac{\sinh\left(\frac{mx}{2}\right)}{x}$$

and $\sinh(x) \le xe^{\frac{x^2}{6}}$ yields the upper bound

$$\frac{\sinh\left(\frac{mx}{2}\right)}{\sinh(x)} \le \frac{\frac{mx}{2} e^{\frac{Mx^2}{24}}}{x} = \frac{m}{2} e^{\frac{Mx^2}{24}}.$$

Thus $\gamma < \frac{M}{12}$. Since x will be small, the inequality will be evaluated at small values and can be improved. Expanding

$$\sinh\left(\frac{mx}{2}\right) = \frac{mx}{2} + \frac{(m)^3 x^3}{48} + \dots$$

and

$$e^{\frac{\gamma x^2}{2}} = 1 + \frac{\gamma x^2}{2} + \dots$$

yields

$$\frac{m}{2} e^{\frac{\gamma x^2}{2}} \sinh(x) = \frac{mx}{2} + \left(\frac{\gamma}{2} + \frac{1}{6}\right) \frac{mx^3}{2} + \dots$$

Considering the linear and cubic term only yields

$$\frac{(m)^3}{48} \le \left(\frac{\gamma}{2} + \frac{1}{6}\right) \frac{m}{2}$$

and hence

$$\gamma \ge \frac{M}{12} - \frac{1}{3}.$$

Using the lower bound on γ we have the following inequality chain

$$\sinh\left(\frac{mx}{2}\right) \le \frac{m}{2} \sinh(x) e^{\frac{\gamma_1 x^2}{2}} + \epsilon'(x) \le \frac{m}{2} \sinh(x) e^{\frac{\gamma_2 x^2}{2}}$$

where $\frac{M}{12} - \frac{1}{3} = \gamma_1 \leq \gamma_2 = \frac{M}{12}$ and $\epsilon'(x)$ is an error term. Since the coefficients of the power series are all positive, the inequality will hold for each coefficient. Hence, the error will be upper bounded by

$$\epsilon'(x) = \frac{m}{2} \sinh(x) e^{\frac{\gamma_2}{2}x^2} - \frac{mx}{2} - \left(\frac{\gamma_2}{2} + \frac{1}{6}\right) \frac{mx^3}{2}$$
$$- \left(\frac{m}{2} \sinh(x) e^{\frac{\gamma_1}{2}x^2} - \frac{mx}{2} - \left(\frac{\gamma_1}{2} + \frac{1}{6}\right) \frac{mx^3}{2}\right).$$

Dividing by $\frac{m}{2} \sinh(x)$ and $e^{\left(\frac{M}{12} - \frac{1}{3}\right)\frac{x^2}{2}}$ yields

$$\epsilon_1(x) = 1 + \frac{\epsilon'(x)}{\frac{m}{2} \sinh(x) e^{\left(\frac{M}{12} - \frac{1}{3}\right)\frac{x^2}{2}}}.$$

The term $\epsilon'(x)$ satisfies $\epsilon'(x) = O(x^5)$ as $x \to 0$ and thus $\epsilon_1(x) = 1 + O(x^5/x)$ as $x \to 0$. In order to get the final result we scale

$$x \to x\sqrt{\frac{3}{2}(M-1)}.$$

$\square$

Theorem 5.3 *Let $\mathcal{Q} = M\text{-}PSK$. Then the following upper bound holds:*

$$\Pr\left(\max_{t \in [0,1)} |F_a(t)| > \lambda\sqrt{n}\right) \leq \min_{\varepsilon > 0} \min_{r > 1, rn \in \mathbb{N}} \min_{h > 2, h \in \mathbb{N}} rhn \cdot e^{-\varepsilon\mu + \frac{\varepsilon^2 n}{4}} \cdot \epsilon_2^n(\varepsilon),$$

where $\mu = \frac{\lambda\sqrt{n}}{C_r \hat{C}_h}$, C_r and $\hat{C}_h$ are defined in (5.1) and (5.2), and

$$\epsilon_2(x) = 1 + \frac{e^x - 1 - x - \frac{x^2}{2} - \left(e^{\frac{x^4}{4}} - 1 - \frac{x^4}{4}\right)}{e^{\frac{x^4}{2}}},$$

satisfying $\epsilon_2(x) \to (1 + O(x^3))$ as $x \to 0$.

Proof We will prove that for $a_k \in M\text{-}PSK$ the following upper bound holds:

$$E\left(e^{x\Re(a_k e^{i(2\pi kt + \alpha)})}\right) \leq e^{\frac{x^2}{4}} \epsilon_2(x).$$

Expanding the exponential function yields

$$\frac{1}{M} \sum_{s=0}^{M-1} e^{x \cos \frac{2\pi s}{M}} = 1 + \frac{\sum_{s=0}^{M-1} \cos\left(\alpha + \frac{2\pi s}{M}\right) x}{M} + \frac{\sum_{s=0}^{M-1} \cos^2\left(\alpha + \frac{2\pi s}{M}\right) x^2}{2M} + \dots$$

$$= 1 + \frac{x^2}{4} + \dots$$

where we used (see Theorems 3.14 and 3.15)

$$\sum_{s=0}^{M-1} \cos\left(\alpha + \frac{2\pi s}{M}\right) = 0$$

and

$$\sum_{s=0}^{M-1} \cos^2\left(\alpha + \frac{2\pi s}{M}\right) = \frac{M}{2},$$

valid for any α. The coefficients are trivially upper bounded by the coefficients of e^x and we can write

$$\frac{1}{M} \sum_{s=0}^{M-1} e^{x\cos\left(\alpha + \frac{2\pi s}{M}\right)} \le e^{\frac{x^2}{4}} + \epsilon'(x),$$

where the error is upper bounded by

$$\epsilon'(x) = e^x - 1 - x - \frac{x^2}{2} - \left(e^{\frac{x^4}{4}} - 1 - \frac{x^4}{4}\right).$$

The term $\epsilon'(x)$ satisfies $\epsilon'(x) = O(x^3)$ as $x \to 0$ and thus $\epsilon_2(x) = 1 + O\left(\frac{x^3}{1+x^2}\right) = 1 + O(x^3)$ as $x \to 0$.   □

Corollary 5.4   *For M-QAM,*

$$\Pr(\mathrm{PMEPR}(\mathbf{a}) > \lambda) \le \min_{r>1, h>2} rhn\, e^{-\frac{\lambda}{C_r^2 \hat{C}_h^2}} \cdot \epsilon_1^{2n}\left(\frac{2\sqrt{\lambda}}{C_r \hat{C}_h \sqrt{n}} \sqrt{\frac{3}{2(M-1)}}\right) \quad (5.8)$$

and for M-PSK,

$$\Pr(\mathrm{PMEPR}(\mathbf{a}) > \lambda) \le \min_{r>1, h>2} rhn\, e^{-\frac{\lambda}{C_r^2 \hat{C}_h^2}} \cdot \epsilon_2^{n}\left(\frac{2\sqrt{\lambda}}{C_r \hat{C}_h \sqrt{n}}\right). \quad (5.9)$$

Proof   Set $\varepsilon = \frac{2\sqrt{\lambda}}{C_r \hat{C}_h \sqrt{n}}$ in Theorems 5.2 and 5.3 and use the definition of PMEPR.   □

Simulations show that the derived bounds compare favorably to other approximations in the low probability region. Let us consider the behavior of the bounds when $n \to \infty$.

Theorem 5.5   *For growing n and* **a** *randomly chosen from either M-QAM or M-PSK,*

$$\Pr(\mathrm{PMEPR} > \ln n + \ln \ln n + c) \le e^{-c}(1 + o(1)).$$

Proof   In this case we may choose growing r and h, pushing the error terms ϵ_1 and ϵ_2 to tend to 1. Therefore we face only minimization of $p = rhn\, e^{-\frac{\lambda}{C_r^2 \hat{C}_h^2}}$. Restricting ourselves to even h, we obtain, using $\cos^2 x > 1 - x^2$,

$$\frac{\lambda}{C_r^2 \hat{C}_h^2} \geq \lambda \cdot \cos^2 \frac{\pi}{2r} \cdot \cos^2 \frac{\pi}{h} > \lambda \left(1 - \frac{\pi^2}{4r^2} - \frac{\pi^2}{h^2} \right).$$

Setting $h = 2r$, and $a = \frac{\pi^2}{2r^2}$, we conclude that

$$p < \frac{\pi^2}{a} n \, e^{-\lambda(1-a)}.$$

Furthermore, setting $a = \frac{\epsilon(n)}{\ln n}$ for a slowly growing function $\epsilon(n)$, and

$$\lambda = \ln n - \ln \epsilon(n) - 2 \ln \pi + \ln \ln n + c,$$

we obtain

$$p < e^{-c + \epsilon(n) + o(n)},$$

giving the claim. $\qquad\square$

5.2 Lower bounds for PMEPR distribution

We start with the QAM case.

Theorem 5.6 *Let $F_{\mathbf{a}}(t) = \sum_{k=0}^{n-1} a_k e^{2\pi \imath k t}$ be an MC signal, where a_k are chosen equiprobably from M-QAM with $E_{av} = 1$. Then*

$$\Pr\left(\text{PMEPR}(\mathbf{a}) \leq \ln n - 6.5 \ln \ln n\right) \leq O\left(\frac{1}{\ln^4 n}\right).$$

Proof We will construct a lower bound on $\Pr(\text{PMEPR}(\mathbf{a}) > \lambda)$. Instead of analyzing the continuous maximum of $|F_{\mathbf{a}}(t)|$, it is enough to consider the maximum of $\Re(F_{\mathbf{a}}(t))$ over its n equidistributed samples at $t_j = \frac{j}{n}$, $j = 0, 1, \ldots, n-1$. Define a function $u(x)$, $0 \leq u(x) \leq 1$, satisfying

$$u(x) = \begin{cases} 0, & |x| \leq L, \\ 1, & |x| \geq L + \Delta, \end{cases}$$

where $\Delta = \sqrt{\frac{n}{\ln n}}$ and $L = \sqrt{n \ln n - 6.5 n \ln \ln n} - \Delta$. We also assume that $u(x)$ is a function that is ten-times differentiable such that $u^{(r)}(x) = O(\Delta^{-r})$ for $1 \leq r \leq 10$.

We will need the following technical lemma.

Lemma 5.7 *Let*

$$u(x) = \int_{-\infty}^{\infty} e^{\imath s x} v(s) \, ds, \qquad (5.10)$$

then

$$v(s) = \frac{1}{2\pi} \int_{-\infty}^{\infty} (u(x) - 1)e^{-\imath s x} \, dx, \tag{5.11}$$

and the following properties hold:

$$|s^r v(s)| = O\left(\frac{1}{\Delta^{r-1}}\right), \quad 1 \le r \le 10, \tag{5.12}$$

$$\int_{-\infty}^{\infty} |v(s)| \, ds = O\left(\frac{L}{\Delta}\right), \tag{5.13}$$

$$\int_{-\infty}^{\infty} |s^p v(s)| \, ds = O\left(\frac{1}{\Delta^p}\right), \quad 1 \le p \le 8, \tag{5.14}$$

$$\int_{|s|>\ell_0} |v(s)| \, ds = O\left(\frac{1}{\Delta^9}\right), \quad \text{for any constant } \ell_0 > 0, \tag{5.15}$$

$$\left| \int_{-\infty}^{\infty} e^{-\frac{ns^2}{4}} s^p v(s) \, ds \right| = O\left(\frac{\sqrt{n}\, e^{-\frac{L^2}{n}}}{L\Delta^p}\right), \quad 1 \le p \le 8, \tag{5.16}$$

$$\int_0^1 \int_0^1 \int_{-\infty}^{\infty} e^{-\frac{s^2}{4}\left(\sum_{k=0}^{n-1} \cos^2(2\pi kt + \gamma) + \sin^2(2\pi kt + \gamma)\right)} s^p v(s) \, ds \, d\gamma \, dt = O\left(\frac{\sqrt{n}\, e^{-\frac{L^2}{n}}}{L\Delta^p}\right). \tag{5.17}$$

Proof Since $u(x)$ is 1 everywhere except in the interval $[-L - \Delta, L + \Delta]$, $v(s) \, ds$ has a jump of unity at $s = 0$. Apart from this point it is evenly continuous with the derivative $v(s)$, given by (5.11).

(5.12): Since $u(x) - 1$ is nonzero only in $[-L - \Delta, L + \Delta]$, we have

$$|v(s)| = \left| \frac{1}{2\pi} \int_{-L-\Delta}^{L+\Delta} (u(x) - 1)e^{-\imath s x} \, dx \right| \le \frac{2(L+\Delta)}{2\pi} = O(L). \tag{5.18}$$

By partial integration we obtain

$$|s^r v(s)| \le 2\Delta \max |u^{(r)}(x)| = O\left(\frac{1}{\Delta^{r-1}}\right) \quad \text{for } r = 1, \dots, 10. \tag{5.19}$$

(5.13): Use (5.18) for $|s| \le \frac{1}{\Delta}$ and (5.19) with $r = 2$ otherwise.
(5.14): Use (5.19) with $r = p$ for $|s| \le \frac{1}{\Delta}$ and with $r = p + 2$ otherwise.
(5.15): Use (5.12) with $r = 10$ to show that $|v(s)| = O\left(\frac{1}{s^{10}\Delta^9}\right)$.
(5.16): We use Parseval's theorem and the properties of Fourier transform to obtain

$$\left| \int_{-\infty}^{\infty} e^{-\frac{ns^2}{4}} s^p v(s) \, ds \right| = \left| \frac{1}{\sqrt{\pi n}} \int_{-\infty}^{\infty} e^{-\frac{x^2}{n}} u^{(p)}(x) \, ds \right|. \tag{5.20}$$

Now we can use the fact that $u^{(p)}(x)$ is zero for $|x| < L$ and equals $O\left(\frac{1}{\Delta^r}\right)$ for $|x| > L$ to rewrite the integral as

$$O\left(\left|\int_{-\infty}^{\infty} e^{-\frac{ns^2}{4}} s^p v(s)\, ds\right|\right) = O\left(\frac{1}{\sqrt{n}\,\Delta^p} \int_{|x|>L} e^{-\frac{x^2}{n}}\, dx\right)$$

$$= O\left(\frac{Q\left(\frac{L}{\sqrt{n/2}}\right)}{\Delta^p}\right), \qquad (5.21)$$

where $Q(x) = \frac{1}{\sqrt{2\pi}} \int_x^{\infty} e^{-\frac{x^2}{2}}\, dx$. Using the asymptotic expansion

$$Q(x) = \frac{e^{-\frac{x^2}{2}}}{x\sqrt{2\pi}}\left(1 - O\left(\frac{1}{x^2}\right)\right),$$

we get

$$O\left(\frac{Q\left(\frac{L}{\sqrt{n/2}}\right)}{\Delta^p}\right) = O\left(\frac{\sqrt{n}\,e^{-\frac{L^2}{n}}}{L\Delta^p}\right). \qquad (5.22)$$

Now (5.16) follows from (5.21) and (5.22).

(5.17): We use (5.16) to write the inner integral in (5.17) as

$$\int_{-\infty}^{\infty} e^{-\frac{s^2}{4}\left(\sum_{k=0}^{n-1}\cos^2(2\pi kt + \gamma) + \sum_{k=0}^{n-1}\sin^2(2\pi kt + \gamma)\right)} s^p v(s)\, ds = O\left(\frac{\sqrt{n}}{L\Delta^p} \cdot e^{-\frac{L^2}{n}}\right). \qquad (5.23)$$

$\square$

Continuation of the proof of Theorem 5.6   Define the random variable η as

$$\eta = \sum_{m=0}^{n-1} u\left(\Re(F_{\mathbf{a}}(t_m))\right) = \sum_{m=0}^{n-1} \int_{-\infty}^{\infty} e^{is\Re(F_{\mathbf{a}}(t))} v(s)\, ds, \qquad (5.24)$$

where we have replaced $u(x)$ by its Fourier transform (5.10). To find a lower bound, we proceed as follows;

$$\Pr\left(\max_{t \in [0,1)} |F_{\mathbf{a}}(t)| \geq L\right) \geq \Pr\left(\max_{m=0,1,\ldots,n-1} |\Re(F_{\mathbf{a}}(t_m))|\right)$$

$$= 1 - \Pr(\eta = 0) \geq 1 - \Pr(\eta = 0 \vee \eta \geq E(\eta))$$

$$= 1 - \Pr(|\eta - E(\eta)| \geq E(\eta)) \geq 1 - \frac{\sigma_\eta}{E^2(\eta)}, \qquad (5.25)$$

where the last inequality is Chebyshev's inequality (Theorem 3.28). Let us analyze the first and second moments of η.

Lemma 5.8

$$E(\eta) \geq O(\ln^6 n), \quad \sigma_\eta^2 \leq O(E(\eta) \ln^2 n + \ln^5 n).$$

Proof The complex coefficients a_k can be presented as

$$a_k = a_k^R + \iota b_k^R,$$

where $a_k^R = \Re(a_k)$, $a_k^I = \Im(a_k)$, and a_k^R and a_k^I are independent identically distributed symmetric random variables. By (5.24),

$$\eta = \sum_{m=0}^{n-1} \int_{-\infty}^{\infty} e^{\iota s \sum_{k=0}^{n-1}(a_k^R \cos ktm - a_k^I \sin ktm)} v(s) \, ds. \tag{5.26}$$

Using the independence of a_k^R and a_k^I, we obtain,

$$\Phi(s, t_m) = E\left(e^{\iota s \sum_{k=0}^{n-1}(a_k^R \cos ktm - a_k^I \sin ktm)}\right)$$

$$= \prod_{k=0}^{n-1} E\left(e^{\iota s a_k^R \cos ktm}\right) E\left(e^{\iota s a_k^I \cos ktm}\right). \tag{5.27}$$

Notice that $\Phi(s, t_m)$ is an even, infinitely differentiable for $|s| < 1$, function in s, and $\Phi(0, t_m) = 1$. Thus, computing the Taylor series for $\log \Phi(s, t_m)$ we conclude that, for $|s| < 1$,

$$E\left(e^{\iota s a_k^R \cos ktm}\right) = e^{-\frac{t^2}{4} - \alpha_1 t^4 + O(t^6)}, \ E\left(e^{\iota s a_k^I \cos ktm}\right) = e^{-\frac{t^2}{4} - \alpha_2 t^4 + O(t^6)}. \tag{5.28}$$

Now using $e^{-a} = e^{-b} + O(|b - a|)$ for $a, b \geq 0$, we can write (5.27) for $|s| < 1$ as

$$\Phi(s, t_m) = e^{-\frac{ns^4}{4}} + O(ns^4). \tag{5.29}$$

Therefore (5.26) can be rewritten as

$$E(\eta) = n \int_{-1}^{1} e^{-\frac{ns^2}{4}} v(s) \, ds + O\left(n^2 \int_{-1}^{1} s^4 |v(s)| \, ds\right) + O\left(n \int_{|s|>1} |v(s)| \, ds\right). \tag{5.30}$$

The first integral may be extended to infinity and the resulting error may be included in the third term. Also, by extending the second integral to infinity the third term can be included in the second integral. Thus (5.30) simplifies to

$$E(\eta) = n \int_{-\infty}^{\infty} e^{-\frac{ns^2}{4}} v(s) \, ds + O\left(n^2 \int_{-\infty}^{\infty} s^4 |v(s)| \, ds\right). \tag{5.31}$$

Using (5.14), we may replace the second term with $O\left(\frac{n^2}{\Delta^4}\right)$, which, by $\Delta = \sqrt{\frac{n}{\ln n}}$, yields

$$E(\eta) = n \int_{-\infty}^{\infty} e^{-\frac{ns^2}{4}} v(s) \, ds + O(\ln^2 n). \tag{5.32}$$

In order to find the second moment of η, we may write η^2 as

$$\eta^2 \sum_{m=0}^{n-1} \sum_{\ell=0}^{n-1} u\left(\Re\left(F_{\mathbf{a}}(t_m)\right)\right) u\left(\Re\left(F_{\mathbf{a}}(t_\ell)\right)\right). \tag{5.33}$$

Therefore, after substituting the Fourier transform of $u(x)$ in (5.33), to evaluate each term of the double summation, we should compute

$$u\left(\Re\left(F_{\mathbf{a}}(t_m)\right)\right) u\left(\Re\left(F_{\mathbf{a}}(t_\ell)\right)\right)$$
$$= \int_{-\infty}^{\infty} \int_{\infty}^{\infty} E\left(e^{\iota \sum_{k=0}^{n-1}\left(a_k^R(s\cos kt_m + h\cos kt_\ell) - a_k^I(s\sin kt_m + h\sin kt_\ell)\right)}\right) v(s)v(h)\, ds\, dh. \tag{5.34}$$

The inner expectation in (5.34) can be split using the independence of a_k^R and b_k^R to calculate

$$\Psi(s, h; t_m, t_\ell) = E\left(e^{\iota \sum_{k=0}^{n-1}\left(a_k^R(s\cos kt_m + h\cos kt_\ell) - a_k^I(s\sin kt_m + h\sin kt_\ell)\right)}\right)$$
$$= \prod_{k=0}^{n-1} E\left(e^{\iota a_k^R(s\cos kt_m + h\cos kt_\ell)}\right) \cdot \prod_{k=0}^{n-1} E\left(e^{-\iota a_k^I(s\sin kt_m + h\sin kt_\ell)}\right). \tag{5.35}$$

Using (5.28) we may rewrite, for $|s|, |h| < \frac{1}{2}$, each expectation as

$$E\left(e^{\iota a_k^R(s\cos kt_m + h\cos kt_\ell)}\right) = e^{-\frac{1}{4}\sum_{k=0}^{n-1}\left((s\cos kt_m + h\cos kt_\ell)^2 + 2\alpha_1(s\cos kt_m + h\cos kt_\ell)^4\right)}$$
$$+ O(n(|s| + |h|)^6). \tag{5.36}$$

Here, for the last term of the exponent we used

$$O((s\cos kt_m + h\cos kt_\ell)^6) = O((|s| + |h|)^6).$$

We can also write a similar equation for a_k^I. After substituting (5.36) into (5.35), we can use the second order approximation

$$e^{-a} = e^{-b} + (b - a)e^{-b} + O((b - a)^2),$$

valid for $a, b > 0$, to write (5.35) as

$$\Psi(s, h; t_m, t_\ell) = e^{-\frac{1}{4}\sum_{k=0}^{n-1}\left((s\cos kt_m + h\cos kt_\ell)^2 + (s\sin kt_m + h\sin kt_\ell)^2\right)}$$
$$+ \sum_{k=0}^{n-1} \left(\alpha_1(s\cos kt_m + h\cos kt_\ell)^4 + \alpha_2(s\sin kt_m + h\sin kt_\ell)^4 + O(n(|s| + |h|)^6)\right)$$
$$\cdot e^{-\frac{1}{4}\sum_{k=0}^{n-1}\left((s\cos kt_m + h\cos kt_\ell)^2 + (s\sin kt_m + h\sin kt_\ell)^2\right)} + O\left(n^2(|s| + |h|)^8\right), \tag{5.37}$$

valid for $|s|, |h| < \frac{1}{2}$. We can further simplify (5.37) by using the identities

$$\sum_{k=0}^{n-1}(s\cos kt_m + h\cos kt_\ell)^2 = \sum_{k=0}^{n-1}(s\sin kt_m + h\sin kt_\ell)^2 = n\frac{s^2 + h^2}{2},$$

for $m \neq \ell$ and $m + \ell \neq n - 1$, to get

$$\Psi(s, h; t_m, t_\ell) = e^{-\frac{n(s^2 + h^2)}{2}} + O(n(|s| + |h|)^4) e^{-\frac{n(s^2 + h^2)}{2}}$$
$$+ O\left(n(|s| + |h|)^6\right) + O\left(n^2(|s| + |h|)^8\right), \qquad (5.38)$$

for $|s|, |h| < \frac{1}{2}$, $m \neq \ell$ and $m + \ell \neq n - 1$. For the other $2n$ terms (i.e., $m = \ell$ or $m + \ell = n - 1$) in (5.26), we can use the following inequality:

$$2 \sum_{m=0}^{n-1} u\left(\Re\left(F_{\mathbf{a}}(t_m)\right)\right) u\left(\Re\left(F_{\mathbf{a}}(t_\ell)\right)\right) \leq 2 \sum_{m=0}^{n-1} u\left(\Re\left(F_{\mathbf{a}}(t_m)\right)\right) = 2\eta, \qquad (5.39)$$

since $0 \leq u(x) \leq 1$. Now (5.33) simplifies to

$$E(\eta^2) \leq (n^2 - n) \int_{-\frac{1}{2}}^{\frac{1}{2}} \int_{-\frac{1}{2}}^{\frac{1}{2}} e^{-\frac{n(s^2 + h^2)}{4}} v(s) v(h) \, ds \, dh$$

$$+ O\left(n^3 \sum_{k=0}^{n-1} \int_{-\frac{1}{2}}^{\frac{1}{2}} \int_{-\frac{1}{2}}^{\frac{1}{2}} (|s| + |h|)^4 \, e^{-\frac{n(s^2 + h^2)}{4}} v(s) v(h) \, ds \, dh\right)$$

$$+ O\left(n^3 \int_{-\frac{1}{2}}^{\frac{1}{2}} \int_{-\frac{1}{2}}^{\frac{1}{2}} (|s| + |h|)^6 \, |v(s)| \, |v(h)| \, ds \, dh\right)$$

$$+ O\left(n^4 \int_{-\frac{1}{2}}^{\frac{1}{2}} \int_{-\frac{1}{2}}^{\frac{1}{2}} (|s| + |h|)^8 \, |v(s)| \, |v(h)| \, ds \, dh\right)$$

$$+ O\left(n^2 \int_{|s| \geq \frac{1}{2}} \int_{|h| \geq \frac{1}{2}} |v(s)| \, |v(h)| \, ds \, dh\right) + 2E(\eta). \qquad (5.40)$$

To evaluate (5.40), we may extend the integrals in the first four terms from $-\infty$ to ∞ to find an upper bound for $E(\eta^2)$. So we may rewrite (5.40) as

$$E(\eta^2) \leq (n^2 - n) \left(\int_{-\infty}^{\infty} e^{-\frac{n(s^2 + h^2)}{4}} v(s) v(h) \, ds \, dh\right)^2$$

$$+ O\left(n^3 \int_{-\infty}^{\infty} \int_{-\infty}^{\infty} (|s| + |h|)^4 \, e^{-\frac{n(s^2 + h^2)}{4}} |v(s)||v(h)| \, ds \, dh\right)$$

$$+ O\left(n^3 \int_{-\infty}^{\infty} \int_{-\frac{1}{2}}^{\frac{1}{2}} (|s| + |h|)^6 \, |v(s)| \, |v(h)| \, ds \, dh\right)$$

$$+ O\left(n^4 \int_{-\infty}^{\infty} \int_{-\frac{1}{2}}^{\frac{1}{2}} (|s| + |h|)^8 \, |v(s)| \, |v(h)| \, ds \, dh\right)$$

$$+ O\left(n \int_{|s| \geq \frac{1}{2}} |v(s)| \, ds\right)^2 + 2E(\eta). \qquad (5.41)$$

Now we can use (5.15) to write the fourth term in (5.41) as $O(\frac{n^2}{\Delta^{18}})$. The second term in (5.41) can also be simplified to

$$
O\left(n^3 \sum_{p=0}^{4} \int_{-\infty}^{\infty} s^p \mathrm{e}^{-\frac{ns^2}{4}} \,\mathrm{d}s \int_{-\infty}^{\infty} h^{4-p} \mathrm{e}^{-\frac{nh^2}{4}} \,\mathrm{d}h\right) = O\left(\frac{n^4 \mathrm{e}^{-\frac{2L^2}{n}}}{\Delta^4 L^2}\right). \qquad (5.42)
$$

In the last equality we used identities (5.16) with $p = k$ and $p = k - 4$. The third term can similarly be evaluated as

$$
O\left(n^3 \left(\sum_{k=1}^{5} \int_{-\infty}^{\infty} |s^k v(s)| \,\mathrm{d}s \int_{\infty}^{\infty} |s^{6-k} v(s)| \,\mathrm{d}s + 2 \int_{-\infty}^{\infty} |v(h)| \,\mathrm{d}h \int_{-\infty}^{\infty} |s^6 v(s)| \,\mathrm{d}s\right)\right)
$$
$$
= O\left(\frac{n^3}{\Delta^6}\right) + O\left(n^3 \frac{1}{\Delta^6} \frac{L}{\Delta}\right) = O\left(\frac{n^3}{\Delta^6}\right) + O\left(\frac{n^3 L}{\Delta^7}\right), \qquad (5.43)
$$

where we again used (5.13) and (5.14) to evaluate both terms in (5.43). Similarly to the third term, the fourth term can be also shown to be $O(\frac{n^4 L}{\Delta^9})$. Therefore, setting the value of L and Δ, we have

$$
E(\eta) \le 2E(\eta) + n^2 \left(\int_{-\infty}^{\infty} \mathrm{e}^{-\frac{ns^2}{4}} v(s) \,\mathrm{d}s\right)^2
$$
$$
+ O\left(\frac{n^4 \mathrm{e}^{-\frac{2L^2}{n}}}{\Delta^4 L^2}\right) + O\left(\frac{n^3 L}{\Delta^7}\right) + O\left(\frac{n^4 L}{\Delta^9}\right)
$$
$$
= 2E(\eta) + n^2 \left(\int_{-\infty}^{\infty} \mathrm{e}^{-\frac{ns^2}{4}} v(s) \,\mathrm{d}s\right)^2 + O(\ln^5 n) + O(\ln n) + O(\ln^2 n). \quad (5.44)
$$

On the other hand, it is easy to show that $E(\eta) \ge O(\ln^6 n)$. By (5.33) it is enough to show that

$$
\int_{-\infty}^{\infty} \mathrm{e}^{-\frac{ns^2}{4}} v(s) \,\mathrm{d}s \ge \frac{\ln^6 n}{n}.
$$

By Parseval's formula, the Fourier transform of $\mathrm{e}^{-\frac{ns^2}{4}}$ is $\sqrt{\frac{4\pi}{n}} \mathrm{e}^{-\frac{x^2}{n}}$, and the integral can be rewritten as

$$
\frac{1}{\sqrt{\pi n}} \int_{-\infty}^{\infty} \mathrm{e}^{-\frac{x^2}{n}} u(x) \,\mathrm{d}x \ge \frac{1}{\sqrt{\pi n}} \int_{L+\Delta \le |x| \le L+2\Delta} \mathrm{e}^{-\frac{x^2}{n}} \,\mathrm{d}x \ge \frac{2\Delta}{\sqrt{\pi n}} \mathrm{e}^{-\frac{(L+\Delta)^2}{n}},
$$

and the sought inequality holds for the chosen values of L and Δ. This implies the result. $\qquad \square$

Continuation of the proof of Theorem 5.6 Substituting the estimates from the lemma into (5.25) we obtain

$$\Pr\left(\max_{t\in[0,1)}|F_{\mathbf{a}}(t)| \geq \sqrt{n\ln n - 6.5n\ln\ln n} - \sqrt{n}\ln n\right) \geq 1 - O\left(\frac{1}{\ln^4 n}\right),$$

thus accomplishing the proof. □

Now we pass to the M-PSK case, for even M. In contrast to the QAM situation, the real and imaginary parts of a_k are not independent. However, we can still use a similar argument to prove a lower bound on the PMEPR distribution.

Theorem 5.9 *Let* $a_k = e^{2\pi\iota\frac{j_k}{M}}$, $k = 0, 1, \ldots, n-1$, *and* j_k *be independently and equiprobably chosen from* $\{0, 1, \ldots, M-1\}$ *for an even* M. *Then*

$$\Pr\left(\mathrm{PMEPR}(\mathbf{a}) \leq \ln n - 6.5\ln\ln n\right) \leq O\left(\frac{1}{\ln^4 n}\right). \tag{5.45}$$

Proof The characteristic function of $\Re\left(F_{\mathbf{a}}(t)\right)$ can be written as

$$\Phi(s) = E\left(e^{\iota s\Re(F_{\mathbf{a}}(t))}\right) = \prod_{k=0}^{n-1} E\left(e^{\iota s\Re(a_k e^{2\pi\iota t k})}\right)$$

$$= \prod_{k=0}^{n-1} E\left(\cos(s\cos(2\pi\iota kt + t_k))\right) = \prod_{k=0}^{n-1} E(\cos s\tau_k), \tag{5.46}$$

where $t_k = \frac{2\pi j_k}{M}$ and $\tau_k = \cos(2\pi kt + t_k)$. Notice that due to the symmetry of constellation, τ_k has an even distribution. Furthermore, for $|s| < 1$, the characteristic function is positive. Therefore, for $|s| < 1$,

$$E(\cos s\tau_k) = e^{-E((\tau_k)^2)\frac{s^2}{2} + \alpha s^4 + O(s^6)}.$$

Next,

$$E((\tau_k)^2) = E(\cos^2(2\pi kt + t_k)) = \frac{1}{2} + \frac{1}{2}E\left(\cos(4\pi kt + 2t_k)\right) = \frac{1}{2}. \tag{5.47}$$

In the last transformation we used the symmetry of the constellation, guaranteeing that the expectation in the second term is zero. Therefore, from (5.46) we have

$$\Phi(s) = e^{-\frac{ns^2}{4} + n\alpha s^4 + O(ns^6)},$$

for $|s| \leq 1$. Now we can use the same argument as that of Theorem 5.6 to find the mean and variance of η as in (5.32) and (5.44) respectively. □

5.3 Gaussian process models

Modeling the behavior of the MC signals as stochastic processes sometimes yields quite accurate approximations to the PMEPR distributions.

We start with a Gaussian approximation. Let the multicarrier signal,

$$F_{\mathbf{a}}(t) = X_{\mathbf{a}}(t) + \iota Y_{\mathbf{a}}(t) = \sum_{k=0}^{n-1} e^{2\pi \iota kt},$$

be sampled at the Nyquist rate. Denote

$$R_j = \frac{\left| F_{\mathbf{a}}\left(\frac{j}{n}\right) \right|}{\sqrt{n}}.$$

Assuming that $X_{\mathbf{a}}(\frac{j}{n})$, $j = 0, 1, \ldots, n-1$, and $Y_{\mathbf{a}}(\frac{j}{n})$, $j = 0, 1, \ldots, n-1$, are pairwise independent identically distributed (i.i.d.) Gaussian random variables, we conclude that R_j are i.i.d. Rayleigh random variables of which the probability density function is

$$f_R(r) = 2re^{-r^2}.$$

Therefore

$$\Pr\left(\max_{j=0,1,\ldots,n-1} R_j < r \right) = \Pr(R < r)^n = \left(1 - e^{-r^2} \right)^n,$$

and

$$\Pr\left(\text{PMEPR}(\mathbf{a}) > \lambda \right) \approx 1 - (1 - e^{-\lambda})^n. \tag{5.48}$$

We will refer to this bound as the *Gaussian approximation* to the PMEPR distribution. The distribution (5.48), however, does not fit simulations, especially when we consider the low-probability region. There are several possible reasons for this. First, as we have seen in Chapter 4, the maximum over the Nyquist samples of the signal can differ significantly from the continuous maximum. Second, the assumption about the Gaussian distribution of the samples when n is finite is correct only for a restricted interval around the expectation, and gives wrong results when we consider probabilities of high power levels. Third, the assumption about pairwise independence of the Nyquist-rate samples is correct only in the case when $\Re(a_k)$ and $\Im(a_k)$ are i.i.d. Gaussian random variables, which is not the case when we pick a_k from specific constellations, such as PSK and QAM.

To improve on the analysis, we assume that $X_{\mathbf{a}}(t)$ and $Y_{\mathbf{a}}(t)$ are independent stationary band-limited Gaussian processes with zero mean and variance $\sigma_X^2 = \frac{n}{2}$, and consequently

$$R_{\mathbf{a}}(t) = \frac{|F_{\mathbf{a}}(t)|}{\sqrt{n}} = \sqrt{\frac{X_{\mathbf{a}}^2(t) + Y_{\mathbf{a}}^2(t)}{n}}$$

is a stationary band-limited Rayleigh process. Since the derivative operation is linear, the derivatives $X_{\mathbf{a}}(t)$ and $Y_{\mathbf{a}}(t)$ are also Gaussian random processes with zero mean and variance $\sigma_{\dot{X}}^2 = E(\dot{X}_a(t)^2)$. We will need several technical results. Let us define the level-crossing rate $\nu_a(\lambda)$ of the process $R_a(t)$ per unit time as the mean number of positive crossings of the process of the level λ.

Lemma 5.10

$$\nu(\lambda) \approx \sqrt{\frac{\pi}{3}} \cdot n\lambda e^{-\lambda^2}.$$

Proof   Let $X, \dot{X}, Y, \dot{Y}$ denote the samples of the corresponding Gaussian processes at the same time instant. The joint p.d.f. of $X, \dot{X}, Y, \dot{Y}$ is

$$f_{X,\dot{X},Y,\dot{Y}}(\mathbf{X}) = \frac{1}{\sqrt{(2\pi)^m |\mathbf{R}|}} \, e^{-\frac{1}{2}\mathbf{X}\mathbf{R}^{-1}\mathbf{X}^t} \tag{5.49}$$

where $(\mathbf{X}) = (x, \dot{x}, y, \dot{y})$, $m = 4$, $\mathbf{R}$ is the covariance matrix, and $|\mathbf{R}|$ is the determinant of $\mathbf{R}$. Since $X_a(t)$ and $Y_a(t)$ are uncorrelated and

$$E\left(X\dot{X}\right) = \frac{1}{2} \cdot \frac{\mathrm{d}}{\mathrm{d}t} E\left(X^2\right) = 0,$$

the covariance matrix is given by

$$\mathbf{R} = \begin{pmatrix} \sigma_X^2 & 0 & 0 & 0 \\ 0 & \sigma_{\dot{X}}^2 & 0 & 0 \\ 0 & 0 & \sigma_X^2 & 0 \\ 0 & 0 & 0 & \sigma_{\dot{X}}^2 \end{pmatrix}.$$

By changing the variables to polar coordinates as

$$\begin{aligned} X &= \sqrt{2\sigma_X^2} \, R \cos \Theta, \\ Y &= \sqrt{2\sigma_X^2} \, R \sin \Theta, \end{aligned} \tag{5.50}$$

the joint p.d.f. of $R, \dot{R}, \Theta, \dot{\Theta}$ is given by

$$f_{R,\dot{R},\Theta,\dot{\Theta}}(r, \dot{r}, \theta, \dot{\theta}) = \frac{r^2}{\pi^2 \kappa} \, e^{-r^2 - \frac{1}{\kappa}(\dot{r}^2 + r^2\dot{\theta}^2)} \tag{5.51}$$

where

$$\kappa = \frac{\sigma_{\dot{X}}}{\sigma_X^2}. \tag{5.52}$$

Integrating (5.51) by θ from 0 to 2π and by $\dot{\theta}$ from $-\infty$ to ∞, we obtain the joint p.d.f. of R and $\dot{R}$,

$$f_{R,\dot{R}}(r, \dot{r}) = 2re^{-r^2} \frac{1}{\sqrt{\pi\kappa}} \, e^{-\frac{\dot{r}^2}{\kappa}}. \tag{5.53}$$

Let $\rho_X(\tau)$ and $S_X(f)$ be the autocorrelation function and the power spectral density (p.s.d.) of $X_a(t)$, respectively. Then

$$\sigma_X^2 = \rho_X(0) = \int_{-\infty}^{\infty} S_X(f)\,\mathrm{d}f,$$

and

$$\sigma_{\dot{X}}^2 = \rho_{\dot{X}}(0) = \int_{-\infty}^{\infty} (2\pi f)^2 S_X(f)\,\mathrm{d}f.$$

Since the p.s.d. of the band-limited signals is nearly rectangular, $S_X(f)$ can be assumed to be constant over the entire bandwidth, i.e.

$$S_X(f) \approx \begin{cases} \frac{1}{W}\sigma_X^2, & \text{for } f \in [0, W], \\ 0, & \text{otherwise.} \end{cases}$$

Thus

$$\sigma_{\dot{X}}^2 = \frac{\pi^2}{3} W^2 \sigma_X^2,$$

and by (5.52) we have

$$\kappa = \frac{\pi^2}{3} W^2. \tag{5.54}$$

The level-crossing level can be computed using

$$\upsilon(\lambda) = \int_0^{\infty} \dot{r} f_R(\lambda, \dot{r})\,\mathrm{d}\dot{r}.$$

Substituting (5.53) and carrying out the integration, we have

$$\upsilon(\lambda) = \sqrt{\frac{\kappa}{\pi}} \cdot \lambda e^{-\lambda^2}.$$

Substitution of κ from (5.54), and noticing that $W \approx n$, yields the result. $\qquad\square$

Let $N(\lambda)$ denote the mean number of the peaks above the level λ in one MC symbol.

Lemma 5.11

$$N(\lambda) = \frac{4n}{\sqrt{15\pi}} \int_{\lambda}^{\infty} u^2 \int_0^{\infty} e^{-(\phi^2+1)u^2}$$

$$\cdot \left(e^{-\frac{5}{4}(\phi^2-1)u^2} - \frac{\sqrt{5\pi}}{2}(\phi^2-1)u\,\mathrm{erfc}\left(\frac{\sqrt{5\pi}}{2}(\phi^2-1)u\right) \right) \mathrm{d}\phi\,\mathrm{d}u, \tag{5.55}$$

where

$$\mathrm{erfc}(s) = \frac{2}{\sqrt{\pi}} \int_s^{\infty} e^{-z^2}\,\mathrm{d}z. \tag{5.56}$$

Proof Since the derivatives of $\dot{X}_a(t)$ and $\dot{Y}_a(t)$ are also Gaussian with zero mean, variance

$$\sigma_{\ddot{X}}^2 = E(\ddot{X}_a(t)^2),$$

$$E(\dot{X}_a(t)\ddot{X}_a(t)) = \frac{1}{2}\frac{d}{dt}E(\dot{X}_a(t)^2) = 0,$$

and

$$E(X_a(t)\ddot{X}_a(t)) = \frac{d}{dt}E(X_a(t)\dot{X}_a(t)) - E(\dot{X}_a(t)^2),$$

the joint p.d.f. of Gaussian random variables X, $\dot{X}$, $\ddot{X}$, Y, $\dot{Y}$, $\ddot{Y}$, which are the samples of the Gaussian processes $X_a(t)$, $\dot{X}_a(t)$, $\ddot{X}_a(t)$, $Y_a(t)$, $\dot{Y}_a(t)$, $\ddot{Y}_a(t)$, at the same time instant, is given by (5.50) with $\mathbf{X} = (x, \dot{x}, \ddot{x}, y, \dot{y}, \ddot{y})$, $m = 6$, and

$$\mathbf{R} = \begin{pmatrix} \sigma_X^2 & 0 & -\sigma_{\dot{X}}^2 & 0 & 0 & 0 \\ 0 & \sigma_{\dot{X}}^2 & 0 & 0 & 0 & 0 \\ -\sigma_{\dot{X}}^2 & 0 & -\sigma_{\ddot{X}}^2 & 0 & 0 & 0 \\ 0 & 0 & 0 & \sigma_X^2 & 0 & -\sigma_{\dot{X}}^2 \\ 0 & 0 & 0 & 0 & \sigma_{\dot{X}}^2 & 0 \\ 0 & 0 & 0 & -\sigma_{\dot{X}}^2 & 0 & \sigma_{\ddot{X}}^2 \end{pmatrix}.$$

Changing the variables as in (5.50) and integrating out the variables θ, $\dot{\theta}$, $\ddot{\theta}$, we obtain the joint p.d.f. of the envelope

$$f_R(r, \dot{r}, \ddot{r}) = \int_{-\infty}^{\infty} \frac{2\sqrt{\eta}}{(\kappa\pi)^{\frac{3}{2}}} r^2 \cdot \exp\left(-(1 + \eta + (1 - 2\eta)\phi^2 + \eta\phi^4)r^2\right.$$

$$\left. - \frac{2\eta}{\kappa}(1 - \phi^2)\ddot{r}r - \frac{1}{\kappa}\left(\dot{r}^2 + \frac{\eta}{\kappa}\ddot{r}^2\right)\right) d\phi,$$

where

$$\eta = \frac{\sigma_{\dot{X}}^4}{\sigma_X^2\sigma_{\ddot{X}}^2 - \sigma_{\dot{X}}^4}.$$

Notice that we have modified the variable $\phi = \sqrt{\kappa}\theta$. The rate density of peaks, which is the mean number of peaks of the process $R_a(t)$ in the region $r \in [u, u + du]$, per unit time, is given by

$$v(u)\,du = du \int_{-\infty}^0 -\ddot{r} f_{R,\dot{R},\ddot{R}}(u, 0, \ddot{r})\,d\ddot{r}.$$

Therefore,

$$v(u)\,du = du\,\frac{2}{\pi^{\frac{3}{2}}}\sqrt{\frac{\kappa}{\eta}}u^2\int_0^\infty e^{-(\phi^2+1)u^2}$$

$$\cdot\left(e^{-\eta(\phi^2-1)^2u^2} - \sqrt{\pi\eta}(\phi^2 - 1)u\,\mathrm{erfc}(\sqrt{\eta}(\phi^2 - 1)u)\right) d\phi.$$

Taking into account

$$\sigma_{\ddot{X}}^2 = \rho_{\ddot{X}}(0) = \int_{-\infty}^{\infty} (2\pi f)^4 S_X(f)\,\mathrm{d}f = \frac{\pi^4}{5} W^4 \sigma_X^2,$$

from which we obtain $\eta = \frac{5}{4}$. Consequently,

$$N(\lambda) = \int_0^{\infty} v(u)\,\mathrm{d}u,$$

yielding the result. □

Now we are in a position to estimate the PMEPR distribution. We first compute the probability $p(\lambda)$ that an arbitrary peak is above level $\lambda\sqrt{n}$ as the ratio between the mean number of peaks above μ to the mean number of peaks,

$$p(\lambda) = \frac{N(\lambda)}{N(0)}. \tag{5.57}$$

The values involved in this expression can be calculated using (5.55). Further we make a heuristic assumption that the peaks are statistically mutually uncorrelated. There is numerical evidence that $N(0) \approx 0.64n$, and thus

$$\Pr\left(\mathrm{PMEPR}(\mathbf{a}) > \lambda\right) \approx 1 - (1 - p(\sqrt{\lambda}))^{0.64n}. \tag{5.58}$$

This expression is not convenient for computation since it involves double integration in (5.55). Let us pick a level $\mu\sqrt{n}$ that satisfies the following assumption: each positive crossing of the level $\mu\sqrt{n}$ has a single positive peak above $\mu\sqrt{n}$. Then the probability $p(\lambda, \mu)$ of a peak being above the level $\lambda\sqrt{n}$ under the condition that the peak is above $\mu\sqrt{n}$ can be approximated as

$$p(\lambda, \mu) \approx \frac{v(\lambda)}{v(\mu)}. \tag{5.59}$$

Notice that v is much simpler to compute, and it is estimated in Lemma 5.10. Therefore,

$$p(\lambda, \mu) \approx \frac{\lambda e^{-\lambda^2}}{\mu e^{-\mu^2}}. \tag{5.60}$$

Now we make a heuristic assumption that the peaks above level $\mu\sqrt{n}$ are independent, and approximate the probability that the signal exceeds the level $\lambda\sqrt{n}$ as $p(\lambda, \mu)^{N(\mu)}$. For large values of μ we have

$$N(\mu) \approx \sqrt{\frac{\pi}{3}} n\mu e^{-\mu^2}.$$

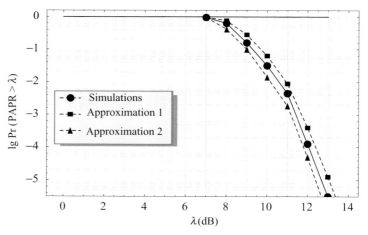

Figure 5.1 Approximations to PMEPR distribution of MC signal with 256 QPSK modulated subcarriers. Approximation 1 – Expression (5.62), Approximation 2 – Expression (5.64)

Finally we arrive at the following approximation:

$$\Pr\left(\text{PMEPR}\left(\mathbf{a}\right) > \lambda\right) \approx \begin{cases} 1 - \left(1 - \dfrac{\sqrt{\lambda}e^{-\lambda}}{\sqrt{\mu}e^{-\mu}}\right)^{\sqrt{\frac{\pi}{3}}n\sqrt{\mu}e^{-\mu}} & \text{for } \lambda > \mu, \\ 1 & \text{for } \lambda \le \mu. \end{cases} \tag{5.61}$$

It remains to choose the parameter μ. This can be done numerically to fit better simulation results in the high-probability region.

For growing n, assuming that $\dfrac{\sqrt{\lambda}e^{-\lambda}}{\sqrt{\mu}e^{-\mu}} \to 0$, and using $\ln(1-x) \approx -x$, for small x, we obtain

$$\Pr\left(\text{PMEPR}\left(\mathbf{a}\right) > \lambda\right) \approx 1 - \exp\left(-\sqrt{\frac{\pi\lambda}{3}}\, n \cdot e^{-\lambda}\right). \tag{5.62}$$

This expression already does not depend on μ, and can be used as an approximation. Taking the first term of the Taylor expansion of (5.62) we obtain

$$\Pr\left(\text{PMEPR}\left(\mathbf{a}\right) > \lambda\right) \approx \sqrt{\frac{\pi\lambda}{3}}\, n \cdot e^{-\lambda}. \tag{5.63}$$

Another good approximation is justified by extreme value theory,

$$\Pr\left(\text{PMEPR}\left(\mathbf{a}\right) > \lambda\right) \approx 1 - \exp\left(-\sqrt{\frac{\pi \ln n}{3}}\, n \cdot e^{-\lambda}\right). \tag{5.64}$$

I have omitted details of its derivation.

In Fig. 5.1, I compare the bounds with the simulation results when $n = 256$ and QPSK modulation is used. The results demonstrate very good accuracy of the derived expressions in the high-probability range.

Here it is worth mentioning again that the Gaussian approach is clearly flawed if one seeks a mathematically rigorous justification. Indeed an approximate joint Gaussian distribution only holds for samples in any fixed finite subset of the symbol interval as n becomes large. Unfortunately, it does not hold for samples in the complete interval (which also increases with n when the occupied bandwidth is kept constant), which would be required for a proof of the claim.

5.4 Lower bound on the number of signals with constant PMEPR

Although I have shown in the previous section that most of the signals have maximum peaks close to $\sqrt{n \ln n}$, it will turn out that signals with maxima of const $\cdot \sqrt{n}$ are not so rare. For the sake of simplicity here, we will consider only BPSK modulated signals.

A *linear form* L in n variables $(x_1, \ldots, x_n)$ is

$$L(x_1, \ldots, x_n) = \sum_{j=0}^{n-1} \eta_j x_j,$$

where all η_j are real. If $|\eta_j| \le \mathcal{A}$ for $j = 0, \ldots, n-1$, and a finite $\mathcal{A} > 0$, the form is said to be bounded by $\mathcal{A}$.

Let us replace the problem of minimizing the continuous maximum of a MC signal with that of minimizing its discrete version. Given $r, r > 1$, the oversampling factor such that rn is an integer, h, and $h > 1$, the number of projection axes, we are facing joint minimization of $r \cdot h \cdot n$ bounded linear forms,

$$L_j(\mathbf{a}) = L_j(a_0, \ldots, a_{n-1}) = \sum_{k=0}^{n-1} \eta_{j,k} \cdot a_k, \quad j = 0, 1, \ldots, rhn - 1,$$

where $t_j = \frac{j}{rn}$ and

$$\eta_{j,k} = \begin{cases} \Re\left(e^{2\pi \iota(t_j k)}\right), & j = 0, 1, \ldots, rn - 1, \\ \Re\left(e^{2\pi \iota(t_j k - \frac{1}{h})}\right), & j = rn+, rn + 1, \ldots, 2rn - 1, \\ \cdots \\ \Re\left(e^{2\pi \iota(t_j k - \frac{h-1}{h})}\right), & j = (h-1)rn, \ldots, rhn - 1. \end{cases} \quad (5.65)$$

Notice that all the linear forms are bounded, $|\eta_{j,k}| \le 1$, and the number of forms exceeds the number of variables.

Clearly, for an $F_\mathbf{a}(t) = \sum_{k=0}^{n-1} a_k e^{2\pi \iota k t}$,

$$\mathcal{M}_c(F_\mathbf{a}) = \max_{t \in [0,1)} |F_\mathbf{a}(t)| \le C_r C_h \cdot \max_j |L_j(\mathbf{a})|, \quad (5.66)$$

where $C_m \le \frac{1}{\cos \frac{\pi}{2m}}$.

Thus we have to prove that there are many vectors **a** with

$$\max_j |L_j(\mathbf{a})| = \text{const} \cdot \sqrt{n}.$$

Lemma 5.12 *Let $\ell \leq m$ and*

$$L_k(x_0, x_1, \ldots, x_{\ell-1}) = \zeta_{k,0}x_0 + \zeta_{k,2}x_1 + \ldots + \zeta_{k,\ell-1}x_{\ell-1}, \quad k = 0, 1, \ldots, m-1,$$

be m linear forms in ℓ variables with all $|\zeta_{k,j}| \leq 1$. Then, if ℓ is sufficiently large, there exist $\epsilon_0, \ldots, \epsilon_{\ell-1} \in \{-1, 0, 1\}$ with

$$|\{j : \epsilon_j = 0\}| < 6 \cdot 10^{-7} \cdot \ell,$$

$$\max_{k=0,1,\ldots,m-1} |L_k(\epsilon_0, \epsilon_1, \ldots, \epsilon_\ell)| < 10\sqrt{\ell \cdot \ln \frac{2m}{\ell}}. \tag{5.67}$$

Proof Define

$$T(\epsilon_0, \epsilon_1, \ldots, \epsilon_{\ell-1}) = (b_0, b_1, \ldots, b_{m-1}),$$

where $\epsilon_j \in \{-1, 1\}$, $b_k \in \mathbb{Z}$,

$$b_k = \text{int}\left(\frac{L_k(\epsilon_0, \epsilon_1, \ldots, \epsilon_\ell)}{20\sqrt{\ell \cdot \ln \frac{2m}{\ell}}}\right)$$

and int(υ) is the nearest integer to υ. Let B be the set of integer-valued vectors for all $s = 1, 2, \ldots$ satisfying

$$|\{k : |b_k| \geq s\}| \leq m \cdot \left(\frac{2m}{\ell}\right)^{-50(2s-1)^2} \cdot 2^{s+1}.$$

I will show that more than half of 2^ℓ possible vectors $(\epsilon_0, \ldots, \epsilon_{\ell-1})$ are mapped by T to B. Moreover, it will be proved that

$$|B| < 2^{250 \cdot 2^{-50} \cdot \ell}. \tag{5.68}$$

Let $\epsilon_0, \epsilon_1, \ldots, \epsilon_{\ell-1}$, be independent and uniform, and let $L_0, \ldots, L_{m-1}$, $b_0, \ldots, b_{m-1}$, be the corresponding values they generate. By the Chernoff bound (Theorem 3.29)

$$\Pr(|b_k| \geq s) = \Pr\left(|L_k| \geq 10s\sqrt{\ell \cdot \ln \frac{2m}{\ell}}\right) < \left(\frac{2m}{\ell}\right)^{-50(2s-1)^2}.$$

Therefore,

$$E(|\{k : |b_k| \geq s\}|) < m \cdot \left(\frac{2m}{\ell}\right)^{-50(2s-1)^2}.$$

Using Markov's inequality (Theorem 3.27) we obtain

$$\Pr\left(|\{k : |b_k| \geq s\}| > 2^{s+1} \cdot m \cdot \left(\frac{2m}{\ell}\right)^{-50(2s-1)^2}\right) < \frac{1}{2^{s+1}}.$$

Thus,

$$\Pr((b_0, \ldots, b_{m-1}) \notin B) < \sum_{s=1}^{\infty} \frac{1}{2^{s+1}} = \frac{1}{2},$$

yielding the claim about the number of vectors $(\epsilon_0, \ldots, \epsilon_{\ell-1})$ mapping to B.

To estimate the size of B, we use the following argument. Let

$$\alpha_s = 2^{s+1} \cdot \left(\frac{2m}{\ell}\right)^{-50(2s-1)^2}. \tag{5.69}$$

Notice that

$$\frac{1}{2} > \alpha_1 > \alpha_2 > \ldots$$

and

$$B = \left\{(b_0, \ldots, b_{m-1}) \in \mathbb{Z}^m : |\{k : |b_k| > s\}| \leq \alpha_s m, \ s = 1, 2, \ldots\right\}.$$

Then

$$|B| \leq \prod_{s=1}^{\infty} \left(\left(\sum_{k=0}^{\alpha_s m} \binom{m}{k}\right) 2^{\alpha_s m}\right).$$

Indeed, $\{k : |b_k| = s\}$ can be chosen in, at most, $\sum_{k=0}^{\alpha_s m} \binom{m}{k}$ ways, and, having been selected, can be split into $\{k : b_k = s\}$ and $\{k : b_k = -s\}$ in, at most, $2^{\alpha_s m}$ ways. We bound

$$\sum_{k=0}^{\alpha m} \binom{m}{k} \leq 2^{m H(\alpha)},$$

where $H(\alpha)$ is the entropy function,

$$H(\alpha) = -\alpha \log_2 \alpha - (1 - \alpha) \log_2(1 - \alpha).$$

Therefore, $|B| \leq 2^{\beta m}$, where

$$\beta = \sum_{s=0}^{\infty} (H(\alpha_s) + \alpha_s),$$

and α_s are defined in (5.69). Since $\frac{2m}{\ell} \geq 2$, we have $\alpha_{s+1} \leq 2^{-49} \alpha_s$ for all s, and all $\alpha_s \leq \alpha_1 \leq 2^{-48}$. Therefore,

$$H(\alpha_{s+1}) + \alpha_{s+1} \leq 2^{-47} (H(\alpha_s) + \alpha_s).$$

Thus β is dominated by the first term,

$$|B| < (1 + 2^{-46})(H(\alpha_1) + \alpha_1) < 1.1\alpha_1(-\log_2 \alpha_1)$$

$$< 5\left(\frac{2m}{\ell}\right)^{-50}\left(48 + 50\log_2 \frac{m}{\ell}\right).$$

The second inequality follows from $\alpha_1 < 2^{-48}$.
Then $|B| \leq 2^{\beta m} \leq 2^{\gamma \ell}$, where

$$\gamma = \frac{m}{\ell} \cdot 5 \cdot \left(\frac{2m}{\ell}\right)^{-50}\left(48 + 50\log_2 \frac{m}{\ell}\right)$$

$$= 5 \cdot 2^{-50} \cdot \left(\frac{m}{\ell}\right)^{-49}\left(48 + 50\log_2 \frac{m}{\ell}\right) < 250 \cdot 2^{-50},$$

since letting $y = \frac{m}{\ell}$, the inequality $y^{-49}(48 + 50\log_2 y) \leq 50$ is valid for all $y \geq 1$.
This proves (5.68).

Let $A_{\mathbf{b}}$ be the set of $(\epsilon_1, \ldots, \epsilon_{\ell-1})$ that are mapped by T to $\mathbf{b} = (b_0, \ldots, b_{m-1})$.
Therefore, there exists a value of $\mathbf{b} \in B$ such that

$$|A_{\mathbf{b}}| \geq 2^{\ell(1-\gamma)-1}. \tag{5.70}$$

Let p_0 be such that $H\left(\frac{1}{2} - p_0\right) = 1 - \gamma$ and let p be such that $p > p_0$. In our case

$$p_0 \approx \left(\frac{\ln 2}{2}\gamma\right)^{\frac{1}{2}} < 3 \cdot 10^{-7},$$

so we take $p = 3 \cdot 10^{-7}$. Let ℓ be sufficiently large so that

$$\ell(1 - \gamma) - 1 \geq \ell H\left(\frac{1}{2} - p\right).$$

Then

$$A_{\mathbf{b}} \geq 2^{\ell H\left(\frac{1}{2}-p\right)}.$$

Let $\text{diam}(A_{\mathbf{b}})$ stand for the maximum Hamming distance between any two vectors of $A_{\mathbf{b}}$. By the Kleitman theorem (see Theorem 3.46),

$$\text{diam}(A_{\mathbf{b}}) \geq (1 - 2p)\ell.$$

Let $\mathbf{x}, \mathbf{y} \in A_{\mathbf{b}}$ such that the Hamming distance between the vectors

$$d_H(\mathbf{x}, \mathbf{y}) = \text{diam}(A_{\mathbf{b}}).$$

Then

$$\mathbf{z} = (\epsilon_0, \ldots, \epsilon_{\ell-1}) = \frac{1}{2}(\mathbf{x} - \mathbf{y})$$

will have zeros on the positions where **x** and **y** coincide, and -1 or 1 in the rest of the coordinates. It is straightforward to check that **z** satisfies the claims of the lemma. □

Lemma 5.13 *Let $\ell \leq m$ and*

$$L_k(x_0, x_1, \ldots, x_{\ell-1}) = \zeta_{k,0}x_0 + \zeta_{k,2}x_1 + \ldots + \zeta_{k,\ell-1}x_{\ell-1}, \quad k = 0, 1, \ldots, m-1,$$

be m linear forms in ℓ variables with all $|\zeta_{k,j}| \leq 1$. Then, if ℓ is sufficiently large, there exist $\epsilon_0, \ldots, \epsilon_{\ell-1} \in \{-1, 1\}$ with

$$\max_{k=0,1,\ldots,m-1} |L_k(\epsilon_0, \epsilon_1, \ldots, \epsilon_\ell)| < 11\sqrt{\ell \cdot \ln \frac{2m}{\ell}}. \tag{5.71}$$

Proof Let $c = 6 \cdot 10^{-7}$. Set $\ell_0 = \ell$ and apply Lemma 5.12 to find values $\epsilon_j \in \{-1, 1\}$ to all but ℓ_1 variables with $\ell_1 \leq c\ell_0$. Iterate this process, at each stage applying it to the still equal zero variables, having ℓ_2 undetermined variables after the second step, etc. Let w be an absolute constant so that Lemma 5.12 applies for all $\ell \geq w$. Then the process will terminate when $\ell_{u+1} < w$. At this stage the undetermined variables could be arbitrarily set to $+1$ or -1. Therefore,

$$\max_{k=0,1,\ldots,m-1} |L_k(\epsilon_0, \epsilon_1, \ldots, \epsilon_\ell)| \leq \ell_{u+1} + \sum_{t=0}^{u} 10\sqrt{\ell_t \cdot \ln \frac{2m}{\ell_t}}$$

$$\leq p + \sum_{t=0}^{\infty} 10\sqrt{\ell c^t \cdot \ln \frac{2m}{\ell c^t}}.$$

Set $\varsigma = \ln \frac{2m}{\ell}$, so that $\varsigma \geq \ln 2$. We use the inequality

$$\sqrt{x + y} \leq \sqrt{x} + \sqrt{y},$$

valid for all $x, y \geq 0$. Then

$$\sum_{t=0}^{\infty} \sqrt{c^t \ln(\varsigma c^{-t})} \leq \sum_{t=0}^{\infty} \sqrt{c^t}\left(\sqrt{\ln \varsigma} + \sqrt{t \ln c^{-1}}\right),$$

$$\sum_{t=0}^{\infty} \sqrt{c^t} = \frac{1}{1 - \sqrt{c}} < 1 + 10^{-3},$$

$$\sum_{t=0}^{\infty} \sqrt{c^t \cdot t \ln c^{-1}} < 3 \cdot 10^{-3},$$

$$\sum_{t=0}^{\infty} \sqrt{c^t \cdot \ln(\varsigma c^{-t})} < \sqrt{\ln \varsigma}(1 + 10^{-3}) + 3 \cdot 10^{-3} < 1.005\sqrt{\ln \varsigma},$$

and

$$\max_{k=0,1,\dots,m-1} |L_k(\epsilon_0, \epsilon_1, \dots, \epsilon_\ell)| \leq w + 10\sqrt{\ell} \sum_{t=0}^{\infty} \sqrt{c^t \ln(\varsigma c^{-t})}$$

$$\leq w + 10 \cdot 1.005 \cdot \sqrt{\ell \ln \varsigma}.$$

This result holds for all ℓ. When ℓ is sufficiently large the constant w may be absorbed into the main term, giving the claim.   □

Corollary 5.14 *For n large enough, there exists $\mathbf{a} = (a_0, \dots, a_{n-1}) \in \{-1, 1\}^n$ such that*

$$\mathrm{PMEPR}(\mathbf{a}) \leq 4026.$$

Proof Applying Lemma 5.13 to the linear forms (5.65), and employing (5.66), we obtain

$$\mathcal{M}_c(F_{\mathbf{a}}) \leq \min_{r,h} \frac{1}{\cos \frac{\pi}{2k}} \frac{1}{\cos \frac{\pi}{2h}} 11\sqrt{rhn \cdot \ln(2rh)},$$

where the minimum is over $r > 1$ such that rn is an integer, and $h > 2$ is an integer. Choosing $r = h = 2$ we obtain the result.   □

Notice that the constant here is quite large. We will see in the next chapter that it can be made as small as 2. However, the techniques used allow us to prove that there are exponentially many sequences with constant PMEPR.

Theorem 5.15 *There are at least $(2 - \delta_K)^n$ coefficient vectors $\mathbf{a} \in \{-1, 1\}^n$, such that*

$$\mathrm{PMEPR}(\mathbf{a}) \leq K,$$

where $\delta_K \in (0, 1)$, and

$$\lim_{K \to \infty} \delta_K = 0.$$

Proof We will show that actually many vectors satisfy Lemma 5.12. In (5.70) we determined the size of $A_{\mathbf{b}}$ when the maximum is restricted to $h\sqrt{n \ln \frac{2m}{n}}, h = 10$. For $h \geq 10$ we have

$$|A_{\mathbf{b}}| \geq 2^{n(1-c(h))(1+o(1))},$$

where $c(h) \to 0$ as $h \to \infty$. Let $p(h) > 0$ satisfy $H\left(\frac{1}{2} - p(h)\right) = 1 - c(h)$, so that $p(h) \to 0$ as $L \to \infty$. Set $A_{\mathbf{b}}^{(0)} = A_{\mathbf{b}}$. For $t = 1, 2, \dots, \frac{|A_{\mathbf{b}}|}{4}$, having defined $A_{\mathbf{b}}^{(t-1)}$, let $\mathbf{x}^{(t)}, \mathbf{y}^{(t)} \in A_{\mathbf{b}}^{(t-1)}$ be a pair of vectors at maximal distance and set

$$A_{\mathbf{b}}^{(t)} = A_{\mathbf{b}}^{(t-1)} - \{\mathbf{x}^{(t)}, \mathbf{y}^{(t)}\}.$$

Since $|A_{\mathbf{b}}^{(t)}| \geq \frac{A_{\mathbf{b}}}{2}$, we have

$$d_H\left(\mathbf{x}^{(t)}, \mathbf{y}^{(t)}\right) \geq n(1 - 2p(h) - o(1)).$$

Set $\mathbf{z}^{(t)} = \frac{1}{2}(\mathbf{x}^{(t)} - \mathbf{y}^{(t)})$, so that $\mathbf{z}^{(t)} \in \{-1, 0, 1\}^n$, then at most $n(2p(h) + o(1))$ of its coefficients are zero,

$$\max_{k=0,1,\ldots,n-1} |L_k(\mathbf{z})| \leq h\sqrt{n \ln \frac{2m}{n}}.$$

By Lemma 5.13 we transform $\mathbf{z}^{(t)}$ to $\mathbf{w}^{(t)} \in \{-1, 1\}^n$ with

$$\max_{k=0,1,\ldots,n-1} |L_k(\mathbf{w})| \leq h'\sqrt{n \ln \frac{2m}{n}},$$

where h' may be chosen to equal $h + 1$ for $h \geq 10$. On those coefficients where $\mathbf{x}^{(t)}$ and $\mathbf{y}^{(t)}$ differ, the vectors $\mathbf{x}^{(t)}, \mathbf{y}^{(t)}$, hence $\mathbf{x}^{(t)}$ and $\mathbf{w}^{(t)}$ are the same. Thus

$$d_H\left(\mathbf{x}^{(t)}, \mathbf{w}^{(t)}\right) \leq n(4p(h) + o(1)).$$

Let $q(h) = H(4p(h))$ so that $q(h) \to 0$ as $h \to \infty$. For each s there are at most $2^{n(q(h)+o(1))}$ indices t with $\mathbf{w}^{(s)} = \mathbf{w}^{(t)}$. There are $\frac{|A_{\mathbf{b}}|}{4} \geq 2^{n(1-c(h)+o(1))}$ indices t. Let δ_h be defined by the equation

$$2 - \delta_h = 2^{1-c(h)-q(h)},$$

so that $\lim_{h\to\infty} \delta_h = 0$. The number of distinct $\mathbf{w}^{(t)} \in \{-1, 1\}^n$ with

$$\max_{k=0,1,\ldots,n-1} |L_k(\mathbf{z})| \leq h'\sqrt{n \ln \frac{2m}{n}}$$

is at least

$$\frac{2^{n(1-c(h)+o(1))}}{2^{n(q(h)+o(1))}} = (2 - \delta_h + o(1))^n.$$

Now assuming $m = \mathrm{const} \cdot n$ and choosing $K = h^2$, we obtain the claim. $\qquad\square$

5.5 BPSK signals with essentially high peaks

The characterization of signals with high peaks is based on the following result.

Theorem 5.16 *Let* $F_{\mathbf{a}}(t) = \sum_{k=0}^{n-1} a_k e^{2\pi \imath k t}$, $a_k \in \{\pm 1\}$, *and*

$$\left|\sum_{k=0}^{n-1} a_k\right| \leq \frac{n}{2}, \qquad \left|\sum_{k=0}^{n-1} (-1)^k a_k\right| \leq \frac{n}{2}.$$

Then, whenever $n > 251$,

$$\max_{t \in [0,1)} |F_{\mathbf{a}}(t)| \le \frac{3}{4} n.$$

Proof   We start with demonstrating that the functions $F_{\mathbf{a}}(t)$ with maximum of the absolute value greater than $3/4$ may assume this maximum only in the vicinity of $t = 0$ and $t = 1/2$, and, moreover, the values of $F_{\mathbf{a}}(t)$ in these points have a very restricted argument.

Let us introduce three vectors:

$$\mathbf{a} = (a_0, \ldots, a_{n-1}),$$

$$\mathbf{e}_+ = \left(1, e^{2\pi \imath t}, e^{2\pi \imath 2t}, \ldots, e^{2\pi \imath (n-1)t}\right),$$

$$\mathbf{e}_- = \left(1, e^{-2\pi \imath t}, e^{-2\pi \imath 2t}, \ldots, e^{-2\pi \imath (n-1)t}\right).$$

Consider the Gram matrix for the three vectors:

$$G = \begin{pmatrix} (\mathbf{a}, \mathbf{a}) & (\mathbf{e}_+, \mathbf{a}) & (\mathbf{e}_-, \mathbf{a}) \\ (\mathbf{a}, \mathbf{e}_+) & (\mathbf{e}_+, \mathbf{e}_+) & (\mathbf{e}_-, \mathbf{e}_+) \\ (\mathbf{a}, \mathbf{e}_-) & (\mathbf{e}_+, \mathbf{e}_-) & (\mathbf{e}_-, \mathbf{e}_-) \end{pmatrix},$$

where, as usual,

$$(\mathbf{w}, \mathbf{v}) = \sum_{k=0}^{n-1} w_k v_k^*,$$

and v_k^* is the complex conjugate of v_k. Clearly,

$$G = \begin{pmatrix} n & F_{\mathbf{a}}(t) & F_{\mathbf{a}}(-t) \\ F_{\mathbf{a}}(-t) & n & D(-2t) \\ F_{\mathbf{a}}(t) & D(2t) & n \end{pmatrix},$$

where

$$D(x) = \sum_{k=0}^{n-1} e^{2\pi \imath k x}$$

is the Dirichlet kernel, see (4.7). We calculate

$$\det G = n^3 + F_{\mathbf{a}}^2(t) D(-2t) + F_{\mathbf{a}}^2(-t) D(2t) - n |F_{\mathbf{a}}(t)|^2 - n |F_{\mathbf{a}}(t)|^2 - n |D(2t)|^2.$$

Since $F_{\mathbf{a}}^2(t) D(-2t) = \left(F_{\mathbf{a}}^2(-t) D(2t)\right)^*$ we conclude that $\det G \in \mathbb{R}$.

It is known that $\det G \ge 0$ always, and if $\det G = 0$ then the vectors $\mathbf{a}, \mathbf{e}_+, \mathbf{e}_-$ are linearly dependent. Thus we have inequality

$$n^3 + F_{\mathbf{a}}^2(t) D(-2t) + F_{\mathbf{a}}^2(-t) D(2t) \ge n \left(2 |F_{\mathbf{a}}(t)|^2 + |D(2t)|^2\right). \qquad (5.72)$$

Denoting by φ the argument of $F_{\mathbf{a}}^2(t)D(-2t)$ (i.e., $e^{i\varphi} = F_{\mathbf{a}}^2(t)D(-2t)$), we have

$$F_{\mathbf{a}}^2(t)D(-2t) + F_{\mathbf{a}}^2(-t)D(2t) = |F_{\mathbf{a}}(t)|^2|D(2t)|e^{i\varphi} + |F_{\mathbf{a}}(t)|^2|D(2t)|e^{-i\varphi}$$
$$= 2|F_{\mathbf{a}}(t)|^2|D(2t)| \cdot \cos\varphi.$$

Then (5.72) may be rewritten as

$$n^3 + 2|D(2t)| \cdot |F_{\mathbf{a}}(t)|^2 \cdot \cos\varphi \geq n\left(2|F_{\mathbf{a}}(t)|^2 + |D(2t)|^2\right). \tag{5.73}$$

From now on let us assume that t is such that $|F_{\mathbf{a}}(t)| \geq \frac{3}{4}n$. Since

$$2|D(2t)| \cdot \cos\varphi \leq 2n,$$

the inequality will stay true if we replace $|F_{\mathbf{a}}(t)|$ with $\frac{3}{4}n$. We obtain

$$n^3 + 2|D(2t)| \cdot \frac{9}{16}n^2 \cdot \cos\varphi \geq n\left(2\frac{9}{16}n^2 + |D(2t)|^2\right)$$

or

$$9n \cdot |D(2t)| \cdot \cos\varphi \geq n^2 + 8|D(2t)|^2. \tag{5.74}$$

Let us now study the behavior of $|D(2t)| = \frac{\sin 2\pi nt}{\sin 2\pi t}$. A direct check shows that this is an oscillating function with zeros at points $t = \frac{k}{n}, k = 1, \ldots, n-1$, and decreasing maxima between the zeros. A weaker version of (5.74) gives

$$9n|D(2t)| \geq n^2 + 8|D(2t)|^2$$

or

$$|D(2t)| \geq \frac{1}{8}n.$$

Comparing the consecutive maxima of $|D(2t)|$, we verify that the fourth maximum (between $\frac{3}{n}$ and $\frac{4}{n}$) is less than $\frac{n}{8}$, and conclude that

$$\text{dist}(2t) < \frac{3}{n}. \tag{5.75}$$

Here, $\text{dist}(x)$ stands for the distance from x to the closest integer.

Let us choose γ in such a way that $F_{\mathbf{a}}(t)e^{-2\pi i\gamma}$ is positive and real. Then

$$0 \leq |F_{\mathbf{a}}(t)| = F_{\mathbf{a}}(t)e^{-2\pi i\gamma}$$
$$= \sum_{k=0}^{n-1} a_k e^{2\pi i(kt-\gamma)} = \sum_{k=0}^{n-1} a_k \cos 2\pi(kt-\gamma).$$

For the sequence

$$\{\cos 2\pi(kt-\gamma)\} \tag{5.76}$$

we consider two possibilities:

1. The sequence (5.76) has only nonnegative entries;
2. The sequence (5.76) has at least one change of sign.

In the first case,

$$|F_{\mathbf{a}}(t)| = \sum_{k=0}^{n-1} a_k \cos 2\pi(kt - \gamma) \le \sum_{k:a_k=1} a_k \cos 2\pi(kt - \gamma)$$

$$\le \sum_{k:a_k=1} 1 \le \max\left\{\sum_{k:a_k=1} 1, \sum_{k:a_k=-1} 1\right\} = \frac{n + F_{\mathbf{a}}(0)}{2}.$$

Let us consider now the second case. Since $\text{dist}(2t) < \frac{3}{n}$, then

$$t \in \left[-\frac{3}{2}\frac{1}{n}, \frac{3}{2}\frac{1}{n}\right] \bigcup \left[\frac{1}{2} - \frac{3}{2}\frac{1}{n}, \frac{1}{2} + \frac{3}{2}\frac{1}{n}\right] = I_1 \bigcup I_2.$$

Notice also that

$$F_{\mathbf{a}}\left(t + \frac{1}{2}\right) = \sum_{k=0}^{n-1} a_k e^{2\pi\imath k(\frac{1}{2}+t)}$$

$$= \sum_{k=0}^{n-1} a_k(-1)^k e^{2\pi\imath kt} = \sum_{k=0}^{n-1} \eta_k e^{2\pi\imath kt},$$

where for all k we have $\eta_k = a_k(-1)^k$. Therefore, any conditions on $\{a_k\}_{k=0}^{n-1}$ guaranteeing $|F_{\mathbf{a}}(t)| \le 3n/4$, for any $t \in I_1$, will be sufficient for $|F_{\mathbf{a}}(t)| \le 3n/4$, for any $t \in I_2$, where $F_{\mathbf{a}}(t)$ is defined by the sequence $\{\eta_k\}_{k=0}^{n-1}$.

Let $t_1 \in I_1$, i.e. $|t_1| \le \frac{3}{2}\frac{1}{n}$. Evidently,

$$|F_{\mathbf{a}}(t)| = \sum_{k=0}^{n-1} a_k \cos 2\pi(kt - \gamma)$$

$$\le \sum_{k=0}^{n-1} |\cos 2\pi(kt - \gamma)| \le \max_x \sum_{k=0}^{n-1} |\sin 2\pi(kt + x)|,$$

where the maximum is taken over such x that the sequence $\{\cos 2\pi(kt + x)\}_{k=0}^{n-1}$ has at least one change of sign. Using integral approximation for the sum we get:

$$J_n = \frac{1}{n}\sum_{k=0}^{n-1} |\sin 2\pi(kt + x)| = \frac{1}{(n-1)t}\int_x^{(n-1)t+x} |\sin 2\pi u|\, du + E.$$

Since

$$|E| \le \frac{1}{n} \max_u |(\sin 2\pi u)'| = \frac{1}{n} \max_u |\cos 2\pi \cos 2\pi u| \le \frac{2\pi}{n},$$

we have, for any $t \in I_1$,

$$|F_\mathbf{a}(t)| \le n J_n \le \frac{n}{(n-1)t} \int_x^{(n-1)t+x} |\sin 2\pi u| \, du + 2\pi,$$

and x is such that in the interval of integration the function $\sin 2\pi u$ has at least one change of sign. Moreover, the size of the interval is, at most, $3/2$. Computations show that

$$\kappa = \max_{\beta, \alpha} \int_\alpha^\beta |\sin 2\pi u| \, du,$$

for $\alpha \in [0, 1/2]$, $\beta \in [1/2, 1]$, $\beta - \alpha \le \frac{3}{2}$, satisfies $\kappa < 0.725$. Therefore,

$$|F_\mathbf{a}(t)| \le \kappa n + 2\pi \le 0.75n$$

whenever $n > 251$. Thus we have arrived at the claimed result.   □

To find how many sequences do not satisfy this pair of inequalities, recall that by Lemma 3.41, $\sum_{0 \le s \le \lambda n} \binom{n}{s} \le 2^{n H(\lambda)}$, where $H(\lambda) = -\lambda \log_2 \lambda - (1 - \lambda) \log_2(1 - \lambda)$, and thus the number of bad sequences is at most (here we use union bound):

$$4 \cdot 2^{n H(0.25)} < 2^{0.812n},$$

which is much less than 2^{n-1}. Thus we expect to lose at most one bit in encoding good sequences.

Let us now describe the encoding algorithm. For simplicity, let n be a multiple of 8. We will substitute the condition of Theorem 5.16 by the following stronger ones:

$$\left| \sum_{k \text{ even}} a_k \right| \le \frac{n}{4}, \qquad \left| \sum_{k \text{ odd}} a_k \right| \le \frac{n}{4},$$

and our purpose is to generate sequences satisfying these conditions. Let us consider the first $\frac{n}{2} - 1$ components of the sequence $\{a_{2j}\}_{j=0}^{n/2-2}$, and assume that they are chosen arbitrarily. Let $\{b_{2j}\}_{j=0}^{n/2-2}$ be an arbitrary sequence having $\frac{n}{4} - 1$ ones. If the sequence $\{a_{2j}\}_{j=0}^{n/2-2}$ contains the number of ones in the interval $\left[\frac{n}{8}, \frac{3n}{8}\right]$, we put $a_{n-2} = -1$, otherwise, we produce the new sequence $\{a_{2j}\} = \{a_{2j}b_{2j}\}_{j=0}^{n/2-2}$ and define $a_{n-2} = 1$. It is easy to check that the resulting sequence satisfies the defined condition. We do the same for the odd coordinates. Decoding is simple, since the last symbol indicates whether the sequence has been modified, and the inverse modification is straightforward.

5.6 Notes

Section 5.1 Initial study of the distribution of peaks in MC signals is by van Eetvelt *et al.* [100] and Shepherd *et al.* [379]. The idea of using Chernoff bounds was first exploited by Wunder and Boche [443]; see also [439]. The presented versions of Theorems 5.2 and 5.3 are by Litsyn and Wunder [252]. The theorems actually hold for more general constellations. In particular the following result was proved by Litsyn and Wunder [252].

Theorem 5.17 *Let* a_k, $k = 0, \ldots, n - 1$, *be random variables with* $E(|a_k|^2) = 1$ *(they need not be necessarily independent). Suppose that*

$$\frac{1}{rhn} \sum_{l_1=0}^{rn-1} \sum_{l_2=0}^{h-1} E\left(\exp\left(\iota\omega\Re\left(\sum_{k=0}^{n-1} a_k e^{\iota\left(2\pi k t_{l_1,r} + \alpha_{l_2,h}\right)}\right)\right)\right) = e^{-\frac{n\omega^2}{4} + n\sum_{m=3}^{5} a_m\omega^m + O(n\omega^6)},$$

holds for any natural numbers $r > 1$, $h > 2$ *and for any real number* ω *in the nonempty interval* $[-d, d]$, *and, furthermore,*

$$\frac{1}{n^2} \sum_{l_1=0}^{n-1} \sum_{l_2=0, l_2 \neq l_1}^{n-1} E\left(\exp\left(\iota\Re\left(\sum_{k=0}^{n-1} a_k \left(\omega_1 e^{\iota 2\pi k t_{l_1}} + \omega_2 e^{\iota 2\pi k t_{l_2}}\right)\right)\right)\right)$$
$$= e^{-\frac{n\omega_1^2}{4} - \frac{n\omega_2^2}{4} + n\sum_{m=3}^{5} A_m(|\omega_1| + |\omega_2|)^m + O(n(|\omega_1| + |\omega_2|)^6)},$$

holds for all real numbers ω_1, ω_2 *in the nonempty interval* $[-d, d]$, *where* t_l, $l = 0, \ldots, O(n) - 1$, *is any sampling set of cardinality* $O(n)$ *and* A_3, A_4, A_5 *are complex numbers independent of* $\omega, \omega_1, \omega_2$ *(finite sixth moment of the random variables!).*

Then there is a constant $\gamma \geq 5$ *independent of n so that for growing n,*

$$\Pr\left(\ln n - \gamma \ln \ln n < \text{PMEPR}(\mathbf{a}) < \ln n + \gamma \ln \ln n\right) = 1 - O\left(\frac{1}{\ln^4 n}\right).$$

□

The theorem holds for any modulation scheme where the real and imaginary parts are independent and identically distributed, and the corresponding probability functions have vanishing first moments and finite support.

Section 5.2 Halász [148] proved the concentration of the peak distribution for real trigonometric polynomials with coefficients chosen from $\{-1, 1\}$. Gersho *et al.* [128] generalized the result of Halász to complex trigonometric polynomials with the real coefficients satisfying some restrictions on the characteristic function. Sharif and Hassibi [369, 370, 374] generalized the earlier results on concentration to the coefficients chosen from PSK and QPSK constellations. I have followed [374] in this section.

Section 5.3 Gaussian approximation is a folklore. It appears, e.g., in Dinur and Wulich [95]. However this approximation gives reasonable accuracy only for small numbers of carriers. Van Nee and de Wild [287] suggested using an empirical approximation,

$$\Pr\left(\text{PMEPR}(\mathbf{a}) > \lambda\right) \approx 1 - (1 - e^{-\lambda})^{2.8n},$$

to improve on its accuracy.

The basic analysis of peaks in Gaussian and Rayleigh processes is from Rice [339, 340]. Cartwright and Longuet-Higgins [58] extended the analysis to extreme peaks in a Gaussian process. In this section, I have followed Ochiai and Imai [303]. The estimate (5.62) is by Ochiai and Imai. The expression (5.63) was obtained by Dinur and Wulich [95], who used analysis of the number of level crossings of stochastic processes, while (5.64) was derived by Wei *et al.* [429] using modern extremal value theory.

Section 5.4 The results of this section are from Spencer [387].

Section 5.5 The results of the section are by Freiman *et al.* [119].

6

Coded MC signals

The peak power distributions considered in the previous chapter are derived under the assumption that the coefficients are chosen independently from a constellation. In this chapter, I will consider a more complicated situation when there exists a dependence between subcarriers. In Section 6.1 the PMEPR distribution in spherical codes is considered. The only restriction on these signals is that they have constant energy. I prove results about concentration of the PMEPR distribution. In Section 6.2, I study the existence of spherical codes with a given minimum (Euclidean or Hamming) distance and PMEPR. In Section 6.3, I relate the PMEPR distribution of coded signals to the distance distribution of codes. In Section 6.4, I specify the previous analysis to the case of BCH codes. I show that when the length grows, the distance distribution of BCH codes approaches the normalized binomial distribution. This allows analysis of the PMEPR distribution for a subclass of BCH codes. In Section 6.5, I show how the PMEPR of a code can be computed in an efficient way if the code possesses a fast maximum likelihood decoding.

6.1 Spherical codes

An n-dimensional constellation has a spherical distribution if the points of the constellation are uniformly distributed over the n-dimensional complex sphere with radius $\sqrt{n}$ denoted $\mathbb{S}^{N-1}$.

Theorem 6.1 *Let $\mathbf{c}$ be spherically distributed. Then*

$$\Pr(\text{PMEPR}(\mathbf{c}) > \lambda) \geq \sum_{\ell=1}^{\lfloor \frac{n}{\lambda} \rfloor} \binom{n}{\ell} (-1)^{\ell+1} \left(1 - \frac{\ell\lambda}{n}\right)^{n-1}, \tag{6.1}$$

$$\Pr(\text{PMEPR}(\mathbf{c}) > \lambda) \leq r \sum_{\ell=1}^{\lfloor \frac{nC_r^2}{\lambda} \rfloor} \binom{n}{\ell} (-1)^{\ell+1} \left(1 - \frac{\ell\lambda}{nC_r^2}\right)^{n-1}, \tag{6.2}$$

where $r > 1$ is an integer.

138

Proof We will study first the situation when the coefficients are circularly Gaussian distributed with unit power. Let $\mathbf{c}$ be a random vector from such an ensemble. The magnitude of the corresponding signal is Rayleigh distributed, i.e., $\Pr\left(|F_{\mathbf{c}}(t)| \le \lambda n\right) = 1 - e^{-\lambda^2}$, and the Nyquist-rate samples $F_{\mathbf{c}}\left(\frac{j}{n}\right)$, $j = 0, 1, \ldots, n-1$, are independent. Denoting

$$\mathcal{M}_d = \mathcal{M}_d(F_{\mathbf{c}}) = \max_{j=0,1,\ldots,n-1} \left| F_{\mathbf{c}}\left(\frac{j}{n}\right) \right|,$$

we conclude that

$$\Pr(\mathcal{M}_d > \lambda\sqrt{n}) = 1 - \left(1 - e^{-\lambda^2}\right)^n.$$

Now,

$$\Pr(\mathcal{M}_d > \lambda\sqrt{n}) = \int_0^\infty f_P(x) \cdot \Pr\left(\mathcal{M}_d > \lambda\sqrt{n} \;\middle|\; \sum_{j=0}^{n-1} |c_j|^2 = x\right) dx,$$

and f_P is the probability density function of $P = \sum_{j=0}^{n-1} |c_j|^2$. This function, see, e.g., [335, Section 1.1.4], is the central χ^2 distribution with $2n$ degrees of freedom,

$$f_P(x) = \frac{x^{n-1}}{(n-1)!} \cdot e^{-x}.$$

Denoting

$$\mathcal{F}(\lambda, x) = \Pr\left(\mathcal{M}_d > \lambda\sqrt{n} \;\middle|\; \sum_{j=0}^{n-1} |c_j|^2 = x\right),$$

we thus arrive at (a Fredholm integral equation of the first kind)

$$1 - \left(1 - e^{-\lambda^2}\right)^n = \int_0^\infty \frac{x^{n-1}}{(n-1)!} \cdot e^{-x} \cdot \mathcal{F}(\lambda, x)\, dx.$$

Notice that $\mathcal{F}(\lambda, x)$ depends only on the ratio $\tau = \frac{x}{\lambda^2 n}$, and thus

$$\frac{(n-1)!}{(\lambda^2 n)^n}\left(1 - \left(1 - e^{-\lambda^2}\right)^n\right) = \int_0^\infty \tau^{n-1} \cdot e^{-\tau\lambda^2 n} \cdot \mathcal{F}(\lambda, \tau\lambda^2 n)\, d\tau.$$

Denoting

$$y(\tau) = \tau^{n-1} \cdot \mathcal{F}(\lambda, \tau\lambda^2 n),$$

we obtain

$$\frac{(n-1)!}{\left(\lambda^2 n\right)^n}\left(1 - \left(1 - e^{-\lambda^2}\right)^n\right) = \int_0^\infty y(\tau) \cdot e^{-\tau\lambda^2 n}\, d\tau.$$

The right-hand side can be seen as the Laplace transform of $y(\tau)$, if $\tau\lambda^2 n$ is considered as a complex frequency variable. Consequently the sought function, $y(\tau)$, can be obtained by inverse Laplace transform. Using

$$1 - \left(1 - e^{-\lambda^2}\right)^n = \sum_{\ell=1}^{n} \binom{n}{\ell} \cdot (-1)^{\ell+1} \cdot e^{-\ell\lambda^2},$$

and dividing by τ^{n-1}, we obtain

$$\mathcal{F}(\lambda, \tau\lambda^2 n) = \sum_{\ell=1}^{n} \binom{n}{\ell} \cdot (-1)^{\ell+1} \cdot u\left(\tau - \frac{\ell}{n}\right) \cdot \left(1 - \frac{\ell}{\tau n}\right)^{n-1},$$

where $u(\cdot)$ is the unit step function. Noticing that

$$\Pr\left(\mathcal{M}_d > \lambda\sqrt{n} \,\Big|\, \sum_{j=0}^{n-1} |c_j|^2 = n\right) = \mathcal{F}(\lambda, n),$$

and thus choosing $\tau = \frac{1}{\lambda^2}$, and the fact that

$$\text{PMEPR}(\mathbf{c}) \geq \frac{\mathcal{M}_d^2(F_{\mathbf{c}})}{n},$$

we conclude that

$$\Pr(\text{PMEPR} > \lambda) \geq \sum_{1 \leq \ell < \lfloor\frac{n}{\lambda}\rfloor} \binom{n}{\ell} (-1)^{\ell+1} \left(1 - \frac{l\lambda}{n}\right)^{n-1}.$$

Clearly the derivation of the bound holds for any shifted (in t) Nyquist-rate sampling set. Hence, using the union bound for $r > 1$ shifted versions of (6.1) and scaling λ by C_r, the upper bound (6.2) is obtained. $\square$

Having the upper and lower bound, we now ask for the asymptotic behavior of these expressions. We need the following technical lemma.

Lemma 6.2   *Let the distribution of c_k, $k = 0, \ldots, n-1$, be circularly complex Gaussian with unit power. Then, for any real $\gamma > 0.5$ and increasing n, we have*

$$\Pr\left(\ln n - 1.1\gamma \ln\ln n < \text{PMEPR}(\mathbf{c}) < \ln n + 1.1\gamma \ln\ln n\right) = 1 - \frac{1.1}{(\ln n)^{\gamma-0.5}}.$$

Proof   Since the magnitude of the absolute values of the signal sampled at the Nyquist frequency is Rayleigh distributed, we have, for any $t \in [0, 1)$,

$$\Pr\left(|F_{\mathbf{c}}(t)| \leq \lambda\sqrt{n}\right) = 1 - e^{-\lambda^2}.$$

Therefore,

$$\Pr\left(\text{PMEPR}(\mathbf{c}) \leq \lambda\right) \leq \left(1 - e^{-\lambda}\right)^n.$$

Setting $\lambda = (1 - \varepsilon)\ln n$ for some $\varepsilon > 0$ we have

$$\Pr\left(\text{PMEPR}(\mathbf{c}) \leq (1 - \varepsilon)\ln n\right) \leq \left(1 - \frac{n^{\varepsilon}}{n}\right)^{n} \leq e^{-n^{\varepsilon}},$$

which is the desired lower bound if we set $\varepsilon = \frac{\ln \ln n}{\ln n}$.

The upper bound is obtained by observing that for any $r \geq 1$ we have

$$\Pr\left(\max_{j=0,1,\dots,rn-1}\left|F_{\mathbf{c}}\left(\frac{j}{rn}\right)\right| > \frac{\lambda}{C_r} \cdot \sqrt{n}\right) \leq r\left(1 - \left(1 - e^{-\frac{\lambda^2}{C_r^2}}\right)^{n}\right).$$

Setting this time $\lambda = \sqrt{(1 + \varepsilon)\ln n}$ for some $\varepsilon > 0$ yields

$$\Pr\left(\frac{1}{n} \cdot \max_{j=0,1,\dots,rn-1}\left|F_{\mathbf{c}}\left(\frac{j}{rn}\right)\right|^{2} > \frac{(1 + \varepsilon)}{C_r^2}\ln n\right) \leq r\left(1 - \left(1 - n^{-\frac{(1+\varepsilon)}{C_r^2}}\right)^{n}\right).$$

Since $\frac{1}{C_r^2} \geq 1 - \frac{2\pi^2}{8r^2}$, we can choose r so that $\frac{(1+\varepsilon)}{C_r^2} = 1 + \varepsilon' > 1$, and hence

$$\Pr\left(\frac{1}{n} \cdot \max_{j=0,1,\dots,rn-1}\left|F_{\mathbf{c}}\left(\frac{j}{rn}\right)\right|^{2} > (1 + \varepsilon')\ln n\right) \leq r\left(1 - \left(1 - \frac{n^{-\varepsilon'}}{n}\right)^{n}\right).$$

Using (this time) $\ln(1 - x) \geq -1.1x$ for x small enough, we have

$$\Pr\left(\text{PMEPR}(\mathbf{c}) > (1 + \varepsilon)\ln n\right) \leq r\left(1 - e^{-1.1n^{-\varepsilon'}}\right),$$

for n large enough. Setting $r = \sqrt{\ln n}$ and $\varepsilon = 1.1\gamma\frac{\ln \ln n}{\ln n}$ we obtain $\varepsilon' \geq \gamma\frac{\ln \ln n}{\ln n}$. Thus for n large enough

$$\Pr\left(\text{PMEPR}(\mathbf{c}) > \ln n + 1.1\gamma\ln \ln n\right) \leq O\left(\frac{1}{(\ln n)^{\gamma - 0.5}}\right),$$

which is the desired upper bound for $\gamma > 0$. Combining this with the lower bound yields the result. □

The following theorem proves that the asymptotic behavior of the Gaussian and spherical distributions is identical.

Theorem 6.3   *Let the cumulative distribution of the c_k, $k = 0, \dots, n - 1$, be such that the code words are uniformly distributed on $\mathbb{S}^{n-1}$. Then for any real $\gamma > 0.5$,*

$$\Pr\left(\ln n - 1.1\gamma\ln \ln n < \text{PMEPR}(\mathbf{c}) < \ln n + 1.1\gamma\ln \ln n\right) = 1 - \frac{1.1}{(\ln n)^{\gamma - 0.5}}.$$

Proof   Since $\mathbf{c}$ is uniformly distributed on $\mathbb{S}^{n-1}$ it can be represented as

$$\mathbf{c} = \frac{\sqrt{n}\mathbf{c}'}{\|\mathbf{c}'\|},$$

where $\mathbf{c}'$ has an n-variate spherical distribution, i.e., it is invariant under multiplication by a unitary matrix. The distribution of $\mathbf{c}'$ is not unique and we can assume that c_k' are circularly Gaussian distributed with unit power. In the following let $\varepsilon > 0$ be arbitrary and $\varepsilon_1 > 0$ be such that $(1 + \varepsilon_1)\sqrt{(1 - \varepsilon)} < 1$. By Lemma 6.2,

$$\lim_{n\to\infty} \Pr\left((1 - \varepsilon)\ln n < \text{PMEPR}(\mathbf{c}') < (1 + \varepsilon)\ln n\right) = 0.$$

I will now show that this yields the main result. The proof is sketched as follows: define an ε-neighborhood around the sphere of radius $\sqrt{n}$ and partition the event space into a set with $\|\mathbf{c}\| \approx \sqrt{n}$ (within the ε-neighborhood) and its complementary set. The latter event is indeed very unlikely for large n, while the first event can be tackled using standard Gaussian distribution analysis obeying the claimed asymptotic behavior.

Let us consider now the first inequality of the theorem. By the law of total probability

$$\Pr\left(\text{PMEPR}(\mathbf{c}) < (1 - \varepsilon)\ln n\right)$$
$$= \Pr\left(\{\text{PMEPR}(\mathbf{c}) < (1 - \varepsilon)\ln n\} \cap \{\|\mathbf{c}\| < (1 - \varepsilon_1)\sqrt{n}\}\right)$$
$$+ \Pr\left(\{\text{PMEPR}(\mathbf{c}) < (1 - \varepsilon)\ln n\} \cap \{(1 - \varepsilon_1)\sqrt{n} < \|\mathbf{c}\| < (1 + \varepsilon_1)\sqrt{n}\}\right)$$
$$+ \Pr\left(\{\text{PMEPR}(\mathbf{c}) < (1 - \varepsilon)\ln n\} \cap \{\|\mathbf{c}\| > (1 + \varepsilon_1)\sqrt{n}\}\right).$$

I will show that

$$\lim_{n\to\infty} \Pr\left(\{\text{PMEPR}(\mathbf{c}) < (1 - \varepsilon)\ln n\} \cap \{(1 - \varepsilon_1)\sqrt{n} < \|\mathbf{c}\| < (1 + \varepsilon_1)\sqrt{n}\}\right) = 0$$

with the same rate of convergence as in Lemma 6.2. We have

$$\Pr\left(\{\text{PMEPR}(\mathbf{c}) < (1 - \varepsilon)\ln n\} \cap \{(1 - \varepsilon_1)\sqrt{n} < \|\mathbf{c}\| < (1 + \varepsilon_1)\sqrt{n}\}\right)$$
$$= \Pr\left(\left\{ \max_{t\in[0,1)} \frac{1}{\|\mathbf{c}\|} |F_{\mathbf{c}}(t)| < \sqrt{(1 - \varepsilon)\ln n} \right\} \right.$$
$$\left. \cap \{(1 - \varepsilon_1)\sqrt{n} < \|\mathbf{c}\| < (1 + \varepsilon_1)\sqrt{n}\} \right)$$
$$\leq \Pr\left(\frac{1}{(1 + \varepsilon_1)\sqrt{n}} \cdot \max_{t\in[0,1)} |F_{\mathbf{c}}(t)| < \sqrt{(1 - \varepsilon)\ln n} \right)$$
$$= \Pr\left(\frac{1}{\sqrt{n}} \cdot \max_{t\in[0,1)} |F_{\mathbf{c}}(t)| < (1 + \varepsilon_1)\sqrt{(1 - \varepsilon)\ln n} \right).$$

Further,

$$\Pr\left(\|\mathbf{c}\| < (1 - \varepsilon_1)\sqrt{n}\right) \leq \frac{1}{\varepsilon_1 n}$$

and

$$\Pr\left(\|\mathbf{c}\| \geq (1 + \varepsilon_1)\sqrt{n}\right) \leq \frac{1}{\varepsilon_1 n}.$$

Setting $\varepsilon_1 = \frac{1}{\sqrt{\ln n}}$ and $\varepsilon = \ln \ln n$ we obtain $\varepsilon > \varepsilon_1$ for n large enough, and both expressions tend to zero faster than the remainder term of Lemma 6.2. The upper bound follows the same derivations. □

6.2 Bounds on PAPR of codes

Let us consider signals defined by

$$F_{\mathbf{a}}(t, \zeta) = \sum_{k=0}^{n-1} a_k e^{2\pi \iota (\zeta + k)t},$$

where the coefficients a_k are from a code word $\mathbf{a} \in \mathcal{C}$, and ζ is the ratio between the carrier frequency and the bandwidth of the tones (see Section 2.2 for the definitions). For the sake of simplicity, we consider the signals of constant energy, equal n. Then

$$\text{PAPR}(\mathcal{C}) = \frac{1}{n} \cdot \max_{t \in [0,1)} \Re\left(F_{\mathbf{a}}(t, \zeta)\right).$$

The error-correcting capability of $\mathcal{C}$ is characterized by the minimum of the Euclidean distance between any two distinct code vectors,

$$d_E(\mathcal{C}) = \min_{\mathbf{a}, \mathbf{b} \in \mathcal{C}, \mathbf{a} \neq \mathbf{b}} d_E(\mathbf{a}, \mathbf{b}).$$

We will establish here bounds on the PAPR as a function of the code's rate $R = \frac{1}{n} \log_2 |\mathcal{C}|$, and minimum Euclidean distance d_E.

Let $\mathcal{C}$ denote such a code. Then the code words $\mathbf{a}$ of $\mathcal{C}$ are points on the n-dimensional complex sphere of radius $\sqrt{n}$. We define the curve Ω by

$$\Omega = \{\mathbf{w}(t, \zeta), \quad t \in [0, 1)\},$$

where

$$\mathbf{w}(t, \zeta) = \left(e^{2\pi \iota \zeta t}, e^{2\pi \iota (\zeta + 1)t}, \ldots, e^{2\pi \iota (\zeta + n - 1)t}\right).$$

This curve lies on the same sphere as the code words of $\mathcal{C}$. We define the curve $-\Omega$ to consist of all the points $\{-\mathbf{w}(t, \zeta), t \in [0, 1)\}$. The following result shows that the closer a code word is to the curve $\Omega \cup (-\Omega)$, the larger is the PAPR.

Theorem 6.4 *Let $\mathcal{C}$ be a code of length n, rate R, and minimum Euclidean distance d_E. Let $\hat{d}$ denote the minimum Euclidean distance between the code words of $\mathcal{C}$ and* *the points of* $\Omega \cup (-\Omega)$. *Then* $\hat{d} \leq \sqrt{2n}$ *and*

$$\text{PAPR}(\mathcal{C}) \geq n \left(1 - \frac{\hat{d}^2}{2n} \right)^2.$$

Proof We first prove that $\hat{d} \leq \sqrt{2n}$. For any $t \in [0, 1)$, $\mathbf{w}(t, \zeta)$ and $-\mathbf{w}(t, \zeta)$ are antipodal points on the n-dimensional complex sphere of radius $\sqrt{n}$. It follows that for an arbitrary $\mathbf{a} \in \mathcal{C}$,

$$\|\mathbf{w}(t, \zeta) - \mathbf{a}\|^2 + \| - \mathbf{w}(t, \zeta) - \mathbf{a}\|^2 = 4n. \tag{6.3}$$

This means that either

$$\|\mathbf{w}(t, \zeta) - \mathbf{a}\| \leq \sqrt{2n},$$

or

$$\| - \mathbf{w}(t, \zeta) - \mathbf{a}\| \leq \sqrt{2n}.$$

It follows that $\hat{d} \leq \sqrt{2n}$. One of the following two cases can occur.

Case 1: In this case, there exist $\mathbf{a} \in \mathcal{C}$ and $t' \in [0, 1)$, such that $\|\mathbf{a} - \mathbf{w}(t', \zeta)\| \leq \hat{d}$. Then

$$2\Re \left(\sum_{k=0}^{n-1} a_k e^{2\pi \iota (\zeta + k) t'} \right) = \|\mathbf{a}\|^2 + \|\mathbf{w}(t', \zeta)\|^2 - \|\mathbf{a} - \mathbf{w}(t', \zeta)\|^2$$

$$= 2n - \|\mathbf{a} - \mathbf{w}(t', \zeta)\|^2 \geq 2n - \hat{d}^2.$$

Thus

$$\text{PAPR}(\mathbf{a}) \geq \frac{1}{n} \left| \Re \left(\sum_{k=0}^{n-1} a_k e^{2\pi \iota (\zeta + k) t'} \right) \right|^2 \geq \frac{1}{n} \left(n - \frac{\hat{d}^2}{2} \right)^2.$$

Case 2: In this case, there exist $\mathbf{a} \in \mathcal{C}$ and $t' \in [0, 1)$, such that $\|\mathbf{a} - (-\mathbf{w}(t', \zeta))\| \leq \hat{d}$. Thus

$$2\Re \left(\sum_{k=0}^{n-1} a_k e^{2\pi \iota (\zeta + k) t'} \right) = 2n - \|\mathbf{a} + \mathbf{w}(t', \zeta)\|^2 \geq 2n - \hat{d}^2.$$

Thus

$$\text{PAPR}(\mathbf{a}) \geq \frac{1}{n} \left| -\Re \left(\sum_{k=0}^{n-1} a_k e^{2\pi \iota (\zeta + k) t'} \right) \right|^2 \geq \frac{1}{n} \left(n - \frac{\hat{d}^2}{2} \right)^2.$$

$\square$

For any subset, S, of the n-dimensional complex sphere of radius $\sqrt{n}$ and any $r \geq 0$, we define $H(S, r)$ to be the surface consisting of all those points of the sphere that are within distance r of S. Let $A(S, r)$ denote the area of $H(S, r)$. So for any point, $\mathbf{x}$, on the sphere, $H(\mathbf{x}, r)$ is a spherical cap having surface (see, e.g., [364])

$$A(r) = A(\mathbf{x}, r) = \frac{2(\pi n)^{n - \frac{1}{2}}}{(n - \frac{3}{2})!} \int_0^{2 \arcsin \frac{r}{2\sqrt{n}}} \sin^{2n-2} \theta \, d\theta, \tag{6.4}$$

where we define

$$\left(\frac{2s + 1}{2} \right)! = \frac{(2s + 2)!}{2^{2s+2}(s + 1)!} \cdot \sqrt{\pi}. \tag{6.5}$$

Theorem 6.5 *Let C be a code of length n, rate R, and minimum Euclidean distance d_E. Suppose that for some $\hat{d}$ with $\frac{d}{2} \leq \hat{d} \leq \sqrt{2n}$,*

$$A\left(\Omega \cup -\Omega, \hat{d} - \frac{d_E}{2} \right) + 2^{nR} A\left(\frac{d_E}{2} \right) \geq \frac{2\pi^n n^{n - \frac{1}{2}}}{(n - 1)!}. \tag{6.6}$$

Then

$$\mathrm{PAPR}(C) \geq n \left(1 - \frac{\hat{d}^2}{2n} \right)^2.$$

Proof The spherical caps $H\left(\mathbf{a}, \frac{d_E}{2} \right)$, $\mathbf{a} \in C$, and the surface $H\left(\Omega \cup -\Omega, \hat{d} - \frac{d_E}{2} \right)$ must meet or overlap since the sum of their areas is at least that of the area of the n-dimensional complex sphere appearing in the right-hand side of (6.6). Therefore, the minimum distance of C from the points of the curve $\Omega \cup -\Omega$ is at most $\hat{d}$. Now the result follows from Theorem 6.4. $\qquad \square$

To make the inequality into a usable one, we need to obtain a lower bound on $A\left(\Omega \cup -\Omega, \hat{d} - \frac{d_E}{2} \right)$. Clearly, the curve Ω twists around the sphere many times since ζ is assumed to be large. To obtain a lower bound we will restrict our attention to almost one complete rotation of the curve Ω around the complex sphere. Let

$$\tilde{\Omega} = \left\{ \mathbf{w}(t, \zeta) : t \in \left[0, \frac{1}{\zeta} \right) \right\}.$$

Clearly,

$$H(\tilde{\Omega} \cup -\tilde{\Omega}, r) \subseteq H(\Omega \cup -\Omega, r),$$

and

$$A(\tilde{\Omega} \cup -\tilde{\Omega}, r) \leq A(\Omega \cup -\Omega, r),$$

for any $r \geq 0$.

We need the following technical result.

Lemma 6.6 *Let $\mathcal{D}$ be the curve on the n-dimensional complex sphere of radius $\sqrt{n}$ given by the set of points*

$$\mathcal{D} = \left\{ (e^{2\pi i \zeta t}, e^{2\pi i \zeta t}, \ldots, e^{2\pi i \zeta t}), \quad t \in \left[0, \frac{1}{\zeta}\right) \right\}.$$

Let

$$\hat{r} = r - \frac{2\pi n^{\frac{3}{2}}}{\sqrt{3}\zeta},$$

then, provided that $\hat{r} \geq 0$, we have

$$H(\mathcal{D}, \hat{r}) \subseteq H\left(\tilde{\Omega} \cup -\tilde{\Omega}, r\right). \tag{6.7}$$

Proof Let $\mathbf{x} \in H(\mathcal{D}, \hat{r})$. We aim to show that

$$\mathbf{x} \in H\left(\tilde{\Omega} \cup -\tilde{\Omega}, r\right).$$

Now there exists $\hat{t} \in \left[0, \frac{1}{\zeta}\right)$ such that $\|\mathbf{x} - \mathbf{y}\| \leq \hat{r}$, where

$$\mathbf{y} = (e^{2\pi i \zeta \hat{t}}, e^{2\pi i \zeta \hat{t}}, \ldots, e^{2\pi i \zeta \hat{t}}).$$

Consider the point $\mathbf{w}(\hat{t}, \zeta)$ on $H(\tilde{\Omega})$. We have (see Theorem 3.14)

$$\|\mathbf{w}(\hat{t}, \zeta) - \mathbf{y}\|^2 = \|\mathbf{w}(\hat{t}, \zeta)\|^2 + \|\mathbf{y}\|^2 - 2\Re(\mathbf{w}(\hat{t}, \zeta), \mathbf{y}^*)$$

$$= 2n - 2\sum_{k=0}^{n-1} \cos(2\pi k \hat{t}) = \begin{cases} (2n-1) - \frac{\sin((2n-1)\pi\hat{t})}{\sin \pi \hat{t}}, & \text{if } \hat{t} \neq 0, \\ 0, & \text{if } \hat{t} = 0. \end{cases}$$

Using the inequalities (see the proof of Theorem 3.17)

$$x - \frac{x^3}{6} \leq \sin x \leq x, \quad x \geq 0,$$

we have, for $\hat{t} \neq 0$,

$$\|\mathbf{w}(\hat{t}, \zeta) - \mathbf{y}\|^2 = (2n-1) - \frac{\sin((2n-1)\pi\hat{t})}{\sin \pi \hat{t}}$$

$$\leq (2n-1) - \frac{\sin((2n-1)\pi\hat{t}) - \frac{1}{6}((2n-1)\pi\hat{t})^3}{\pi \hat{t}}$$

$$= \frac{(2n-1)^3(\pi\hat{t})^2}{6} \leq \frac{4\pi^2 n^3}{3\zeta^2}.$$

It follows that

$$\|\mathbf{w}(\hat{t}, \zeta) - \mathbf{y}\| \leq \frac{2\pi n^{\frac{3}{2}}}{\sqrt{3}\zeta}.$$

Thus

$$\|\mathbf{x} - \mathbf{w}(\hat{t}, \zeta)\| \le \|\mathbf{x} - \mathbf{y}\| + \|\mathbf{w}(\hat{t}, \zeta) - \mathbf{y}\|$$

$$\le \hat{r} + \frac{2\pi n^{\frac{3}{2}}}{\sqrt{3}\zeta} = r.$$

Hence $\mathbf{x} \in H(\tilde{\Omega} \cup -\tilde{\Omega})$. □

Lemma 6.7 *For any $\hat{r} \ge 0$, we have*

$$A(\mathcal{D}, \hat{r}) \ge \max\left\{1, \left\lfloor \frac{\pi}{\arcsin \frac{\hat{r}}{\sqrt{n}}} \right\rfloor\right\} \cdot \frac{2(\pi n)^{n-\frac{1}{2}}}{(n-\frac{3}{2})!} \int_0^{2\arcsin \frac{\hat{r}}{2\sqrt{n}}} \sin^{2n-2}\theta \, d\theta,$$

where we define

$$\left\lfloor \frac{\pi}{\arcsin \frac{\hat{r}}{\sqrt{n}}} \right\rfloor = 0, \quad \text{for } \hat{r} \ge \sqrt{n}.$$

Proof   Let

$$\ell = \max\left\{1, \left\lfloor \frac{\pi}{\arcsin \frac{\hat{r}}{\sqrt{n}}} \right\rfloor\right\},$$

when $\hat{r} \in [0, \sqrt{n}]$, and $\ell = 1$ otherwise. Consider any ℓ points $\mathbf{x}_k, k = 0, 1, \dots, \ell - 1$, on the curve $\mathcal{D}$ having circular angular distance $2\arcsin \frac{\hat{r}}{\sqrt{n}}$ from one another. Then the spherical caps $H(\mathbf{x}_k, \hat{r})$ do not overlap. Each of these caps is also contained in $H(\mathcal{D}, \hat{r})$. Application of (6.4) accomplishes the proof. □

We are now in a position to state Theorem 6.5 in an effective way.

Theorem 6.8   *Let $\mathcal{C}$ denote a code of length n, rate R, and minimum Euclidean distance d_E. Let $\hat{d}$ denote any value of x that is greater than*

$$\frac{d_E}{2} + \frac{2\pi n^{\frac{3}{2}}}{\sqrt{3}\zeta},$$

for which the inequality

$$\max\left\{1, \left\lfloor \frac{\pi}{\theta_2(x)} \right\rfloor\right\} \cdot \int_0^{2\theta_1(x)} \sin^{2n-2}\theta \, d\theta$$

$$+ 2^{nR} \int_0^{2\arcsin \frac{d_E}{4\sqrt{n}}} \sin^{2n-2}\theta \, d\theta \ge \frac{\sqrt{\pi}\,(n-\frac{3}{2})!}{(n-1)!} \qquad (6.8)$$

is satisfied. Here

$$\theta_1(x) = \arcsin\left(\frac{x - \frac{d_E}{2} - \frac{2\pi n^{\frac{3}{2}}}{\sqrt{3}\zeta}}{2\sqrt{n}}\right),$$

$$\theta_2(x) = \arcsin\left(\frac{x - \frac{d_E}{2} - \frac{2\pi n^{\frac{3}{2}}}{\sqrt{3}\zeta}}{\sqrt{n}}\right).$$

Suppose further that $\hat{d} \leq \sqrt{2n}$. Then

$$\text{PAPR}(\mathcal{C}) \geq n\left(1 - \frac{\hat{d}^2}{2n}\right)^2.$$

Proof   We first establish that inequality (6.8) always has solutions. Notice that on rescaling the inequality throughout by a factor of

$$\frac{2(\pi n)^{n-\frac{1}{2}}}{(n - \frac{3}{2})!},$$

the right-hand side becomes equal to the area of the n-dimensional complex sphere of radius $\sqrt{n}$, while the first term on the left-hand side is lower-bounded by the area of a single spherical cap of radius

$$x - \frac{d_E}{2} - \frac{2\pi n^{\frac{3}{2}}}{\sqrt{3}\zeta}.$$

Putting

$$x = \frac{d_E}{2} + \frac{2\pi n^{\frac{3}{2}}}{\sqrt{3}\zeta}$$

makes this area equal to zero, while putting

$$x = 2\sqrt{n} + \frac{d_E}{2} + \frac{2\pi n^{\frac{3}{2}}}{\sqrt{3}\zeta}$$

ensures that the cap encompasses the whole sphere. The second term on the left-hand side becomes equal to $2^{nR}A\left(\frac{d_E}{2}\right)$ after rescaling, and is nonnegative. It follows that the inequality is satisfied for at least some values of x that are greater than or equal to

$$\frac{d_E}{2} + \frac{2\pi n^{\frac{3}{2}}}{\sqrt{3}\zeta}.$$

From Lemma 6.6, we know that

$$A\left(\Omega \cup -\Omega, \hat{d} - \frac{d_E}{2}\right) \geq A\left(\tilde{\Omega} \cup -\tilde{\Omega}, \hat{d} - \frac{d_E}{2}\right) \geq A\left(\mathcal{D}, \hat{d} - \frac{d_E}{2} - \frac{2\pi n^{\frac{3}{2}}}{\sqrt{3}\zeta}\right),$$

while from Lemma 6.7, we know that

$$A\left(\mathcal{D}, \hat{d} - \frac{d_E}{2} - \frac{2\pi n^{\frac{3}{2}}}{\sqrt{3}\zeta}\right) \geq \max\left\{1, \left\lfloor \frac{\pi}{\theta_2(\hat{d})} \right\rfloor\right\} \cdot \frac{2(\pi n)^{n-\frac{1}{2}}}{(n - \frac{3}{2})!} \int_0^{2\theta_1(\hat{d})} \sin^{2n-2}\theta \, d\theta,$$

where we again define $\left\lfloor \frac{\pi}{\theta_2(\hat{d})} \right\rfloor$ to equal zero when the arcsin function in θ_2 is undefined. So from (7.31), which holds for $x = \hat{d}$, we obtain the inequality

$$A\left(\Omega \cup -\Omega, \hat{d} - \frac{d_E}{2}\right) + 2^{nR} A\left(\frac{d_E}{2}\right) \geq \frac{2\pi^n n^{n-\frac{1}{2}}}{(n-1)!}.$$

The theorem now follows from Theorem 6.5. □

The above bound proves that there is a trade-off between the rate, minimum distance, and PAPR of a code. It shows in a strict sense that redundancy introduced by considering only those code words with low PAPR cannot all be exploited to provide error correction. Informally, this is because the words of low PAPR are restricted to lie in a certain region of the sphere, and this region shrinks as the PAPR decreases.

Now we pass to existence bounds of codes with a given minimum Euclidean distance and PAPR.

Theorem 6.9 *Let nonnegative $d, \hat{d} \in \mathbb{R}$, and $n \in \mathbb{N}$ be given. For $r \in \mathbb{R}$, such that $rn \in \mathbb{N}$, and $\ell \in \mathbb{N}$, let 2^{nR} be the largest integer for which*

$$2\ell r n A(\hat{d}) + 2^{Rn} A(d) \leq \frac{2n^{n-\frac{1}{2}}\pi^n}{(n-1)!} \tag{6.9}$$

holds. Then there exists a constant energy code $\mathcal{C}$ of rate R with minimum Euclidean distance at least d and

$$\text{PAPR}(\mathcal{C}) \leq \frac{1}{\cos\frac{\pi}{2r}} \cdot \frac{1}{\cos\frac{\pi}{2\ell}} \cdot \left(n - \frac{\hat{d}^2}{2}\right).$$

Proof For $k = 0, 1, \ldots, rn - 1$, and $j = 0, 1, \ldots, \ell - 1$, define

$$\mathbf{b}^{(k,j)} = \left(e^{-2\pi\iota\frac{j}{\ell}}, e^{-2\pi\iota\left(\frac{k}{rn}+\frac{j}{\ell}\right)}, e^{-2\pi\iota 2\left(\frac{k}{rn}+\frac{j}{\ell}\right)}, \ldots, e^{-2\pi\iota 2\left(\frac{k}{rn}+\frac{j}{\ell}\right)}\right).$$

Notice that if, for a vector $\mathbf{a} \in \mathbb{R}^n$,

$$\left\|\mathbf{a} - \mathbf{b}^{(k,j)}\right\| \geq \hat{d} \tag{6.10}$$

and

$$\|\mathbf{a} + \mathbf{b}^{(k,j)}\| \geq \hat{d}, \tag{6.11}$$

then

$$2 \left| \Re \left(\sum_{m=0}^{n-1} a_m e^{2\pi \imath m \left(\frac{k}{rn} + \frac{j}{\ell} \right)} \right) \right| = 2 \left| F_\mathbf{a} \left(\frac{k}{rn} \right) \cdot e^{2\pi \imath \frac{j}{\ell}} \right| \leq 2n - \hat{d}^2.$$

Applying Theorems 4.8 and 4.14 we obtain that if $\mathbf{a}$ satisfies (6.10) and (6.11),

$$\mathrm{PMEPR}(\mathbf{a}) \leq \frac{1}{\cos \frac{\pi}{2k}} \cdot \frac{1}{\cos \frac{\pi}{2\ell}} \cdot \left(n - \frac{\hat{d}^2}{2} \right).$$

Now we will construct the code recursively, point by point. Assume that we have chosen s code words $\mathbf{a}^{(0)}, \mathbf{a}^{(1)}, \ldots, \mathbf{a}^{(s-1)}$. Then if the union of

$$H\left((\pm \mathbf{b}^{(k,j)}, \hat{d}), \quad k = 0, 1, \ldots, rn - 1, \ j = 0, 1, \ldots, \ell - 1, \right.$$

and

$$H\left(\mathbf{a}^{(m)}, d \right), \quad m = 0, 1, \ldots, s - 1,$$

does not cover the whole complex sphere, an extra code word, $\mathbf{c}^{(s)}$, can be added to the code. Since the area of the union is less than or equal to the sum of the areas, this can be done as long as (6.9) holds, and we are done.   □

Let us analyze the asymptotics of the established bound. For this, we will use the following inequality: for any $n \geq 2$ and $\theta_1 \in \left[0, \frac{\pi}{2} \right)$,

$$0 \leq \int_0^{\theta_1} \sin^{2n-2} \theta \, d\theta \leq \frac{\sin^{2n-1} \theta_1}{(2n-1) \cos \theta_1}. \tag{6.12}$$

The left inequality is straightforward, the right one follows from

$$\sin^{2n-1} \theta_1 = \int_0^{\theta_1} \frac{d \sin^{2n-1} \theta}{d\theta} \, d\theta$$

$$= (2n-1) \int_0^{\theta_1} \sin^{2n-2} \theta \, \cos \theta \, d\theta$$

$$\geq (2n-1) \cos \theta_1 \int_0^{\theta_1} \sin^{2n-2} \theta \, d\theta,$$

where we have used the fact that

$$\min_{\theta \in [0, \frac{\pi}{2}]} \cos \theta = \cos \theta_1.$$

Theorem 6.10   *Let $R \geq 0$, and $\Delta \geq 0$, be such that*

$$2^R \sqrt{2\Delta \left(1 - \frac{\Delta}{2}\right)} < 1. \qquad (6.13)$$

Then, for all sufficiently large n, there exists a constant energy code $\mathcal{C}$ of length n, rate R, and minimum Euclidean distance $d_E \geq \sqrt{2\Delta n}$ with

$$\mathrm{PAPR}(\mathcal{C}) \leq \ln n (1 + o(1)).$$

Proof   Choose in the previous theorem $\ell = \sqrt{\ln \ln n}$, $r = \sqrt{\ln \ln n}$, and

$$\delta = \frac{\hat{d}^2}{n} = 1 - \sqrt{\frac{\ln n}{n}}.$$

Using the following easy-to-check inequality

$$\frac{1}{\cos x} \leq 1 + \frac{x^2}{2} + \frac{x^4}{2},$$

valid for $|x| \leq 1$, we conclude that

$$\mathrm{PMEPR}(\mathcal{C}) \leq \left(1 + \frac{\pi^2}{8 \ln \ln n} + \frac{\pi^4}{32(\ln \ln n)^2}\right) \ln n = \ln n (1 + o(1)).$$

To confirm that such a code exists we have to prove (6.9), which, by (6.4), reduces to

$$2n \ln \ln n \int_0^{2 \arcsin \frac{\sqrt{2\delta}}{2}} \sin^{2n-2} \theta \, d\theta + 2^{Rn} \int_0^{2 \arcsin \frac{\sqrt{2\Delta}}{2}} \sin^{2n-2} \theta \, d\theta \leq \frac{2\sqrt{\pi} \left(n - \frac{3}{2}\right)!}{(n-1)!}.$$

By (6.5),

$$\frac{\left(n - \frac{3}{2}\right)!}{(n-1)!} = \frac{\sqrt{\pi}}{2^{2n-2}} \binom{2n-2}{n-1} \leq \sqrt{\frac{1}{n-1}} < \frac{2}{\sqrt{n}},$$

where the inequality before the last follows from the upper bound in Lemma 3.40. Thus, using (6.12), we arrive at the equivalent inequality

$$\frac{2n^{\frac{3}{2}} \ln \ln n}{(2n-1)} \cdot \frac{\sin^{2n-1}(\theta_1)}{\cos \theta_1} + 2^{Rn} \frac{\sqrt{n}}{(2n-1)} \cdot \frac{\sin^{2n-1}(\theta_2)}{\cos \theta_2} \leq 4\sqrt{\pi}, \qquad (6.14)$$

where

$$\theta_1 = 2 \arcsin \frac{\sqrt{2\delta}}{2}, \qquad \theta_2 = 2 \arcsin \frac{\sqrt{2\Delta}}{2}.$$

I will prove that the inequality holds for large enough n. With the choice of δ, we have

$$\sin \theta_1 = \sqrt{1 - \frac{\ln n}{n}}, \quad \sin^2 \theta_1 = 1 - \frac{\ln n}{n}, \quad \cos \theta_1 = \sqrt{\frac{\ln n}{n}}.$$

Thus,

$$\frac{2n^{\frac{3}{2}} \ln \ln n}{(2n-1)} \cdot \frac{\sin^{2n-1}(\theta_1)}{\cos \theta_1} = \frac{2n \ln \ln n}{(2n-1)\left(1 - \frac{\ln n}{n}\right)\sqrt{\ln n}} \cdot n \left(1 - \frac{\ln n}{n}\right)^n.$$

Since

$$\lim_{n \to \infty} n \left(1 - \frac{\ln n}{n}\right)^n = 1,$$

we conclude that

$$\lim_{n \to \infty} \frac{2n \ln \ln n}{(2n-1)\left(1 - \frac{\ln n}{n}\right)\sqrt{\ln n}} \cdot n \left(1 - \frac{\ln n}{n}\right)^n = 0.$$

Therefore, the first summand in the left-hand side of (6.14) tends to 0 when n increases. The same is true for the second summand under the condition (6.13). Thus, the left-hand side of (6.14) tends to 0 while the right-hand side is a constant. This yields that starting from some n the inequality is valid, and we are done. □

For the BPSK modulation and Hamming distance a similar analysis can be applied. Asymptotically this will have the following form.

Theorem 6.11 *Let $R \geq 0$, and $\Delta \geq 0$, be such that*

$$R \leq 1 - H(\Delta), \tag{6.15}$$

where

$$H(\Delta) = -\Delta \log_2 \Delta - (1 - \Delta)\log_2(1 - \Delta),$$

is the binary entropy function. Then for all sufficiently large n, there exists a code $\mathcal{C}$ of length n, rate R, and minimum Hamming distance $d_H \geq \Delta n$, with

$$\mathrm{PAPR}(\mathcal{C}) \leq \ln n(1 + o(1)).$$

□

6.3 Codes with known distance distribution

In this section, I will relate the peak power distribution to distance distributions of codes. We assume here underlying binary codes with the mapping $0 \to 1$ and $1 \to -1$, i.e., the code $\mathcal{C}$ consists of n-tuples of BPSK symbols. Let the cardinality of $\mathcal{C}$ equal $|\mathcal{C}| = M_1$. We shall give a bound in terms of the distance distribution

$$B_k = \frac{1}{M_1} \left|\left\{\left(\mathbf{c}^{(1)}, \mathbf{c}^{(2)}\right) : d\left(\mathbf{c}^{(1)}, \mathbf{c}^{(2)}\right) = k, \mathbf{c}^{(1)} \in \mathcal{C}, \mathbf{c}^{(2)} \in \mathcal{C}\right\}\right|, \tag{6.16}$$

where $k = 0, 1, \ldots, n$, and $d(\mathbf{x}, \mathbf{y})$ is the Hamming distance between $\mathbf{x}$ and $\mathbf{y}$, i.e., the number of positions where the code words differ. Note that the distance distribution coincides with the weight distribution for linear codes, i.e.,

$$W_k = |\{\mathbf{c} : w(\mathbf{c}) = k, \mathbf{c} \in \mathcal{C}\}| . \tag{6.17}$$

where $w(\mathbf{c})$ is the weight of the code word, corresponding to the number of -1 components in $\mathbf{c}$. Furthermore, if the code contains the code word consisting of only -1s we have $W_k = W_{n-k}$, i.e., the weight distribution is symmetric with respect to $\frac{n}{2}$.

Given an integer j, let $\mathbf{i} = (i_0, i_1, \ldots, i_{n-1})$ be a nonnegative, integer-valued vector of length n satisfying $i_0 + i_1 + \ldots + i_{n-1} = j$, and let $\mathcal{I}_j$ denote the set of all such vectors. Define the moments

$$m_{\mathbf{i}} = E\left(c_0^{i_0}, c_1^{i_1}, \ldots, c_{n-1}^{i_{n-1}}\right), \tag{6.18}$$

where the expectation is over the vectors $\mathbf{c} = (c_0, c_1, \ldots, c_{n-1})$ of code $\mathcal{C}$. Let us introduce the following notations:

$$b_{\mathbf{i}}^{(j)} = \frac{j}{i_0! i_1! \cdots i_{n-1}!}, \tag{6.19}$$

and

$$k_{\mathbf{i}}(t, \alpha)$$
$$= \left(\cos^{i_0}(\alpha) \cos^{i_1}(2\pi t + \alpha) \cos^{i_2}(2 \cdot 2\pi t + \alpha), \ldots, \cos^{i_{n-1}}((n-1)2\pi t + \alpha)\right).$$

We need the following technical lemma on the moments of linear, binary codes.

Lemma 6.12 *The moments of linear, binary codes are either 0 or n.*

Proof The sum $\sum_{\mathbf{c} \in \mathcal{C}} \prod_{k=0}^{n-1} c_k^{i_k}$ is just the sum over all code words of $(-1)^p$ where p is the parity of the subvectors on the corresponding positions of the code words where i_k is odd. Whatever subset of positions one picks, the parity of the subvectors is either all even, or half of the subvectors are of even and half the subvectors are of odd parity. This is because the sum of two subvectors with different (equal) parities yields odd (even) parity, and the code is linear. □

We will use the notations and bound from Theorem 5.1. Expanding the exponential function in the Chernoff bound yields

$$E\left(e^{\varepsilon \Re(F_{\mathbf{c}}(t_{l_1,r}) e^{i\alpha_{l_2,h}})}\right) = \sum_{j=0}^{N} \frac{\varepsilon^j}{j!} \sum_{\mathbf{i} \in \mathcal{I}_j} b_{\mathbf{i}}^{(j)} m_{\mathbf{i}} k_{\mathbf{i}}\left(t_{l_1,r}, \alpha_{l_2,h}\right) + \frac{\varepsilon^{N+1} n^{N+1} e^{\varepsilon n}}{(N+1)!},$$

where the error term of the right-hand side depends on the natural number $N > 0$ and is given by Taylor's theorem (and can be made arbitrarily small by choosing N large enough).

Then, exchanging integration and summation, for any $\varepsilon > 0$ we obtain

$$\Pr\left(\max_{t\in[0,1)}|F_{\mathbf{c}}(t)| > \lambda\sqrt{n}\right)$$

$$\leq \sum_{l_1=0}^{rn-1}\sum_{l_2=0}^{h-1} e^{-\frac{\varepsilon\sqrt{n}\lambda}{C_r\tilde{C}_h}} \sum_{j=0}^{N}\frac{\varepsilon^j}{j!}\sum_{\mathbf{i}\in\mathcal{I}_j}b_{\mathbf{i}}^{(j)}m_{\mathbf{i}}k_{\mathbf{i}}(t_{l_1,r},\alpha_{l_2,h}) + \frac{\varepsilon^{N+1}n^{N+1}e^{\varepsilon n}}{(N+1)!}$$

for any r and h. However, from the practical point of view, the moments are rather unwieldy to evaluate. On the other hand, we can obtain simpler expressions for linear, binary codes. The following theorem relies on the fact that moments of linear codes are nonnegative.

Theorem 6.13 *Let $\mathcal{C}$ be a linear binary code with symmetric weight distribution. Then*

$$\Pr\left(\text{PMEPR}(\mathbf{c}) > \lambda\right) \leq \min_{\varepsilon>0}\sum_{k=0}^{n}\frac{f_n(\varepsilon,\lambda)\,W_k\cosh\left(\varepsilon(n-2k)\right)}{M_1} \qquad (6.20)$$

with

$$f_n(\varepsilon,\lambda) = \min_{r>1,h>3} rhn\cdot e^{-\frac{\varepsilon\sqrt{n}\lambda}{C_r\tilde{C}_h}}. \qquad (6.21)$$

Proof   Fixing $\varepsilon > 0$ we know that

$$\Pr\left(\max_{t\in[0,1)}|F_{\mathbf{c}}| > \lambda\right) \leq \min_{r>1,h>3}\sum_{l_1=0}^{rn-1}\sum_{l_2=0}^{h-1} e^{-\frac{\varepsilon\sqrt{n}\lambda}{C_r\tilde{C}_h}} \sum_{j=0}^{N}\frac{\varepsilon^j}{j!}\sum_{\mathbf{i}\in\mathcal{I}_j}b_{\mathbf{i}}^{(j)}m_{\mathbf{i}}k_{\mathbf{i}}(t_{l_1,r},\alpha_{l_2,h})$$

$$+ \frac{\varepsilon^{N+1}n^{N+1}e^{\varepsilon n}}{(N+1)!}.$$

By the Cauchy–Schwartz inequality

$$\Pr\left(\max_{t\in[0,1)}|F_{\mathbf{c}}| > \lambda\right) \leq \min_{r>1,h>3}\sum_{l_1=0}^{rn-1}\sum_{l_2=0}^{h-1} e^{-\frac{\varepsilon\sqrt{n}\lambda}{C_r\tilde{C}_h}} \sum_{j=0}^{N}\frac{\varepsilon^j}{j!}\max_{\mathbf{i}\in I_j}\left|k_{\mathbf{i}}\left(t_{l_1,r},\alpha_{l_2,h}\right)\right|\sum_{\mathbf{i}\in\mathcal{I}_j}b_{\mathbf{i}}^{(j)}m_{\mathbf{i}}$$

$$+ \frac{\varepsilon^{N+1}n^{N+1}e^{\varepsilon n}}{(N+1)!}$$

$$= \sum_{j=0}^{N}\frac{f_n(\varepsilon,\lambda)\,\varepsilon^j}{j!}\sum_{\mathbf{i}\in\mathcal{I}_j}b_{\mathbf{i}}^{(j)}m_{\mathbf{i}} + \frac{\varepsilon^{N+1}n^{N+1}e^{\varepsilon n}}{(N+1)!}$$

provided that $m_{\mathbf{i}} \geq 0$, $i \in \mathcal{I}_j$, which is indeed the case for linear codes by Lemma 6.12.

Next, we need to represent the term

$$m_{\mathbf{i}} = E\left(c_0^{i_0}c_1^{i_1}\ldots c_{n-1}^{i_{n-1}}\right) = \frac{1}{M_1}\sum_{\mathbf{c}\in\mathcal{C}}\prod_{k=0}^{n-1}c_k^{i_k}$$

in terms of the weight distribution of the linear code. Note that for $\mathbf{c} \in \{\pm 1\}^n$, we have

$$\sum_{k=0}^{n-1} c_k = n - 2w\left(\mathbf{c}\right),$$

where $w(\mathbf{c})$ is the weight of the code word $\mathbf{c}$. Consider the following sum

$$s_j = \sum_{\mathbf{c} \in \mathcal{C}} \left(\sum_{k=0}^{N-1} c_k \right)^j.$$

On the one hand, it is

$$s_j = \sum_{\mathbf{c} \in \mathcal{C}} \left(n - 2w\left(\mathbf{c}\right) \right)^j = \sum_{k=0}^{n} \left(n - 2k \right)^j W_k.$$

On the other hand, we have

$$s_j = \sum_{\mathbf{c} \in \mathcal{C}} \left(\sum_{k=0}^{n-1} c_k \right)^j = \sum_{\mathbf{c} \in \mathcal{C}} \sum_{\mathbf{i} \in I_j} b_{\mathbf{i}}^{(j)} \prod_{k=0}^{n-1} c_k^{i_k} = \sum_{\mathbf{i} \in \mathcal{I}_j} b_{\mathbf{i}}^{(j)} \sum_{\mathbf{c} \in \mathcal{C}} \prod_{k=0}^{n-1} c_k^{i_k}.$$

Thus

$$\sum_{\mathbf{i} \in \mathcal{I}_j} b_{\mathbf{i}}^{(j)} \sum_{\mathbf{c} \in \mathcal{C}} \prod_{k=0}^{n-1} c_k^{i_k} = \sum_{k=0}^{n} \left(n - 2k \right)^j W_k$$

and

$$\Pr\left(\max_{t \in [0,1)} |F_\mathbf{c}| > \lambda \right) \leq \sum_{j=0}^{N} \frac{f_n\left(\varepsilon, \lambda\right) \varepsilon^j}{j!} \sum_{\mathbf{i} \in \mathcal{I}_j} b_{\mathbf{i}}^{(j)} m_{\mathbf{i}} + \frac{\varepsilon^{N+1} n^{N+1} e^{\varepsilon n}}{(N+1)!}$$

$$= \sum_{j=0}^{N} \frac{f_n\left(\varepsilon, \lambda\right) \varepsilon^j}{M_1 j!} \sum_{k=0}^{n} \left(n - 2k \right)^j W_k + \frac{\varepsilon^{N+1} n^{N+1} e^{\varepsilon n}}{(N+1)!}.$$

Observing that the sum is zero for odd j and exchanging the order of summation

$$\Pr\left(\max_{t \in [0,1)} |F_\mathbf{c}| > \lambda \right) \leq \sum_{k=0}^{n} \frac{f_n\left(\varepsilon, \lambda\right) W_k}{M_1} \sum_{j=0}^{N} \frac{\varepsilon^{2j} \left(n - 2k \right)^{2j}}{(2j)!} + \frac{\varepsilon^{N+1} n^{N+1} e^{\varepsilon n}}{(N+1)!}$$

$$\leq \limsup\nolimits_{n \to \infty} \sum_{k=0}^{n} \frac{f_n\left(\varepsilon, \lambda\right) W_k}{M_1} \sum_{j=0}^{N} \frac{\varepsilon^{2j} \left(n - 2k \right)^{2j}}{(2j)!}$$

$$+ \frac{\varepsilon^{N+1} n^{N+1} e^{\varepsilon n}}{(N+1)!}.$$

Since N is arbitrary,

$$\sum_{j=0}^{N} \frac{\varepsilon^{2j}(n-2k)^{2j}}{(2j)!} \rightarrow \cosh(\varepsilon(n-2k))$$

and

$$\frac{\varepsilon^{N+1}n^{N+1}e^{\varepsilon n}}{(N+1)!} \rightarrow 0$$

as $N \rightarrow \infty$, the latter yields the desired result.   □

In Theorem 6.13, it is assumed that the code possesses a symmetric weight distribution. This assumption can be dropped if the dual code is considered as follows. Let

$$K_k(x) = \sum_{j=0}^{k}(-1)^j \binom{x}{j}\binom{n-x}{k-j}$$

be binary Krawtchouk polynomials. For properties of Krawtchouk polynomials see Section 3.4.4. The MacWilliams identity is

$$W_j^{\perp} = \frac{1}{M_1}\sum_{k=0}^{n} W_k K_j(k)$$

where $W_j^{\perp}$ stands for the distance distribution of the dual code. Let $d^{\perp}$, the dual distance, be the minimum nonzero index of a strictly positive component in the dual distance distribution.

Theorem 6.14 *Let $\mathcal{C}$ be a linear binary code. Then*

$$\Pr(\mathrm{PMEPR}(\mathbf{c}) > \lambda) \leq \min_{\varepsilon>0}\left[\sum_{j=0,\,j\text{ even}}^{N} \frac{f_n(\varepsilon,\lambda)\varepsilon^j}{j!}\sum_{k=0}^{n}(n-2k)^j\binom{n}{k} + \frac{(\varepsilon n)^{N+1}e^{\varepsilon n}}{(N+1)!}\right]$$

$$(6.22)$$

where $N \leq d^{\perp}$ and f_n is defined in (6.21).

Proof We will need the following simple argument. Let $j < d^{\perp}$, and

$$(n-2k)^j = \sum_{\ell=0}^{n} a_{\ell}K_{\ell}(k)$$

(such an expansion exists by orthogonality and completeness of the system of Krawtchouk polynomials). To estimate a_0 notice that

$$a_0 = 2^{-n}\sum_{k=0}^{n}(n-2k)^j K_0(k) = \sum_{k=0}^{n}(n-2k)^j\binom{n}{k},$$

and

$$a_0 = 0 \text{ when } j \text{ is odd,}$$
$$a_0 = \sum_{k=0}^{n}(n - 2k)^j \binom{n}{k} \text{ when } j \text{ is even.}$$

Moreover, by orthogonality of Krawtchouk polynomials, for $\ell > j$,

$$a_\ell = 2^{-n} \sum_{k=0}^{n}(n - 2k)^j K_\ell(k) = 0.$$

Thus, if $j < d^\perp$,

$$\sum_{k=0}^{n}(n - 2k)^j W_k = \sum_{\ell=0}^{n} a_\ell \sum_{k=0}^{n} W_k K_\ell(k) = M_1 \sum_{\ell=0}^{j} a_\ell W'_\ell = M_1 a_0.$$

We assume that we could choose $N < d^\perp$ (which is always possible when $d^\perp$ is large enough) yielding the result. $\qquad\square$

The optimization problem in (6.21) can be easily solved using simple line search procedures. The optimization parameter in (6.20) must also be numerically computed. Here, we choose a good starting point from the BPSK case where the optimal ε can be analytically obtained. Note that the remainder term in (6.22) is small as long as the dual distance scales with n.

I will now discuss the asymptotic behavior of the derived bounds.

Theorem 6.15 *Let C be a linear $[n, a]$ code. Suppose that, for all W_j, there is a constant E_s independent of j so that*

$$W_j = 2^{a-n} \binom{n}{j}(1 + E_s).$$

Then

$$\Pr(\text{PMEPR}(\mathbf{c}) \leq \lambda) \leq (1 + E_s)\, rhn \cdot e^{-\frac{\lambda^2}{2\hat{C}_r^2 \hat{C}_h^2}}.$$

Proof Using the identity

$$\sum_{k=0}^{n} \binom{n}{k} \frac{\cosh(\varepsilon(n - 2k))}{2^n} = \cosh^n(\varepsilon),$$

which can be obtained from the BPSK case and the inequality $\cosh(\varepsilon) \leq e^{\frac{\varepsilon^2}{2}}$, yields

$$\Pr\left(\max_{t \in [0,1)} |F_{\mathbf{c}}(t)| > \lambda\sqrt{n}\right) \leq (1 + E_s)\, rhn \cdot e^{-\frac{\varepsilon\lambda\sqrt{n}}{C_r \hat{C}_h}} e^{\frac{\varepsilon^2 n}{2}}.$$

Setting $\varepsilon = \frac{\lambda}{(C_r \hat{C}_h \sqrt{N})}$ gives the final result. $\qquad\square$

From the latter theorem we can conclude that if the weight distribution can be approximated by the binomial distribution up to a constant $(1 + E_s)$ where E_s satisfies

$$\limsup_{n \to \infty} |E_s| < \infty.$$

Then the PMEPR distribution is of the same order as that of BPSK.

Since the proof of the bound relies on the fact that the moments are nonnegative, the bound does not apply to general nonlinear codes. However, a simple observation shows that the bound can also be extended to the general case.

Theorem 6.16 *For any binary code $\mathcal{C}$ of size M_1 and with distance distribution $B_k, k = 0, 1, \ldots, n$, we have*

$$\Pr\left(\text{PMEPR}(\mathbf{c}) > \lambda\right)$$

$$\leq \min_{\varepsilon > 0} \left(\sum_{j=0}^{N} \frac{f_n(\varepsilon, \lambda) \varepsilon^j n^{\frac{j}{2}}}{M_1 j!} \left(\sum_{k=0}^{n} (n - 2k)^j B_k \right)^{\frac{1}{2}} + \frac{\varepsilon^{N+1} n^{N+1} e^{\varepsilon n}}{(N+1)!} \right). \qquad (6.23)$$

Proof   Indeed

$$s_j = \sum_{\mathbf{c}_1 \in \mathcal{C}} \sum_{\mathbf{c}_2 \in \mathcal{C}} (n - 2d(\mathbf{c}_1, \mathbf{c}_2))^j = \sum_{k=0}^{n} (n - 2k)^j B_k$$

$$= \sum_{\mathbf{i} \in I_j} b_{\mathbf{i}}^{(j)} \left(\sum_{\mathbf{c} \in \mathcal{C}} \prod_{k=0}^{n-1} c_k^{i_k} \right)^2.$$

Applying the Cauchy–Schwartz inequality as in Theorem 6.13 and observing that

$$\sum_{\mathbf{i} \in \mathcal{I}_j} k_{\mathbf{i}}^2 \left(t_{l_1, r}, \alpha_{l_2, h} \right) \leq n^j$$

yields the result.   □

Observe that the error term in (6.23) can be made arbitrarily small by simply considering more terms in the sum. If $B_k = B_{n-k}$ then the second sum in (6.23) is zero for odd k.

6.4 BCH codes

Here, I will prove that the weight distribution of a large class of BCH codes is approximately binomial. The idea of the proof is to show that when computing the values of the weight distribution using the MacWilliams transform, they will be mainly determined by the zeroth component of the dual code, and the impact of the other dual codes' weight distribution components is negligible.

We start with an analysis of the behavior of Krawtchouk polynomials. Let us recall some facts from Section 3.4.4. The binary Krawtchouk polynomial $K_k(x) = K_k^n(x)$ (of degree k in x) is defined by the following generating function:

$$\sum_{k=0}^{\infty} K_k(x) z^k = (1 - z)^x (1 + z)^{n-x}. \tag{6.24}$$

We need some particular values of $K_k(x)$, namely,

$$K_k(0) = \binom{n}{k}, \quad K_k(n) = (-1)^k \binom{n}{k}.$$

From Cauchy's integral formula we get, for nonnegative integer x:

$$K_k(x) = \frac{1}{2\pi x} \oint \frac{(1 + z)^{n-x}(1 - z)^x}{z^{k+1}} \, dz$$

$$= \frac{(-\iota)^x}{2\pi} \int_0^{2\pi} e^{\iota(\frac{n}{2} - k)\theta} \left(\cos \frac{\theta}{2} \right)^{n-x} \left(\sin \frac{\theta}{2} \right)^x \, d\theta.$$

Thus, for even n and x, we have

$$|K_k(x)| = \frac{1}{2\pi} \left| \int_0^{2\pi} e^{\iota(\frac{n}{2} - k)\theta} \left(\cos \frac{\theta}{2} \right)^{n-x} \left(\sin \frac{\theta}{2} \right)^x \, d\theta \right|$$

$$\leq \frac{1}{2\pi} \int_0^{2\pi} \left(\cos \frac{\theta}{2} \right)^{n-x} \left(\sin \frac{\theta}{2} \right)^x \, d\theta = \left| K_{\frac{n}{2}}(x) \right|.$$

Hence, the following result holds.

Lemma 6.17 *For even n and j,*

$$|K_k(j)| \leq \left| K_{\frac{n}{2}}(j) \right|.$$

$\square$

Let us find the values $K_{\frac{n}{2}}(j)$ for even n and j. The following symmetry relation holds for integer k and j:

$$\binom{n}{j} K_k(j) = \binom{n}{k} K_j(k). \tag{6.25}$$

From (6.24), we get

$$\sum_{k=0}^{\infty} K_k \left(\frac{n}{2} \right) z^k = (1 - z)^{\frac{n}{2}} (1 + z)^{\frac{n}{2}} = (1 - z^2)^{\frac{n}{2}},$$

thus,

$$K_{2k} \left(\frac{n}{2} \right) = (-1)^k \binom{n/2}{k},$$

and

$$K_{\frac{n}{2}}(j) = (-1)^{\frac{j}{2}} \frac{\binom{n}{n/2}\binom{n/2}{j/2}}{\binom{n}{j}}.$$ (6.26)

So, from (6.25) and Lemma 6.17 we get for even n and j

$$|K_j(k)| = \frac{\binom{n}{j}}{\binom{n}{k}}|K_k(j)| \leq \frac{\binom{n}{j}}{\binom{n}{k}}\left|K_{\frac{n}{2}}(j)\right|.$$

Employing (6.26) we obtain the following lemma.

Lemma 6.18 *For k integer, and n and j even,*

$$|K_j(k)| \leq \frac{\binom{n}{n/2}\binom{n/2}{i/2}}{\binom{n}{k}}.$$

□

Now we are in a position to analyze the weight distribution of BCH codes. Consider the extended BCH code $\mathcal{C}$ of length $n = 2^m$, cardinality 2^{n-mt-1} and minimum distance $d = 2t + 2 \leq 2^{[(m+1)/2]} + 2$. Let the distance distribution of the code be $\mathbf{B} = (B_0, \ldots, B_n)$, $B_0 = B_n = 1$, $B_{2i+1} = 0$ for $i = 0, \ldots, n/2 - 1$, and $B_i = B_{n-i} = 0$ for $i = 1, 2, \ldots, 2t + 1$. For the conventional (nonextended) BCH code of length $n - 1$ and minimum distance $2t + 1$ we denote its spectrum by $\mathbf{b} = (b_0, \ldots, b_{n-1})$. Since the extended BCH code is doubly transitive, we have the following result relating the values of odd and even components of the spectra of BCH codes (see, e.g., Theorems 14 and 16, Section 8.5, in [257]).

Lemma 6.19

$$b_{2j-1} = \frac{2jB_{2j}}{n}, \quad b_{2j} = \frac{(n-2j)B_{2j}}{n}.$$

□

We start with estimating the spectrum of the extended BCH code. Let $\mathbf{B}^\perp = (B_0^\perp, \ldots, B_n^\perp)$ stand for the spectrum of the dual $\mathcal{C}^\perp$ of the extended BCH code, where $\mathbf{B}^\perp$ is determined by the MacWilliams transform of $\mathbf{B}$

$$B_k^\perp = \frac{|\mathcal{C}^\perp|}{2^n}\sum_{j=0}^n B_j K_k(j),$$ (6.27)

and $B_j^\perp = B_{n-j}^\perp$ for $j = 0, \ldots, n$, $B_0^\perp = B_n^\perp = 1$, $B_{2j+1}^\perp = 0$ for $j = 0, \ldots, n/2 - 1$, and for $2t + 1 \leq 2^{[(m+1)/2]} + 1$,

$$B_j^\perp = B_{n-j}^\perp = 0 \text{ for } j = 1, \ldots, d^\perp = [n/2 - (t-1)\sqrt{n}].$$ (6.28)

Denote by $D^\perp$ the segment $[d^\perp, \ldots, n - d^\perp]$. Note that for the considered range of t,

$$\sum_{j=0}^{n} B_j^\perp = |\mathcal{C}^\perp| = 2n^t.$$

Inverting relation (6.27) we have

$$B_j = \frac{1}{2n^t} \sum_{k=0}^{n} B_k^\perp K_j(k) = \frac{1}{2n^t}\left(B_0^\perp K_j(0) + B_n^\perp K_j(n) + \sum_{k \in D^\perp} B_k^\perp K_j(k)\right).$$

We consider only the even j, since, otherwise, $B_j = 0$. Hence,

$$B_j = \frac{1}{2n^t}\left(2\binom{n}{j} + \sum_{k \in D^\perp} B_k^\perp K_j(k)\right),$$

i.e., for even j we have

$$B_j = \frac{\binom{n}{j}}{n^t}(1 + E_j),$$

where

$$|E_j| = \frac{1}{2\binom{n}{j}}\left|\sum_{k \in D^\perp} B_k^\perp K_j(k)\right| \le \frac{n^t}{\binom{n}{j}} \max_{k \in D^\perp} |K_j(k)|.$$

Now since n and j are even we may use Lemma 6.18, thus getting

Theorem 6.20 *In the extended BCH code of length $n = 2^m$ and minimum distance $2t + 2 \le 2^{[m+1]/2} + 2$,*

$$B_j = 0 \text{ for } j \text{ odd},$$

$$B_j = \frac{\binom{n}{j}}{n^t}(1 + E_j) \text{ for } j \text{ even},$$

where

$$|E_j| \le \frac{n^t \binom{n}{n/2}\binom{n/2}{j/2}}{\binom{n}{j}\binom{n}{d^\perp}}. \tag{6.29}$$

$\square$

Using Lemma 6.19, we obtain the following result for conventional BCH codes.

Theorem 6.21 *In the BCH code of length $\hat{n} = n - 1 = 2^m - 1$ and minimum distance $2t + 1 \le 2^{[(m+1)/2]} + 1$,*

$$b_j = \frac{\binom{\hat{n}}{j}}{n^t}(1 + E_{j^*}),$$

where $j^* = j + 1$ for j odd, and $j^* = j$ for j even, and $|E_j|$ is estimated in (6.29).

$\square$

Using the theorems we can analyze some particular cases.

From standard estimates of binomial coefficients (see Lemma 3.42) we have

$$\ln \frac{\binom{n}{n/2}}{\binom{n}{d^\perp}} = 2(t-1)^2 + O\left(\frac{(t-1)^4}{n}\right).$$

Therefore, for $t = o(n^{\frac{1}{4}})$,

$$\ln \frac{\binom{n}{n/2}}{\binom{n}{d^\perp}} = 2(t-1)^2 + o(1),$$

and for $t = o(\sqrt{n})$,

$$\ln \frac{\binom{n}{n/2}}{\binom{n}{d^\perp}} = o(n).$$

This leads to straightforward corollaries.

Corollary 6.22   *If $t = o(\sqrt{n})$, and j grows linearly with n, $\frac{j}{n} = \sigma + o(1)$, then*

$$\frac{1}{n}\log_2 |E_{\sigma n}| \leq -\frac{1}{2}H(\sigma) + o(1).$$

$\square$

Assuming $t = o(n^{\frac{1}{4}})$, we give somehow sharper estimates, particularly good for small j. We use the fact that for $j = o(\sqrt{n})$, from Stirling approximation we have

$$\frac{\binom{n/2}{i/2}}{\binom{n}{j}} = \sqrt{2}\,e^{-\frac{j}{2}}\left(\frac{j}{n}\right)^{\frac{j}{2}}(1 + o(1)).$$

Corollary 6.23   *If $t = o(n^{\frac{1}{4}})$, and $i = o(\sqrt{n})$, then*

$$|E_j| \leq \sqrt{2}\,j^{\frac{j}{2}}\,e^{2(t-1)^2 - \frac{j}{2}}\,n^{t-\frac{j}{2}}(1 + o(1)).$$

$\square$

If t and j are constants then we have $|E_j| = O(n^{t-\frac{j}{2}})$. Since the maximum error occurs in $|E_{2t+2}|$, we always have $|E_j| = O(n^{-1})$.

Finally, for conventional BCH codes we get

Theorem 6.24   *Let $t = o(n^{\frac{1}{4}})$ and $l = [(j+1)/2]$, then in the BCH code of length $n = 2^m - 1$ and with minimum distance $2t + 1$*

$$b_j = \frac{\binom{n}{j}}{(n+1)^t}(1 + E_{2l}),$$

where the error term is upper bounded as follows:

$$|E_{2l}| \leq \sqrt{2}\,(2l)^l\,e^{2(t-1)^2 - \frac{l(n-2l)}{n}}\,n^{t-l}(1 + o(1)).$$

$\square$

Similar, but more technical arguments, yield the following result.

Theorem 6.25 *For all $j > 2t + 1$,*

$$|E_j| \leq |E_{2t+1}|,$$

and

$$|E_{2t+1}| \leq 4(t - 1)^2 t!$$

$\square$

This guarantees that the whole distance distribution is asymptotically binomial, given

$$4(t - 1)^2 t! = O(n).$$

Using the Stirling approximation we check that it suffices for growing n that

$$t \leq \frac{\ln n - 2.5 \ln \ln n}{\ln \ln n}. \tag{6.30}$$

Thus, we obtain the following theorem.

Theorem 6.26 *Let $\mathbf{c}$ be picked at random from the BCH code of length $n = 2^m - 1$ and minimum distance $2t + 1 \leq 2^{\lfloor (m+1)/2 \rfloor} + 1$ where t fulfills (6.30). Then for large n,*

$$\Pr\,(\mathrm{PMEPR}(\mathbf{c}) > \lambda) \to 0,$$

where

$$\lambda = 4 \ln n + 4 \ln \ln n + o(1).$$

Proof By Theorem 6.15 an upper bound is given by

$$\Pr\,(\mathrm{PMEPR}\,(\mathbf{c}) > \lambda) \leq (1 + E_s)\,rhn \cdot e^{-\frac{\lambda}{2C_r^2 \hat{C}_h^2}}.$$

Choosing $r = h = \sqrt{\ln n}$ we have $C_r^{-2}\hat{C}_h^{-2} \geq 1 - \frac{2\pi^2}{\ln n}$ for large n. Setting $E_s = O(n)$ by (6.30) and λ as in the statement of the theorem, we have

$$\Pr\,(\mathrm{PMEPR}(\mathbf{c}) > \lambda) \leq O\left(\frac{1}{\ln^3 n}\right),$$

and the results follows.

$\square$

6.5 Fast computation of PMEPR and PAPR of codes

It is a simple problem to determine the maximum possible PAPR or PMEPR of uncoded signals. However, when we exclude some of the possible combinations of the coefficients, i.e., in the case of coded signals, we have the problem of determining the maximum in a computationally efficient way.

We start with the problem of computing PMEPR.

Theorem 6.27 *Let $\mathcal{C}$ be a code consisting of vectors $\mathbf{a}$, $|a_k| = 1$, $k = 0, 1, \ldots,$ $n - 1$, and $\mathcal{C}$ such that it contains along with every code word $\mathbf{a}$ its negative, $-\mathbf{a}$. Also let for a $\mathbf{b} \in \mathbb{R}^n$,*

$$d_E(\mathbf{b}, \mathcal{C}) = \min_{\mathbf{a} \in \mathcal{C}} d_E(\mathbf{b}, \mathbf{a}),$$

and

$$\mathbf{e}(t, \varphi) = \left(e^{-2\varphi}, e^{-2\pi \imath t - \varphi}, e^{-2\pi \imath 2t - \varphi}, \ldots, e^{-2\pi \imath (n-1)t - \varphi}\right).$$

Then, for any r, $rn \in \mathbb{N}$, $r \geq 1$, and ℓ, $\ell \in \mathbb{N}$, $\ell \geq 2$,

$\mathrm{PMEPR}(\mathcal{C})$

$$\leq \max_{s=0,1,\ldots,rn-1} \max_{j=0,1,\ldots,\ell-1} \frac{1}{n \cdot \cos \frac{\pi}{2r} \cdot \cos \frac{\pi}{2\ell}} \left(\frac{n - 2d_E\left(\mathbf{e}\left(\frac{s}{rn}, \frac{2\pi j}{\ell}\right), \mathcal{C}\right)}{2}\right)^2.$$

Moreover,

$$\mathrm{PMEPR}(\mathcal{C}) \geq \max_{s=0,1,\ldots,rn-1} \max_{j=0,1,\ldots,\ell-1} \frac{1}{n} \left(\frac{n - 2d_E\left(\mathbf{e}\left(\frac{s}{rn}, \frac{2\pi j}{\ell}\right), \mathcal{C}\right)}{2}\right)^2.$$

Proof Let $\mathbf{a} = (a_0, a_1, \ldots, a_{n-1})$ belong to $\mathcal{C}$. Then the problem is to find

$$\mathrm{PMEPR}(\mathcal{C}) = \frac{1}{n} \max_{\mathbf{a} \in \mathcal{C}} \max_{t \in [0,1)} \left| \sum_{k=0}^{n-1} a_k e^{2\pi \imath kt} \right|^2 = \frac{1}{n} \max_{t \in [0,1)} \max_{\mathbf{a} \in \mathcal{C}} \left| \sum_{k=0}^{n-1} a_k e^{2\pi \imath kt} \right|^2.$$

We denote $\mathbf{e}(t, \varphi) = \left(e^{-\varphi}, e^{-2\pi \imath t - \varphi}, e^{-2\pi \imath 2t - \varphi}, \ldots, e^{-2\pi \imath (n-1)t - \varphi}\right)$. Then the Euclidean distance between $\mathbf{a}$ and $\mathbf{e}(t, \varphi)$ is

$$d_E(\mathbf{a}, \mathbf{e}(t, \varphi)) = 2n - 2\Re((\mathbf{a}, \mathbf{e}(t, \varphi))) = 2n - 2\Re\left(\sum_{k=0}^{n-1} a_k e^{2\pi \imath kt + \varphi}\right).$$

Clearly,

$$\left| \sum_{k=0}^{n-1} a_k e^{2\pi \iota k t} \right| = \max_{\varphi \in [0, 2\pi)} \left| \Re \left(\sum_{k=0}^{n-1} a_k e^{2\pi \iota k t + \varphi} \right) \right|$$

$$\leq \frac{1}{\cos \left(\frac{\pi}{2\ell} \right)} \cdot \max_{j=0,1,\dots,\ell-1} \left| \Re \left(\sum_{k=0}^{n-1} a_k e^{2\pi \iota k t + 2\pi \frac{j}{\ell}} \right) \right|$$

$$= \frac{1}{\cos \left(\frac{\pi}{2\ell} \right)} \cdot \max_{j=0,1,\dots,\ell-1} \frac{2n - d_E \left(\mathbf{a}, \mathbf{e} \left(t, \frac{2\pi j}{\ell} \right) \right)}{2}.$$

The last inequality is given by Theorem 4.14 and any integer $\ell \geq 2$ can be used. Furthermore,

$$\max_{t \in [0,1)} \left| \sum_{k=0}^{n-1} a_k e^{2\pi \iota k t} \right| \leq C_r \max_{s=0,1,\dots,rn-1} \left| \sum_{k=0}^{n-1} a_k e^{2\pi \iota k \frac{s}{rn}} \right|,$$

where $r \geq 1$, $rn \in \mathbb{N}$, and C_r is estimated in Theorems 4.8, 4.9, 4.10, and 4.11. For example, we have $C_r \leq \frac{1}{\cos \frac{\pi}{2r}}$. Summarizing, we have

$$\max_{t \in [0,1)} \left| \sum_{k=0}^{n-1} a_k e^{2\pi \iota k t} \right| \leq \frac{C_r}{\cos \frac{\pi}{2\ell}} \max_{s=0,1,\dots,rn-1} \max_{j=0,1,\dots,\ell-1} \frac{2n - d_E \left(\mathbf{a}, \mathbf{e} \left(\frac{s}{rn}, \frac{2\pi j}{\ell} \right) \right)}{2}.$$

Simple algebraic manipulations accomplish the proof of the upper bound. The lower bound is straightforward. □

Therefore, to estimate PMEPR with a maximum relative error of, at most, $\left(\cos \frac{\pi}{2\ell} \cos \frac{\pi}{2r} \right)^{-1}$, one should pick some appropriate r and ℓ, and find ℓrn values of the Euclidean distances between the code and the vectors defined in the theorem. Thus the complexity of this procedure is ℓrn times the complexity of the soft minimum distance decoding of $\mathcal{C}$. Clearly if the code has an efficient decoding procedure, estimating its PMEPR is only a constant times n more complicated.

To deal with the PAPR, one may apply Theorem 4.19, showing that when the carrier frequency is essentially larger than the bandwidth of each tone, bounds for PMEPR allow accurate estimation of PAPR. If the carrier frequency is zero, to estimate PAPR we have to find

$$\frac{1}{n} \cdot \max_{t \in [0,1)} \max_{\mathbf{a} \in \mathcal{C}} \left(\Re \left(\sum_{k=0}^{n-1} a_k e^{2\pi \iota k t} \right) \right)^2,$$

and the same techniques as used for PMEPR (though without projections on axes) apply.

6.6 Notes

Section 6.1   Theorem 6.1 is by Friese [124]. The rest of the section follows Litsyn and Wunder [252]. On the PMEPR distribution for the spherical case, see also Sharif and Hassibi [374].

Section 6.2   The results of this section are by Paterson and Tarokh [328]. Theorems 6.9 and 6.10 are given here with better estimates for the PMEPR than those in [328].

Section 6.3   This section follows Litsyn and Wunder [252]; see also [444]. Quite a lot is known about the distance distributions of codes. Distance distributions of some classes of algebraic codes are discussed in MacWilliams and Sloane [257]; see also references there. The convergence of distance distributions to normalized binomial distributions was analyzed by Ashikhmin *et al.* [14] and Krasikov and Litsyn [213, 214]. The distance distributions of LDPC codes were considered by Burshtein and Miller [56], Di *et al.* [90, 91], Gallager [125], and Litsyn and Shevelev [249, 250]. Yue and Yang [452] discussed the distance distributions of turbo codes.

Section 6.4   The first bound on the weight spectrum of BCH codes showing that it converges to normalized binomial distribution is by Sidel'nikov [380]. This section follows Krasikov and Litsyn [211]. Other papers treating the subject are by Kasami *et al.* [196, 197], Krasikov and Litsyn [211, 212, 215, 216], Keren and Litsyn [201, 202], Solé [386], and Vláduts and Skorobogatov [418, 419]. The result on PMEPR of BCH codes is by Litsyn and Wunder [252].

Section 6.5   The ideas of this section are from Tarokh and Jafarkhani [398, 399], though here I consider PMEPR rather than PAPR estimation. In [399] an estimate for PAPR for high carrier frequencies via lower carrier frequencies is also given.

7

MC signals with constant PMEPR

Although we have seen that most of the MC signals have peaks of value about $\sqrt{n \ln n}$, there are plenty of signals with maxima of order $\sqrt{n}$. This chapter is devoted to methods of constructing such signals. I begin with relating the maxima in signals to the distribution of their aperiodic correlations (Theorem 7.2). Then I describe in Section 7.2 the Rudin–Shapiro sequences over $\{-1, 1\}$, guaranteeing a PMEPR of at most 2 for n being powers of 2. They appear in pairs, where each one of the sequences possesses the claimed property. The Rudin–Shapiro sequences are representatives of a much broader class of complementary sequences discussed in Section 7.3. The signals defined by these sequences also have a PMEPR not exceeding 2, while existing for a wider spectrum of lengths. In Section 7.4, I introduce complementary sets of sequences. The number of sequences in the sets can be more than two, and the corresponding sequences have a PMEPR not exceeding the number of sequences in the set. In Section 7.5, I generalize the earlier derived results to the polyphase case, and describe a general construction of complementary pairs and sets stemming from cosets of the first-order Reed–Muller codes within the second-order Reed–Muller codes. Another idea in constructing sequences with low PMEPR is to use vectors defined by evaluating the trace of a function over finite fields or rings. This topic is explored in Section 7.6 using estimates for exponential sums. Finally, in Sections 7.7 and 7.8, I study two classes of sequences, M-sequences and Legendre sequences, guaranteeing PMEPR of order at most $(\ln n)^2$.

7.1 Peak power and aperiodic correlation

There is an intimate connection between the peak power of MC signals and the values of its aperiodic out-of-phase correlation function. Let a signal in an n-carrier

MC system be defined by

$$F(t) = F_{\mathbf{a}}(t) = \sum_{k=0}^{n-1} a_k e^{2\pi \imath k t},$$

where $\mathbf{a} = (a_0, \ldots, a_{n-1})$ is the vector of complex-valued coefficients. The aperiodic autocorrelation function is

$$\rho(j) = \rho_{\mathbf{a}}(j) = \sum_{k=0}^{n-j-1} a_k a_{k+j}^*, \tag{7.1}$$

where the out-of-phase coefficients correspond to $j \in \{1, 2, \ldots, n-1\}$. Note that $\rho(0) = \sum_{k=0}^{n-1} |a_k|^2$.

Lemma 7.1

$$|F(t)|^2 = \sum_{k=0}^{n-1} |a_k|^2 + 2\Re \left\{ \sum_{j=1}^{n-1} e^{2\pi \imath j t} \rho(j) \right\}.$$

Proof   Indeed,

$$|F(t)|^2 = F(t) F^*(t) = \sum_{k=0}^{n-1} \sum_{j=0}^{n-1} a_k a_j^* e^{2\pi \imath (k-j) t}.$$

Singling out the terms with $j = k$ and since for all $a, b \in \mathbb{C}$,

$$ab^* + a^* b = 2\Re\{ab^*\},$$

we have

$$|F(t)|^2 = \sum_{k=0}^{n-1} |a_k|^2 + 2\Re \left\{ \sum_{k=0}^{n-1} \sum_{j=k+1}^{n-1} a_k a_j^* e^{2\pi \imath (k-j) t} \right\}$$

$$= \sum_{k=0}^{n-1} |a_k|^2 + 2\Re \left\{ \sum_{j=1}^{n-1} e^{2\pi \imath j t} \sum_{k=0}^{n-j-1} a_k a_{k+j}^* \right\}.$$

$\square$

Theorem 7.2

$$\mathrm{PMEPR}(\mathbf{a}) \le \frac{\sum_{k=0}^{n-1} |a_k|^2}{n} + \frac{2}{n} \sum_{j=1}^{n-1} |\rho(j)|. \tag{7.2}$$

Proof For every $a \in \mathbb{C}$, we have $\Re\{a\} \leq |a|$, and thus

$$|F(t)|^2 \leq \sum_{k=0}^{n-1} |a_k|^2 + 2 \left| \sum_{j=1}^{n-1} e^{2\pi \iota jt} \rho(j) \right|$$

$$\leq \sum_{k=0}^{n-1} |a_k|^2 + 2 \sum_{j=1}^{n-1} |\rho(j)|,$$

and the claim follows. $\qquad\square$

Corollary 7.3 *If* $\sum_{k=0}^{n-1} |a_k|^2 = n$, *then*

$$\text{PMEPR}(\mathbf{a}) \leq 1 + \frac{2}{n} \sum_{j=1}^{n-1} |\rho(j)| \leq 1 + \frac{2(n-1)}{n} \max_{j=1,\dots,n-1} |\rho(j)|.$$

$\qquad\square$

Example 7.1 Consider $\mathbf{a} = (1, 1, 1, -1, 1)$. The autocorrelation coefficients are $\rho = (5, 0, 1, 0, 1)$. Hence, the upper bound (7.2) is

$$\text{PMEPR}(\mathbf{a}) \leq 1 + \frac{4}{5}.$$

It coincides with the value $|F(0)|$ or $|F(1/2)|$ and is thus tight. $\qquad\square$

Along with the maximum of the out-of-phase aperiodic correlation, another parameter, called the *merit factor*, is often considered. The merit factor of a vector **a** is

$$\mu(\mathbf{a}) = \frac{n^2}{2 \sum_{j=1}^{n-1} |\rho_\mathbf{a}(j)|^2}. \tag{7.3}$$

The merit factor measures the mean-square deviation of $|F_\mathbf{a}(t)|^2$ from $n = (\sqrt{n})^2$,

$$\frac{1}{\mu(\mathbf{a})} = \frac{1}{n^2} \int_0^1 \left(|F_\mathbf{a}(t)|^2 - n \right)^2 dt.$$

The following relates the merit factor with the PMEPR.

Theorem 7.4

$$\text{PMEPR}(\mathbf{a}) \leq \frac{\sum_{k=0}^{n-1} |a_k|^2}{n} + \sqrt{\frac{2(n-1)}{\mu(\mathbf{a})}}. \tag{7.4}$$

Proof By the Cauchy–Schwartz inequality (see Theorem 3.4),

$$\sum_{j=1}^{n-1} |\rho(j)| = \sum_{j=1}^{n-1} 1 \cdot |\rho(j)| \le \sqrt{(n-1)\sum_{j=1}^{n-1} |\rho(j)|^2}.$$

□

The results above hint that it might be beneficial to expect a low PMEPR from sequences with low maximum out-of-phase autocorrelation. A short account of the current knowledge about the maximum of aperiodic autocorrelation and merit factor appears in the notes (Section 7.9).

7.2 Rudin–Shapiro sequences

In this section, I consider BPSK-modulated MC signals for $n = 2^m$. Let us construct, recursively, a sequence of signals with PMEPR equal to 2. We start with $m = 0$, and set

$$P_0 = 1, \quad Q_0 = 1.$$

The Rudin–Shapiro signals are defined recursively as follows:

$$P_{m+1}(t) = P_m(t) + e^{2\pi \iota \cdot 2^m t} Q_m(t), \tag{7.5}$$

$$Q_{m+1}(t) = P_m(t) - e^{2\pi \iota \cdot 2^m t} Q_m(t). \tag{7.6}$$

Here are the coefficients for the first four lengths:

$P_0 : (1)$	$Q_0 : (1)$
$P_1 : (1, 1)$	$Q_1 : (1, -1)$
$P_2 : (1, 1, 1, -1)$	$Q_2 : (1, 1, -1, 1)$
$P_3 : (1, 1, 1, -1, 1, 1, -1, 1)$	$Q_3 : (1, 1, 1, -1, -1, -1, 1, -1).$

We address the sequence of coefficients of the defined signals as Rudin–Shapiro P_m and Q_m sequences.

Theorem 7.5 *For any nonnegative integer m,*

$$\mathrm{PMEPR}(P_m) \le 2, \quad \mathrm{PMEPR}(Q_m) \le 2.$$

Proof For all $a, b \in \mathbb{C}$, the *parallelogram law* is valid:

$$|a+b|^2 + |a-b|^2 = (a+b)(a^*+b^*) + (a-b)(a^*-b^*) = 2|a|^2 + 2|b|^2.$$

Thus

$$|P_{m+1}(t)|^2 + |Q_{m+1}(t)|^2 = |P_m(t) + e^{2\pi \iota \cdot 2^m t} Q_m(t)|^2 + |P_m(t) - e^{2\pi \iota \cdot 2^m t} Q_m(t)|^2$$

$$= 2(|P_m(t)|^2 + |Q_m(t)|^2)$$

$$= 2^{m+1}(|P_0(t)|^2 + |Q_0(t)|^2) = 2^{m+2}.$$

Therefore,

$$\max_{t \in [0,1)} |P_{m+1}(t)|^2 \le 2^{m+2}, \quad \max_{t \in [0,1)} |Q_{m+1}(t)|^2 \le 2^{m+2}.$$

□

Theorem 7.6

$$P_{2m}(0) = 2^m,\ P_{2m}\left(\frac{1}{2}\right) = 2^m,\ P_{2m+1}(0) = 2^{m+1},\ P_{2m+1}\left(\tfrac{1}{2}\right) = 0,$$

$$Q_{2m}(0) = 2^m,\ Q_{2m}\left(\frac{1}{2}\right) = -2^m,\ Q_{2m+1}(0) = 0,\ Q_{2m+1}\left(\tfrac{1}{2}\right) = 2^{m+1}.$$

Proof We have

$$\begin{aligned}
P_{m+2}(t) &= P_{m+1}(t) + e^{2\pi \imath 2^{m+1} t} Q_{m+1}(t) \\
&= P_m(t) + e^{2\pi \imath 2^m t} Q_m(t) + e^{2\pi \imath 2^{m+1} t} \left(P_m(t) - e^{2\pi \imath 2^m t} Q_m(t) \right) \\
&= \left(1 + e^{2\pi \imath 2^{m+1} t} \right) P_m(t) + e^{2\pi \imath 2^m t} \left(1 - e^{\imath 2^{m+1} t} \right) Q_{m+1}(t).
\end{aligned}$$

Then

$$P_{2m}(0) = (1+1)P_{2m-2}(0) + 0 = \ldots = 2^m P_0(0) = 2^m,$$

$$P_{2m}\left(\frac{1}{2}\right) = 2 P_{2m-2}\left(\frac{1}{2}\right) = \ldots = P_0\left(\frac{1}{2}\right) = 2^m,$$

$$P_{2m+1}(0) = 2 P_{2m-1}(0) = \ldots = 2^m P_1(0) = 2^{m+1},$$

$$P_{2m+1}\left(\frac{1}{2}\right) = 2 P_{2m-1}\left(\frac{1}{2}\right) = \ldots = 2^m P_1\left(\frac{1}{2}\right) = 0.$$

Analogous recursions yield the result for Q-sequences. □

Corollary 7.7 *For any nonnegative m,*

$$\mathrm{PMEPR}(P_{2m+1}) = 2, \quad \mathrm{PMEPR}(Q_{2m+1}) = 2.$$

□

The Rudin–Shapiro sequences possess some symmetry.

Theorem 7.8 *Let P_m and Q_m be two Rudin–Shapiro sequences. Then*

$$|P_m(t)|^2 = 2^{m+1} - \left| P_m\left(\frac{1}{2} - t\right) \right|^2,$$

$$|Q_m(t)|^2 = 2^{m+1} - \left| Q_m\left(\frac{1}{2} - t\right) \right|^2.$$

Proof   The proof is by induction. The claim is clearly true for $m = 0$. Furthermore,

$$|P_{m+1}(t)|^2 = |P_m(t)|^2 + |Q_m(t)|^2 + e^{2\pi i 2^m t} P_m^*(t) Q_m(t) + e^{-2\pi i 2^m t} P_m(t) Q_m^*(t)$$

$$= 2^{m+1} - \left| P_m \left(\frac{1}{2} - t \right) \right|^2 + 2^{m+1} - \left| Q_m \left(\frac{1}{2} - t \right) \right|^2$$

$$+ 2\Re \left\{ e^{2\pi i 2^m t} P_m^*(t) Q_m(t) \right\}$$

$$= 2^{m+2} - \left| P_m \left(\frac{1}{2} - t \right) \right|^2 - \left| Q_m \left(\frac{1}{2} - t \right) \right|^2$$

$$- 2\Re \left\{ e^{2\pi i 2^m (\frac{1}{2} - t)} P_m^* \left(\frac{1}{2} - t \right) Q_m \left(\frac{1}{2} - t \right) \right\}$$

$$= 2^{m+2} - \left| P_{m+1} \left(\frac{1}{2} - t \right) \right|^2 .$$

In the third equality we used $\cos t = -\cos(\pi - t)$. $\qquad\qquad\square$

Although the Rudin–Shapiro sequences are defined only for lengths being a power of 2, one may consider their partial sums. Let $S_n^{(P_m)}$ and $S_n^{(Q_m)}$ stand for the polynomials defined by the first n terms in the corresponding sequences.

Theorem 7.9   *For any nonnegative m and $n \leq 2^m$,*

$$\mathrm{PMEPR}\left(S_n^{(P_m)} \right) \leq (2 + \sqrt{2})^2, \quad \mathrm{PMEPR}\left(S_n^{(P_m)} \right) \leq (2 + \sqrt{2})^2.$$

Proof   We will prove that

$$\max_t \left| S_n^{(P_m)}(t) \right| \leq (2 + \sqrt{2}) 2^{\frac{m}{2}}, \tag{7.7}$$

$$\max_t \left| S_n^{(Q_m)}(t) \right| \leq (2 + \sqrt{2}) 2^{\frac{m}{2}}. \tag{7.8}$$

Indeed, this is true for $m = 0$. Suppose (7.7) and (7.8) hold for some m, and consider $S_n^{(P_{m+1})}$ and $S_n^{(Q_{m+1})}$ with $n \leq 2^{m+1}$. If $n \leq 2^m$, by (7.5) and (7.6), the coefficients of $S_n^{(P_{m+1})}$ and $S_n^{(Q_{m+1})}$ are just the first n coefficients of P_m, and thus

$$\left| S_n^{(P_{m+1})} \right| = \left| S_n^{(Q_{m+1})} \right| = \left| S_n^{(P_m)} \right| \leq (2 + \sqrt{2}) 2^{\frac{m}{2}} < (2 + \sqrt{2}) 2^{\frac{m+1}{2}}.$$

Now let $2^m < n \leq 2^{m+1}$. Then, using (7.5) and Theorem 7.5 we conclude that

$$\left| S_n^{(P_{m+1})} \right| \leq |P_m| + \left| S_{n-2^m}^{(Q_m)} \right| \leq 2^{\frac{m+1}{2}} + (2 + \sqrt{2}) 2^{\frac{m}{2}} = (2 + \sqrt{2}) 2^{\frac{m+1}{2}}.$$

The same estimate holds for $\left| S_n^{(Q_{m+1})} \right|$, and we are done. $\qquad\qquad\square$

7.3 Complementary sequences

The Rudin–Shapiro sequences are representatives of a wider class of sequences called *Golay complementary sequences*, or just complementary sequences.

Let **a** and **b** be two vectors of length n satisfying

$$\sum_{k=0}^{n-1} |a_k|^2 = \sum_{k=0}^{n-1} |b_k|^2 = n.$$

Then the aperiodic correlation functions for these vectors are defined by (7.1). The vectors constitute a *complementary pair* if, for all $j \in \{1, 2, \dots, n-1\}$,

$$\rho_{\mathbf{a}}(j) + \rho_{\mathbf{b}}(j) = 0. \tag{7.9}$$

Every sequence being a member of at least one complementary pair is called a *Golay sequence*.

The relevance of the complementary pairs to the PMEPR problem becomes apparent in the following theorem.

Theorem 7.10   *Let **a** be one of the vectors from a complementary pair. Then*

$$\mathrm{PMEPR}(\mathbf{a}) \leq 2.$$

Proof   Let **a** and **b** constitute a complementary pair. By Lemma 7.1,

$$|F_{\mathbf{a}}(t)|^2 + |F_{\mathbf{b}}(t)|^2 = \sum_{k=0}^{n-1} |a_k|^2 + \sum_{k=0}^{n-1} |b_k|^2 + 2\Re \left\{ \sum_{j=1}^{n-1} e^{2\pi \iota j t}(\rho_{\mathbf{a}}(j) + \rho_{\mathbf{b}}(j)) \right\} = 2n.$$

Therefore,

$$\max_{t \in [0,1)} |F_{\mathbf{a}}(t)|^2 \leq 2n.$$

$\square$

Example 7.2   Let

$$\mathbf{a} = (-1, 1, 1, -1, 1, -1, 1, 1, 1, -1), \quad \mathbf{b} = (-1, 1, 1, 1, 1, 1, 1, -1, -1, 1).$$

Then

$$\rho_{\mathbf{a}}(j) = (10, -3, 0, -1, 0, \quad 1, \quad 2, -1, -2, \quad 1)$$
$$\rho_{\mathbf{a}}(j) = (10, \quad 3, 0, \quad 1, 0, -1, -2, \quad 1, \quad 2, -1)$$

and thus **a** and **b** are a complementary pair. The behavior of $|F_{\mathbf{a}}(t)|$ and $|F_{\mathbf{b}}(t)|$ is presented in Fig. 7.1. $\square$

The lengths of vectors in a complementary pair cannot be arbitrary. Let us consider here the binary case, i.e., $a_i \in \{-1, 1\}$.

Theorem 7.11   *The length, n, of a complementary pair is a sum of two integral squares.*

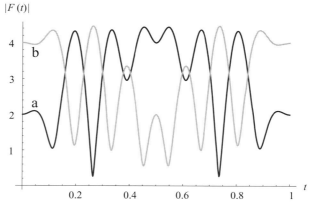

Figure 7.1 $|F_{\mathbf{a}}(t)|$ and $|F_{\mathbf{b}}(t)|$

Proof We have

$$\rho_{\mathbf{a}}(0) = \sum_{k=0}^{n-1} a_k^2 = n.$$

Thus

$$\sum_{j=0}^{n-1} \rho_{\mathbf{a}}(j) = \sum_{j=0}^{n-1} \sum_{k=0}^{n-j-1} a_k a_{k+j} = 2 \left(\sum_{k_1=0}^{n-1} a_{k_1} \right) \left(\sum_{k_2=0}^{n-1} a_{k_2} \right) - \sum_{k=0}^{n-1} a_k^2 = 2 \left(\sum_{k=0}^{n-1} a_k \right)^2 - n.$$

For a complementary pair, $\mathbf{a}$ and $\mathbf{b}$, we have

$$\sum_{j=0}^{n-1} (\rho_{\mathbf{a}}(j) + \rho_{\mathbf{b}}(j)) = \rho_{\mathbf{a}}(0) + \rho_{\mathbf{b}}(0) = 2n.$$

On the other hand,

$$\sum_{j=0}^{n-1} (\rho_{\mathbf{a}}(j) + \rho_{\mathbf{b}}(j)) = 2 \left(\sum_{k=0}^{n-1} a_k \right)^2 + 2 \left(\sum_{k=0}^{n-1} b_k \right)^2 - 2n.$$

Therefore,

$$\left(\sum_{k=0}^{n-1} a_k \right)^2 + \left(\sum_{k=0}^{n-1} b_k \right)^2 = 2n, \tag{7.10}$$

and $\left(\sum_{k=0}^{n-1} a_k \right)^2$ and $\left(\sum_{k=0}^{n-1} b_k \right)^2$ have the same parity. Furthermore,

$$n = \left(\frac{\left(\sum_{k=0}^{n-1} a_k \right) + \left(\sum_{k=0}^{n-1} b_k \right)}{2} \right)^2 + \left(\frac{\left(\sum_{k=0}^{n-1} a_k \right) - \left(\sum_{k=0}^{n-1} b_k \right)}{2} \right)^2,$$

and this proves the claim. $\qquad \square$

Theorem 7.12   *The length n of a complementary pair is even.*

Proof   Let **a** and **b** be a complementary pair. Then (7.9) can be rewritten as the system of quadratic equations:

$$
\begin{aligned}
a_0 a_{n-1} \qquad\qquad &+ \qquad\qquad b_0 b_{n-1} \qquad\qquad &= 0 \\
a_0 a_{n-2} + a_1 a_{n-1} \qquad &+ \qquad b_0 b_{n-2} + b_1 b_{n-1} \qquad &= 0 \\
a_0 a_{n-3} + a_1 a_{n-2} + a_2 a_{n-1} &+ b_0 b_{n-3} + b_1 b_{n-2} + b_2 b_{n-1} &= 0 \qquad (7.11)\\
&\cdots \qquad\qquad\qquad \cdots & \\
a_0 a_1 + a_1 a_2 + \ldots + a_{n-2} a_{n-1} &+ b_0 b_1 + b_1 b_2 + \ldots + b_{n-2} b_{n-1} &= 0.
\end{aligned}
$$

Noticing that for $a, b \in \{-1, 1\}$,

$$ab = a + b - 1 \mod 4,$$

we can reduce the first equation to

$$a_0 + a_{n-1} + b_0 + b_{n-1} = 2 \mod 4. \qquad (7.12)$$

The second equation is

$$a_0 + a_{n-2} + a_1 + a_{n-1} + b_0 + b_{n-2} + b_1 + b_{n-1} = 0 \mod 4.$$

From this, and (7.12), we can conclude that

$$a_1 + a_{n-2} + b_1 + b_{n-2} = 2 \mod 4.$$

Continuing in the same way, we obtain

$$a_j + a_{n-1-j} + b_j + b_{n-1-j} = 2 \mod 4, \qquad (7.13)$$

valid for $0 \le j < \lfloor \frac{n}{2} \rfloor$. If n were odd, the next equation would be

$$2 a_{\frac{n-1}{2}} + 2 b_{\frac{n-1}{2}} = 2 \mod 4,$$

which is impossible for $a_{\frac{n-1}{2}}, b_{\frac{n-1}{2}} \in \{-1, 1\}$.   □

Corollary 7.13   *For any complementary pair* {**a**, **b**}, *and any* j, $0 \le j < \frac{n}{2}$,

$$a_j a_{n-1-j} + b_j b_{n-1-j} = 0.$$

Proof   This follows from (7.13).   □

Thus, for every j, $0 \le j < \frac{n}{2}$, in a complementary pair **a**, **b**, exactly three out of a_j, a_{n-1-j}, b_j and b_{n-1-j} have the same sign.

Still another restriction on the length of complementary pairs deals with its divisibility.

Theorem 7.14   *The length n of a complementary pair has no prime factor congruent to* 3 mod 4.

Proof Let **a** and **b** constitute a complementary pair of length n and assume that p is an odd prime factor of n. Denoting $z = e^{2\pi i t}$, we consider two polynomials of degree $n - 1$,

$$F_{\mathbf{a}}(z) = \sum_{k=0}^{n-1} a_k z^k, \qquad F_{\mathbf{b}}(z) = \sum_{j=0}^{n-1} b_j z^j,$$

with the coefficients $a_k, b_j \in \{-1, 1\}$. The complementarity is equivalent to

$$F_{\mathbf{a}}(z) F_{\mathbf{a}}(z^{-1}) + F_{\mathbf{b}}(z) F_{\mathbf{b}}(z^{-1}) = 2n,$$

for any complex z, $|z| = 1$. This equation, considered modulo p, provides a necessary condition

$$F_{\mathbf{a}}(z) F_{\mathbf{a}}(z^{-1}) + F_{\mathbf{b}}(z) F_{\mathbf{b}}(z^{-1}) = 0 \ \mathrm{mod}\ p.$$

In what follows we consider the polynomials over $\mathbb{F}_p$, i.e., the summations and multiplications are done modulo p. Let $h(z) = \mathrm{g.c.d.}\,(F_{\mathbf{a}}(z), F_{\mathbf{b}}(z))$, and

$$F_{\mathbf{a}}(z) = h(z) f_{\mathbf{a}}(z), \qquad F_{\mathbf{b}}(z) = h(z) f_{\mathbf{b}}(z),$$

where $\mathrm{g.c.d.}\,(f_{\mathbf{a}}(z), f_{\mathbf{b}}(z)) = 1$. Clearly $\deg f_{\mathbf{a}}(z) = \deg f_{\mathbf{b}}(z)$. Let s denote this common degree. This implies

$$f_{\mathbf{a}}(z) \cdot z^s f_{\mathbf{a}}(z^{-1}) + f_{\mathbf{b}}(z) \cdot z^s f_{\mathbf{a}}(z^{-1}) = 0 \ \mathrm{mod}\ p;$$

here $z^s f_{\mathbf{a}}(z^{-1}) = f_{\overleftarrow{\mathbf{a}}}(z)$, where $\overleftarrow{\mathbf{a}}$ is the reversed sequence **a**, and thus is a polynomial containing only nonnegative degrees of z. Furthermore, since $f_{\mathbf{a}}(z)$ and $f_{\mathbf{b}}(z)$ are relatively prime, the polynomial $z^s f_{\mathbf{b}}(z^{-1})$ must be a constant multiple of $f_{\mathbf{a}}(z)$. Hence for some $c \in \mathbb{F}_p^*$, we have

$$z^s f_{\mathbf{b}}(z^{-1}) = c f_{\mathbf{a}}(z), \qquad z^s f_{\mathbf{a}}(z^{-1}) = -c f_{\mathbf{b}}(z).$$

This yields

$$c^2 f_{\mathbf{a}}(z) = c z^s f_{\mathbf{b}}(z^{-1}) = -f_{\mathbf{a}}(z),$$

and, therefore, $c^2 = -1 \ \mathrm{mod}\ p$. However, by Euler's criterion, Corollary 3.34, it is possible only if $p = 1 \ \mathrm{mod}\ 4$. $\qquad\square$

Now let us pass on to constructions of complementary pairs. Analysis of (7.11) allows us to conclude that performing the following operations on a complementary pair, $\{\mathbf{a}, \mathbf{b}\}$, generates other complementary pairs:

- Interchanging the sequences yields the complementary pair $\{\mathbf{b}, \mathbf{a}\}$;
- Let $\overleftarrow{\mathbf{a}} = (a_{n-1}, a_{n-2}, \ldots, a_0)$ stand for the reversed sequence $\mathbf{a} = (a_0, a_1, \ldots, a_{n-1})$. Reversing either or both sequences generates the complementary pairs $\{\overleftarrow{\mathbf{a}}, \mathbf{b}\}$, $\{\mathbf{a}, \overleftarrow{\mathbf{b}}\}$, and $\{\overleftarrow{\mathbf{a}}, \overleftarrow{\mathbf{b}}\}$;
- Let $-\mathbf{a} = (-a_0, -a_1, \ldots, -a_{n-1})$ stand for the negation of **a**. Negation of either or both sequences yields the complementary pairs $\{-\mathbf{a}, \mathbf{b}\}$, $\{\mathbf{a}, -\mathbf{b}\}$, and $\{-\mathbf{a}, -\mathbf{b}\}$;

- Let $\widetilde{\mathbf{a}} = (a_0, -a_1, a_2, -a_3, \ldots, (-1)^j a_j, \ldots, (-1)^{n-1} a_{n-1})$ stand for the sequence with negated even entries. Negation of even entries in $\mathbf{a}$ and $\mathbf{b}$ yields the complementary pair $\{\widetilde{\mathbf{a}}, \widetilde{\mathbf{b}}\}$.

Only the last property requires justification. Indeed, as a result of negation, the values of $\rho_{\mathbf{a}}(j)$ and $\rho_{\mathbf{b}}(j)$ will stay intact for even j, while the signs will alter for odd j. However, this will not violate the complementarity of the sequences.

Now let us pass on to more involved methods of generating complementary pairs. We will need some notation to describe the approaches. Let

$$\mathrm{cat}(\mathbf{a}, \mathbf{b}) = (a_0, a_1, \ldots, a_{n-1}, b_0, b_1, \ldots, b_{n-1})$$

stand for concatenation of two vectors, $\mathbf{a}$ and $\mathbf{b}$, and let

$$\mathrm{int}(\mathbf{a}, \mathbf{b}) = (a_0, b_0, a_1, b_1, \ldots, a_{n-1}, b_{n-1})$$

denote interleaved vectors $\mathbf{a}$ and $\mathbf{b}$.

The first construction is a direct generalization of the method used for the recursive construction of Rudin–Shapiro sequences, see (7.5) and (7.6). It allows the construction of a complementary pair of length $2n$ from a complementary pair of length n.

Theorem 7.15   *Let $\{\mathbf{a}, \mathbf{b}\}$ be a complementary pair. Then*

$$\{\mathbf{c}, \mathbf{d}\} = \{\mathrm{cat}(\mathbf{a}, \mathbf{b}), \mathrm{cat}(\mathbf{a}, -\mathbf{b})\}$$

is a complementary pair.

Proof   Indeed, for $1 \leq j \leq n - 1$,

$$\rho_{\mathbf{c}}(j) = \rho_{\mathbf{a}}(j) + \rho_{\mathbf{b}}(j) + \sum_{k=0}^{j-1} a_{n-j+k} b_k = \sum_{k=0}^{j-1} a_{n-j+k} b_k,$$

while

$$\rho_{\mathbf{d}}(j) = \rho_{\mathbf{a}}(j) + \rho_{-\mathbf{b}}(j) - \sum_{k=0}^{j-1} a_{n-j+k} b_k = -\sum_{k=0}^{j-1} a_{n-j+k} b_k = -\rho_{\mathbf{c}}(j).$$

For $n \leq j \leq 2n - 1$,

$$\rho_{\mathbf{c}}(j) = \sum_{k=0}^{2n-j-1} a_k b_{k+j-n},$$

while

$$\rho_{\mathbf{d}}(j) = -\sum_{k=0}^{2n-j-1} a_k b_{k+j-n} = -\rho_{\mathbf{c}}(j),$$

and we are done. $\square$

Theorem 7.16   *Let* {**a**, **b**} *be a complementary pair. Then*

$$\{\mathbf{c}, \mathbf{d}\} = \{\text{int}(\mathbf{a}, \mathbf{b}), \text{int}(\mathbf{a}, -\mathbf{b})\}$$

is a complementary pair.

Proof   Indeed, for j, $0 \le j \le n - 1$,

$$\rho_{\mathbf{c}}(2j) = \rho_{\mathbf{d}}(2j) = \rho_{\mathbf{a}}(j) + \rho_{\mathbf{b}}(j) = 0,$$

and

$$\rho_{\mathbf{c}}(2j + 1) = \sum_{k=0}^{n-j-1} a_k b_{k+j} + \sum_{k=0}^{n-j-2} b_k a_{k+j+1},$$

$$\rho_{\mathbf{d}}(2j + 1) = -\sum_{k=0}^{n-j-1} a_k b_{k+j} - \sum_{k=0}^{n-j-2} b_k a_{k+j+1},$$

and

$$\rho_{\mathbf{c}}(2j + 1) = -\rho_{\mathbf{d}}(2j + 1).$$

$\square$

Similar ideas yield the following recursive constructions. We will need the definition of the *Kronecker product* of vectors. Let $\mathbf{a} = (a_0, a_1, \ldots, a_{n-1})$ and $\mathbf{b} = (b_0, b_1, \ldots, b_{m-1})$ be two vectors of length n and m, respectively. Then their Kronecker product, $\mathbf{a} * \mathbf{b}$, is a vector of length nm,

$$\mathbf{a} * \mathbf{b} = (b_0\mathbf{a}, b_1\mathbf{a}, \ldots, b_{n-1}\mathbf{a})$$
$$= (b_0 a_0, b_0 a_1, \ldots, b_0 a_{n-1}, b_1 a_0, b_1 a_1, \ldots, b_1 a_{n-1}, \ldots, b_{n-1} a_0,$$
$$b_{n-1} a_1, \ldots, b_{n-1} a_{n-1}).$$

The MC signal corresponding to $\mathbf{a} * \mathbf{b}$ is

$$F_{\mathbf{a}*\mathbf{b}}(t) = F_{\mathbf{a}}(t) F_{\mathbf{b}}(nt). \tag{7.14}$$

Theorem 7.17   *Let* {**a**, **b**} *and* {**c**, **d**} *be two complementary pairs of lengths n and m, respectively. Then*

$$\text{cat}(\mathbf{a} * \mathbf{c}, \mathbf{b} * \mathbf{d}) \text{ and } \text{cat}(\mathbf{a} * \overleftarrow{\mathbf{d}}, \mathbf{b} * (\overleftarrow{-\mathbf{c}})),$$

and

$$\text{int}(\mathbf{a} * \mathbf{c}, \mathbf{b} * \mathbf{d}) \text{ and } \text{int}(\mathbf{a} * \overleftarrow{\mathbf{d}}, \mathbf{b} * (\overleftarrow{-\mathbf{c}})),$$

where the interleaving is block-wise, i.e.,

$$\text{int}(\mathbf{a} * \mathbf{c}, \mathbf{b} * \mathbf{d}) = (c_0\mathbf{a}, d_0\mathbf{b}, c_1\mathbf{a}, d_1\mathbf{b}, \ldots, c_{m-1}\mathbf{a}, d_{m-1}\mathbf{b}),$$
$$\text{int}(\mathbf{a} * \mathbf{c}, \mathbf{b} * \mathbf{d}) = (d_{m-1}\mathbf{a}, -c_{m-1}\mathbf{b}, d_{m-2}\mathbf{a}, -c_{m-2}\mathbf{b}, \ldots, d_0\mathbf{a}, -c_0\mathbf{b}),$$

constitute complementary pairs of length 2mn. □

Let $\mathbf{a} + \mathbf{b}$ stand for the component-wise sum of $\mathbf{a}$ and $\mathbf{b}$,

$$\mathbf{a} + \mathbf{b} = (a_0 + b_0, a_1 + b_1, \ldots, a_{n-1} + b_{n-1}).$$

Theorem 7.18 *Let $\{\mathbf{a}, \mathbf{b}\}$ and $\{\mathbf{c}, \mathbf{d}\}$ be two complementary pairs of lengths n and m, respectively. Then $\{\mathbf{g}, \mathbf{h}\}$, where*

$$\mathbf{g} = \left(\frac{\mathbf{a} + \mathbf{b}}{2}\right) * \mathbf{c} + \left(\frac{\mathbf{a} - \mathbf{b}}{2}\right) * \mathbf{d},$$

and

$$\mathbf{h} = \left(\frac{\overleftarrow{\mathbf{a}} - \overleftarrow{\mathbf{b}}}{2}\right) * \mathbf{c} - \left(\frac{\overleftarrow{\mathbf{a}} + \overleftarrow{\mathbf{b}}}{2}\right) * \mathbf{d},$$

are a complementary pair of length nm. □

Proof Instead of proving complementarity, I will address directly an equivalent statement about PMEPR of $\mathbf{g}$ and $\mathbf{h}$, see Theorem 7.10. Indeed, it is straightforward to check that the obtained vectors, $\mathbf{g}$ and $\mathbf{h}$, have coefficients ± 1. Moreover,

$$F_{\mathbf{g}}(t) = \frac{1}{2}(F_{\mathbf{a}}(t) + F_{\mathbf{b}}(t)) F_{\mathbf{c}}(nt) + \frac{1}{2}(F_{\mathbf{a}}(t) - F_{\mathbf{b}}(t)) F_{\mathbf{d}}(nt),$$
$$F_{\mathbf{h}}(t) = \frac{e^{2\pi \imath (n-1)t}}{2}\left(F_{\mathbf{a}}^*(t) - F_{\mathbf{b}}^*(t)\right) F_{\mathbf{c}}(nt) - \frac{e^{2\pi \imath (n-1)t}}{2}\left(F_{\mathbf{a}}^*(t) + F_{\mathbf{b}}^*(t)\right) F_{\mathbf{d}}(nt),$$

where the second equality follows from (7.14) and, since $\mathbf{a} \in \mathbb{R}^n$,

$$F_{\overleftarrow{\mathbf{a}}}(t) = e^{2\pi \imath (n-1)t} F_{\mathbf{a}}(-t) = e^{2\pi \imath (n-1)t} F_{\mathbf{a}}^*(t).$$

Therefore,

$$\left|F_{\mathbf{g}}(t)\right|^2 + |F_{\mathbf{h}}(t)|^2 = F_{\mathbf{g}}(t)F_{\mathbf{g}}^*(t) + F_{\mathbf{h}}(t)F_{\mathbf{h}}^*(t)$$
$$= \frac{1}{2}(|F_{\mathbf{a}}(t)|^2 + |F_{\mathbf{b}}(t)|^2)(|F_{\mathbf{c}}(t)|^2 + |F_{\mathbf{d}}(t)|^2),$$

and the result follows from Theorem 7.10. □

A *primitive complementary pair* is defined as one that cannot be constructed by means of the described recursive constructions from shorter complementary pairs. The known cases of primitive pairs are of lengths 2, 10 (two pairs), and 26. It is conjectured that there are no more primitive pairs.

Table 7.1 *Numbers of pairs for selected lengths*

n	1	2	4	8	10	16	20	26	32	40	52	64	80
number of pairs	4	8	32	192	128	1536	1088	64	15360	9728	512	184320	102912

$n = 2$ $\mathbf{a}=(1,1)$, $\mathbf{b}=(1,-1)$

$n = 10$ $\mathbf{a}=(1,1,-1,1,-1,1,-1,-1,1,1)$, $\mathbf{b}=(1,1,-1,1,1,1,1,1,-1,-1)$

         $\mathbf{a}=(1,1,1,1,1,-1,1,1,-1,-1,1,1)$, $\mathbf{b}=(1,1,-1,-1,1,1,1,1,-1,1,1,-1)$

$n = 26$ $\mathbf{a}=(1,1,1,1,-1,1,1,1,-1,-1,1,-1,1,-1,1,-1,-1,1,-1,1,1,1,-1,-1,1,1,1,1)$,

         $\mathbf{b}=(1,1,1,1,-1,1,1,1,-1,-1,1,1,-1,1,1,1,1,1,1,-1,1,1,-1,-1,1,1,1,-1,-1,1,1,-1,-1,-1)$

Theorem 7.19   *Complementary pairs exist for all lengths*

$$n = 2^{\alpha} 10^{\beta} 26^{\gamma}, \quad \alpha, \beta, \gamma \geq 0.$$

Proof   This follows from Theorem 7.18. $\square$

The complete enumeration of complementary pairs is known for all possible lengths up to 100 [39]. The number of pairs is given in Table 7.1. Notice, however, that a particular Golay sequence may appear in more than one complementary pair.

7.4 Complementary sets

Complementary pairs can be generalized to sets containing more than two sequences. We say that a set of T sequences over ± 1, $\{\mathbf{a}^{(0)}, \mathbf{a}^{(1)}, \ldots, \mathbf{a}^{(T-1)}\}$, forms a *complementary set* of size T if

$$\sum_{k=0}^{T-1} \rho_{\mathbf{a}^{(k)}}(j) = 0 \quad \text{for } j \neq 0.$$

Theorem 7.20   *For any sequence $\mathbf{a}$ belonging to a complementary set of size T,*

$$\text{PMEPR}(\mathbf{a}) \leq T.$$

Proof   Analogously to Theorem 7.10 the result follows from the identity

$$\sum_{k=0}^{T-1} |F_{\mathbf{a}^{(k)}}(t)|^2 = Tn,$$

where $\{\mathbf{a}^{(0)}, \mathbf{a}^{(1)}, \ldots, \mathbf{a}^{(T-1)}\}$ is a complementary set. $\square$

Theorem 7.21   *The number of sequences in a complementary set is even.*

Table 7.2 *Examples of complementary sets of odd length and size 4*

$n = 3$	$(1, 1, 1)$	$(-1, 1, 1)$	$(1, -1, 1)$	$(1, 1, -1)$
$n = 5$	$(1, -1, -1, -1, -1)$	$(-1, 1, 1, -1, 1)$	$(1, -1, -1, -1, 1)$	$(-1, -1, -1, 1, -1)$
$n = 7$	$(1, 1, 1, -1, 1, 1, 1)$	$(1, -1, 1, 1, 1, -1, -1)$	$(1, -1, -1, 1, -1, 1, 1)$	$(1, 1, -1, 1, -1, -1, -1)$

Proof Let $\{\mathbf{a}^{(0)}, \mathbf{a}^{(1)}, \ldots, \mathbf{a}^{(T-1)}\}$ be a complementary set of length n. For every j, $j = 0, 1, \ldots, T - 1$, $\rho_{\mathbf{a}^{(j)}}(n - 1) \in \{-1, 1\}$. Since the sum of an odd number of summands of the form ± 1 is odd, it cannot be 0. $\square$

Theorem 7.22   *The number of sequences in a complementary set of odd length is divisible by 4.*

Proof   By the previous theorem, T should be even. Composing the system as for (7.11) and reasoning similarly to establishing (7.13), we conclude that

$$2 \sum_{k=1}^{T} a_{\frac{n-1}{2}}^{(k)} = T \bmod 4.$$

However, the last is impossible for $a_{\frac{n-1}{2}}^{(k)} \in \{-1, 1\}$ and $T = 2 \bmod 4$. $\square$

Example 7.3   Table 7.2 presents several examples of complementary sets of odd length and of size 4. Clearly, complementary pairs of such length do not exist. $\square$

To finish with nonexistence results, I give without proof the following result, similar to Theorem 7.11.

Theorem 7.23   *The length n of complementary sets of size 4 is a sum of three squares.* $\square$

Analogous to the case of complementary pairs, the following operations on complementary sets preserve complementarity:

* Reversing any number of the sequences in the set;
* Negating any number of the sequences in the set;
* Negating alternate elements in all sequences in the set.

The following constructive results rely on arguments similar to the case of complementary pairs, and are given without proofs.
    Let $\mathbf{a}_{\text{even}} = (a_0, a_2, \ldots)$, and $\mathbf{a}_{\text{odd}} = (a_1, a_3, \ldots)$, stand for the subvectors consisting of the elements with even and odd indices of $\mathbf{a}$.

Theorem 7.24 *Let n be even and* $\{\mathbf{a}^{(0)}, \mathbf{a}^{(1)}, \ldots, \mathbf{a}^{(T-1)}\}$ *be a complementary set of size T. Then*

$$\left\{ \mathbf{a}_{\text{even}}^{(0)}, \mathbf{a}_{\text{even}}^{(1)}, \ldots, \mathbf{a}_{\text{even}}^{(T-1)}, \mathbf{a}_{\text{odd}}^{(0)}, \mathbf{a}_{\text{odd}}^{(1)}, \ldots, \mathbf{a}_{\text{odd}}^{(T-1)} \right\}$$

is a complementary set of length $\frac{n}{2}$ *and size* $2T$. □

Let $A = \{\mathbf{a}^{(0)}, \mathbf{a}^{(1)}, \ldots, \mathbf{a}^{(T-1)}\}$ and $B = \{\mathbf{b}^{(0)}, \mathbf{b}^{(1)}, \ldots, \mathbf{b}^{(T-1)}\}$ be two complementary sets of length n and size T. Then A and B are called *mates* if

$$\left(\mathbf{a}^{(k)}, \mathbf{b}^{(k)} \right) = 0 \quad \text{for } k = 0, 1, \ldots, T,$$

i.e., the inner product of corresponding pairs of vectors is 0.

Theorem 7.25 *Let A and B be mates of length n and size T. Then*

$$\left\{ \text{int} \left(\mathbf{a}^{(0)}, \mathbf{b}^{(0)} \right), \text{int} \left(\mathbf{a}^{(1)}, \mathbf{b}^{(1)} \right), \ldots, \text{int} \left(\mathbf{a}^{(T-1)}, \mathbf{b}^{(T-1)} \right) \right\}$$

is a complementary set of length $2n$ *and size* T. □

A ± 1 matrix H is *column-orthogonal* if the inner product of any two distinct columns of H is 0. The number of columns in such a matrix does not exceed the number of rows.

Theorem 7.26 *Let H be a* $T \times n$ *column-orthogonal matrix,* $n \leq T$. *Then the rows of H form a complementary set of length n and size T.* □

In particular the theorem holds for $n = T$ when the corresponding square matrix H satisfies $H^{tr} H = n I_n$, I_n is the $n \times n$ identity matrix, and thus is a transposed Hadamard matrix.

Theorem 7.27 *Let H be a* $T \times m$ *column-orthogonal matrix with elements* $h_{k,\ell}$, $k = 0, 1, \ldots, T - 1$; $\ell = 0, 1, \ldots, m - 1$; *and* $A = \{\mathbf{a}^{(0)}, \mathbf{a}^{(1)}, \ldots, \mathbf{a}^{(m-1)}\}$ *be a complementary set of length n and size m. Then*

$$\left\{ \text{cat} \left(h_{j,0} \mathbf{a}^{(0)}, h_{j,1} \mathbf{a}^{(1)}, \ldots, h_{j,m-1} \mathbf{a}^{(m-1)} \right), j = 0, 1, \ldots, T - 1 \right\},$$

is a complementary set of length nm and size T. □

For example, if

$$H = \begin{pmatrix} 1 & 1 \\ 1 & -1 \end{pmatrix},$$

we obtain the Rudin–Shapiro recursion, see Section 7.2. Choosing now

$$H = \begin{pmatrix} 1 & 1 & 1 & 1 \\ 1 & -1 & 1 & -1 \\ 1 & -1 & -1 & 1 \\ 1 & 1 & -1 & -1 \end{pmatrix},$$

we obtain the following result.

Corollary 7.28 *Let* $\left\{\mathbf{a}^{(0)}, \mathbf{a}^{(1)}, \mathbf{a}^{(2)}, \mathbf{a}^{(3)}\right\}$ *be a complementary set of length n and size 4. Then*

$$
\begin{aligned}
&\left\{\operatorname{cat}\left(\mathbf{a}^{(0)}, \mathbf{a}^{(1)}, \mathbf{a}^{(2)}, \mathbf{a}^{(3)}\right), \operatorname{cat}\left(\mathbf{a}^{(0)}, -\mathbf{a}^{(1)}, \mathbf{a}^{(2)}, -\mathbf{a}^{(3)}\right),\right. \\
&\left.\operatorname{cat}\left(\mathbf{a}^{(0)}, -\mathbf{a}^{(1)}, -\mathbf{a}^{(2)}, \mathbf{a}^{(3)}\right), \operatorname{cat}\left(\mathbf{a}^{(0)}, \mathbf{a}^{(1)}, -\mathbf{a}^{(2)}, -\mathbf{a}^{(3)}\right)\right\}
\end{aligned}
$$

is a complementary set of length 4n and size 4. □

One can also construct complementary sets by combining complementary sets of smaller sizes. One such result is given in the next theorem.

Theorem 7.29 *Let* $\{\mathbf{a}^{(0)}, \mathbf{a}^{(1)}\}$ *and* $\{\mathbf{b}^{(0)}, \mathbf{b}^{(1)}\}$ *be two complementary pairs of length n. Then*

$$
\left\{\operatorname{cat}\left(\mathbf{a}^{(0)}, \mathbf{b}^{(0)}\right), \operatorname{cat}\left(\mathbf{a}^{(0)}, -\mathbf{b}^{(0)}\right), \operatorname{cat}\left(\mathbf{a}^{(1)}, \mathbf{b}^{(1)}\right), \operatorname{cat}\left(\mathbf{a}^{(1)}, -\mathbf{b}^{(1)}\right)\right\}
$$

is a complementary set of length 2n and size 4. □

The following result deals with combining complementary sets of different lengths.

Theorem 7.30 *Let there exist m complementary sets of lengths* $n_0, n_1, \ldots, n_{m-1}$, *and sizes* $T_0, T_1, \ldots, T_{m-1}$ *with* $T = \text{l.c.m.}(T_0, T_1, \ldots, T_{m-1})$. *Let there also exist a column-orthogonal matrix of size* $S \times T$. *Then there exists a complementary set of length* $n_0 + n_1 + \ldots + n_{m-1}$ *and size* ST. □

Finally, here is another recursive construction.

Theorem 7.31 *Let there exist a complementary pair of length n and a complementary set of length m and size* 2^r. *Then there exists a complementary set of length* $2^r nm$ *and size* 2^r. □

7.5 Polyphase complementary sequences

As we have seen, the vectors belonging to complementary pairs provide a PMEPR of 2. However, if the vectors are restricted to have entries ± 1, their possible lengths are restricted to specific values. By allowing the entries to take values from a larger set $e^{2\pi i \frac{k}{M}}, k = 0, 1, \ldots, M - 1$, and $M > 2$, we may extend the range of available lengths for complementary pairs and the number of vectors with a low PMEPR. Notice that since the coefficients may now have complex values, the definition (7.1) with conjugacy should be used.

I'll start with an example. By Theorem 7.12, the length of a binary complementary pair is even. Let us check the case $n = 3$. Let $\mathbf{a}$ and $\mathbf{b}$ constitute a complementary pair. We may assume

$$\mathbf{a} = \left(e^{2\pi i \varphi_0}, e^{2\pi i (\varphi_0 + \varphi_1)}, e^{2\pi i (\varphi_0 + \varphi_1 + \varphi_2)} \right),$$

and

$$\mathbf{b} = \left(e^{2\pi i \theta_0}, e^{2\pi i (\theta_0 + \theta_1)}, e^{2\pi i (\theta_0 + \theta_1 + \theta_2)} \right).$$

For complementarity we need

$$e^{2\pi i \varphi_1} + e^{2\pi i \varphi_2} + e^{2\pi i \theta_1} + e^{2\pi i \theta_2} = 0,$$

$$e^{2\pi i (\varphi_1 + \varphi_2)} + e^{2\pi i (\theta_1 + \theta_2)} = 0.$$

Solving this system we obtain

$$\varphi_2 = \varphi_1 + \frac{2k+1}{2}, \qquad \theta_1 = \varphi_1 + \frac{2k - 2m + 2j - 1}{4},$$

$$\theta_2 = \varphi_1 + \frac{2k + 2m + 2j + 1}{4}.$$

For $\mathbf{a}$ to be binary it is necessary that $2\varphi_0, 2(\varphi_0 + \varphi_1), 2(\varphi_0 + \varphi_1 + \varphi_2) \in \mathbb{Z}$. However, it is easy to check that $\mathbf{b}$ will then have entries different from -1 or 1.

Example 7.4 Let $\varphi_0 = \theta_0 = \varphi_1 = k = m = j = 0$, and consequently $\varphi_2 = \frac{1}{2}, \theta_1 = -\frac{1}{4}, \theta_2 = \frac{1}{4}$. We obtain

$$\mathbf{a} = (1, 1, -1), \quad \mathbf{b} = (1, -\imath, 1),$$

which is evidently a complementary pair.   □

Analogous techniques for $n = 4$ give, e.g., the following nonbinary examples:

$$M = 4, n = 4, \quad \mathbf{a} = (1, -\imath, 1, \imath), \quad \mathbf{b} = (1, \imath, 1, -\imath),$$

$$M = 6, n = 4, \quad \mathbf{a} = \left(1, e^{-2\pi i \frac{1}{3}}, e^{-2\pi i \frac{1}{6}}, 1 \right), \quad \mathbf{b} = \left(1, e^{2\pi i \frac{1}{6}}, e^{2\pi i \frac{5}{6}}, -1 \right).$$

There exist quadriphase complementary pairs of lengths 5 and 13:

$$M = 4, n = 5, \quad (1, \imath, -\imath, -1, \imath), \quad (1, 1, 1, \imath, -\imath),$$

and $M = 4, n = 13$,

$$(1, 1, 1, \imath, -1, 1, 1, -\imath, 1, -1, 1, -\imath, \imath),$$

$$(1, \imath, -1, -1, -1, \imath, -1, 1, 1, -\imath, -1, 1, -\imath).$$

Now let us pass to recursive constructions of complementary pairs. Many of the constructions for the two-phase case, like negation, reversion and interleaving of sequences, can be generalized directly to the multiphase case. Moreover, the

Rudin–Shapiro recursion, Theorem 7.15, shows the existence of a complementary pair of length $2n$ constructed from a pair of length n.

Example 7.5   Starting from $\mathbf{a} = (1, 1, -1)$, $\mathbf{b} = (1, \iota, 1)$ we obtain the complementary pair

$$(1, 1, -1, 1, \iota, 1), \quad (1, 1, -1, -1, -\iota, -1). \qquad \Box$$

Theorem 7.17 can also be extended to the polyphase case, and the existence of two pairs of lengths n and m correspondingly implies the existence of a pair of length $2mn$. However, the Turyn construction, Theorem 7.18, does not hold for the polyphase case; this follows, e.g., from the nonexistence of quaternary complementary pairs of length 9.

Let

$$\text{cwp}(\mathbf{a}, \mathbf{b}) = (a_0 b_0, a_1 b_1, \ldots, a_{n-1} b_{n-1})$$

be the component-wise product of two n-dimensional vectors $\mathbf{a}$ and $\mathbf{b}$.

Theorem 7.32   *Let $\{\mathbf{a}, \mathbf{b}\}$ be a complementary pair, and*

$$\mathbf{c} = \left(e^{2\pi \iota \frac{c}{M}}, e^{2\pi \iota \frac{c}{M}}, \ldots, e^{2\pi \iota \frac{c}{M}} \right)$$

for some $c \in \{0, 1, \ldots, M - 1\}$. Then

$$\{\text{cwp}(\mathbf{a}, \mathbf{c}), \text{cwp}(\mathbf{b}, \mathbf{c})\}$$

is also a complementary pair.

Proof   The aperiodic correlation function does not change under constant phase shift of all entries in the sequences. $\qquad \Box$

Theorem 7.33   *Let $\{\mathbf{a}, \mathbf{b}\}$ be a complementary pair of length n, and*

$$\mathbf{c} = \left(e^{2\pi \iota \frac{0 \cdot c}{M}}, e^{2\pi \iota \frac{1 \cdot c}{M}}, \ldots, e^{2\pi \iota \frac{(M-1) \cdot c}{M}} \right)$$

for some $c \in \{0, 1, \ldots, M - 1\}$. Then

$$\{\text{cwp}(\mathbf{a}, \mathbf{c}), \text{cwp}(\mathbf{b}, \mathbf{c})\}$$

is also a complementary pair of length n.

Proof   Let $\mathbf{d} = \text{cwp}(\mathbf{a}, \mathbf{c})$. Then

$$\rho_{\mathbf{d}}(j) = \sum_{k=0}^{n-j-1} a_k e^{2\pi \iota \frac{kc}{M}} \cdot a_{k+j}^* e^{-2\pi \iota \frac{(k+j)c}{M}} = e^{-2\pi \iota \frac{jc}{M}} \rho_{\mathbf{a}}(j).$$

$\qquad \Box$

Notice that when $M = 2$ and $c = 1$, we have $\mathbf{c} = (1, -1, 1, -1, \ldots)$, and we obtain the biphase construction with negation of every other entry.

The next construction is based on Boolean functions, and provides a general framework for construction of complementary pairs. Recall that a *Boolean function* $f = f(x_0, \ldots, x_{m-1})$ is a function from $\mathbb{Z}_2^m$, consisting of the binary m-tuples, $(a_0, \ldots, a_{m-1}), a_j \in \{0, 1\}$, to $\mathbb{Z}_2$. Any Boolean function can be uniquely expressed as a linear combination over $\mathbb{Z}_2$ of monomials $1, x_0, x_1, \ldots, x_{m-1}, x_0x_1, x_0x_2, \ldots, x_0x_{m-1}, x_1x_2, \ldots, x_{m-1}x_m, x_0, x_1x_2, \ldots, x_0x_1 \ldots x_{m-1}$. The function, f, can be as well specified by the list, $\mathbf{f}$, of its values when $(x_0, x_1, \ldots, x_{m-1})$ ranges over all its 2^m values in lexicographic order.

Example 7.6   For $m = 3$ we have

$$\mathbf{f} = (f(0,0,0), f(0,0,1), f(0,1,0), f(0,1,1), f(1,0,0), f(1,0,1), f(1,1,0), f(1,1,1)).$$

If $f = x_0x_1 + x_1x_3$, then

$$\mathbf{f} = (0, 0, 0, 1, 0, 0, 1, 0).$$

$\square$

A *generalized Boolean function* is defined as a function from $\mathbb{Z}_2^m$ to Z_{2h}, where $h \geq 1$. It is straightforward to show that any such function can be uniquely expressed as a linear combination over $\mathbb{Z}_{2h}$ of the monomials, where the coefficient of each monomial belongs to $\mathbb{Z}_{2h}$. As above, we specify a sequence $\mathbf{f}$ of length 2^m corresponding to the generalized Boolean function f.

Example 7.7   Let $h = 2$ and $m = 3$. Then for $f = x_0x_1 + 3x_1x_2 + 2 \cdot 1$ we have $\mathbf{f} = (22212232)$.

$\square$

With a slight abuse of notation we will write $f(j_0, j_1, \ldots, j_{m-1}) = f(j)$, where

$$j = \sum_{k=0}^{m-1} j_k = j,$$

i.e., $(j_0, j_1, \ldots, j_{m-1})_2$ is the binary expansion of j.

Let $M = 2h$, and we consider vectors of length $n = 2^m$, with components of the shape $e^{2\pi i \frac{j}{M}}$, $j = 0, 1, \ldots, M - 1$.

Theorem 7.34   *Let*

$$f = h \sum_{k=0}^{m-2} x_{\pi(k)}x_{\pi(k+1)} + \sum_{k=0}^{m-1} c_k x_k + c, \qquad (7.15)$$

$$g = f + h x_{\pi(1)}, \qquad (7.16)$$

*where c_k, c are arbitrary elements of $\mathbb{Z}_{2h}$ and π is any permutation of $\{0, 1, \ldots,$ $m - 1\}$. Then **a** and **b** are defined by*

$$a_j = e^{2\pi i \frac{f(j)}{M}}, \quad b_j = e^{2\pi i \frac{g(j)}{M}}, \quad j = 0, 1, \ldots, 2^m - 1,$$

and constitute a complementary pair.

Proof The case $m = 1$ is easily checked. So assume $m \geq 2$ and fix $u \neq 0$. By definition $\rho_{\mathbf{a}}(u) + \rho_{\mathbf{b}}(u)$ is the sum over r of terms

$$e^{2\pi i (a_r - a_{r+u})} + e^{2\pi i (b_r - b_{r+u})}.$$

For a given integer, r, set $j = r + u$, and let $(r_0, r_1, \ldots, r_{m-1})_2$ and $(j_0, j_1, \ldots, j_{m-1})_2$ be the binary representation of r and j respectively.

Case 1: $j_{\pi(1)} \neq r_{\pi(1)}$. From (7.16), over $\mathbb{Z}_{2h}$ we have

$$f(r) - f(j) - g(r) + g(j) = h(j_{\pi(1)} - r_{\pi(1)}) = h,$$

so

$$\frac{e^{2\pi i (f(r) - f(j))}}{e^{2\pi i (g(r) - g(j))}} = e^{2\pi i h} = -1.$$

Therefore,

$$e^{2\pi i (f(r) - f(j))} + e^{2\pi i (g(r) - g(j))} = 0.$$

Case 2: $j_{\pi(1)} = r_{\pi(1)}$. Since $j \neq r$, we can define v to be the smallest integer for which $r_{\pi(v)} \neq j_{\pi(v)}$. Let r' be the integer whose binary representation $(r_0, r_1, \ldots, 1 - r_{\pi(v-1)}, \ldots, r_{m-1})_2$ differs from that of r only in position $\pi(v - 1)$. Similarly let j' have binary representation $(j_0, j_1, \ldots, 1 - j_{\pi(v-1)}, \ldots, j_{m-1})_2$. By assumption, $r_{\pi(v-1)} = j_{\pi(v-1)}$ and so $j' = r' + u$. We have therefore defined an invertible map from the ordered pair (r, j) to (r', j'), and both pairs contribute to $\rho_{\mathbf{a}}(u) + \rho_{\mathbf{b}}(u)$. Now substitution for r' in (7.15) gives

$$f(r') = f(r) + hr_{\pi(v-2)} + hr_{\pi(v)} + c_{\pi(v-1)} - 2c_{\pi(v-1)}r_{\pi(v-1)},$$

unless $v = 1$, in which case we just delete terms involving $\pi(v - 2)$ here and in what follows. Therefore,

$$\begin{aligned}
f(r) &- f(j) - f(r') + f(j') \\
&= h\left(j_{\pi(v-2)} - r_{\pi(v-2)}\right) + h\left(j_{\pi(v)} - r_{\pi(v)}\right) - 2c_{\pi(v-1)}\left(j_{\pi(v-1)} - r_{\pi(v-1)}\right) \\
&= h,
\end{aligned}$$

by the definition of v. Then (7.16) implies that

$$g(r) - g(j) - g(r') + g(j') = f(r) - f(j) - f(r') + f(j') = h.$$

Arguing as in Case 1, we obtain

$$e^{2\pi\imath(f(r)-f(j))} + e^{2\pi\imath(f(r')-f(j'))} = 0,$$

and

$$e^{2\pi\imath(g(r)-g(j))} + e^{2\pi\imath(g(r')-g(j'))} = 0.$$

Therefore,

$$\left(e^{2\pi\imath(f(r)-f(j))} + e^{2\pi\imath(g(r)-g(j))}\right) + \left(e^{2\pi\imath(f(r')-f(j'))} + e^{2\pi\imath(g(r')-g(j'))}\right) = 0.$$

Combining these cases we see that $\rho_{\mathbf{a}}(u) + \rho_{\mathbf{b}}(u)$ comprises zero contributions (as in Case 1), and contributions which sum to zero in pairs (as in Case 2). $\square$

Corollary 7.35   *There are* $(2h)^{m+1} \cdot \frac{m!}{2}$ *Golay sequences with entries belonging to the set* $\left\{e^{2\pi\imath\frac{j}{2h}}, j = 0, 1, \ldots, 2h-1\right\}$, *of length* $n = 2^m$ *defined by*

$$f(x_0, \ldots, x_{m-1}) = h \sum_{k=0}^{m-2} x_{\pi(k)}x_{\pi(k+1)} + \sum_{k=0}^{m-1} c_k x_k + c,$$

for any $c, c_k \in \mathbb{Z}_{2h}$ *and any permutation* π *of* $\{0, 1, \ldots, m-1\}$.

Proof   Notice that $\sum_{k=0}^{m-2} x_{\pi(k)}x_{\pi(k+1)}$ is invariant under the mapping $\pi \to \pi'$, where $\pi'(k) = \pi(m-k-1)$. Thus, there are $\frac{m!}{2}$ inequivalent permutations. Moreover, there are $(2h)^{m+1}$ choices of $c, c_0, \ldots, c_{m-1}$. $\square$

Actually, the above construction of complementary pairs is a particular case of a construction of complementary sets based on Boolean functions. We will need the following definitions.

Let Q be the generalized Boolean function acting from $\mathbb{Z}_2^m$ to $\mathbb{Z}_{2h}$, defined by

$$Q(x_0, x_1, \ldots, x_{m-1}) = \sum_{0 \le r < j < m} q_{rj} x_r x_j,$$

where $q_{rj} \in \mathbb{Z}_{2h}$. We associate a labeled graph $G(Q)$ on m vertices with the quadratic form Q as follows. We label the vertices of $G(Q)$ by $0, 1, \ldots, m-1$, and join vertices r and j by an edge labeled q_{rj} if $q_{rj} \ne 0$. In the case $q = 2$, every edge is labeled 1, and thus by convention we will omit edge labels in this case. Of course, from any graph, G, of this type we can recover a quadratic form Q. If f is a generalized Boolean function of degree 2 we define $G(f)$ to be the graph $G(Q)$ where Q is the quadratic part of f. We say that a graph G of the type defined above is a *path* if either

- $m = 1$ (in which case the graph contains a single vertex and no edges), or
- $m \ge 2$ and G has exactly $m - 1$ edges, all labeled h, which form a Hamiltonian path (a path passing through all the vertices exactly once) in G.

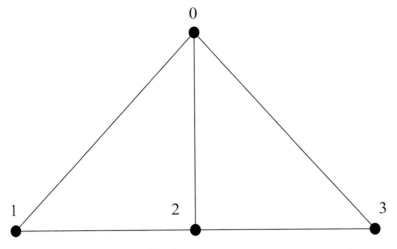

Figure 7.2 Graph $G(Q)$ for $Q = x_0 x_1 + x_0 x_2 + x_0 x_3 + x_1 x_2 + x_2 x_3$

For $m \geq 2$, a path on m vertices corresponds to a quadratic form of the type

$$h \sum_{k=0}^{m-1} x_{\pi(k)} x_{\pi(k+1)}, \tag{7.17}$$

where π is a permutation of $\{0, 1, \ldots, m - 1\}$. I present the following result without proof.

Theorem 7.36 *Let* $Q : \mathbb{Z}_2^m \to \mathbb{Z}_{2h}$ *be a quadratic form in variables* $x_0, x_1,$ $\ldots, x_{m-1}$. *Let* $G(Q)$ *contain a set of* ℓ *distinct vertices labeled* $j_0, j_1, \ldots, j_{\ell-1}$, *with the property that deleting those* ℓ *vertices and all incident edges results in a path. Let* s *be the label of either end vertex in this path (or the single vertex of the graph when* $\ell = m - 1$*). Then for any choice of* c', $c_k \in \mathbb{Z}_{2h}$, *the set of sequence corresponding to*

$$f = Q + \sum_{k=0}^{m-1} c_k x_k + c' + h \left(\sum_{r=0}^{\ell-1} d_r x_{j_r} + d x_s \right),$$

for all choices of d', $d_r \in \{0, 1\}$, *constitutes a complementary set of length* 2^m *and size* $2^{\ell+1}$. $\quad\square$

When $\ell = 0$, we obtain the previous construction of complementary pairs.

Example 7.8 Let $2h = 2$, $m = 4$, and

$$Q = Q(x_0, x_1, x_2, x_3) = x_0 x_1 + x_0 x_2 + x_0 x_3 + x_1 x_2 + x_2 x_3.$$

The graph $G(Q)$ is shown in Fig. 7.2. Deleting the vertex labeled 0 results in a path graph on vertices 1, 2, and 3. Applying Theorem 7.36 with $\ell = 1$, we get for each choice of $c', c_0, c_1, c_2, c_3 \in \mathbb{Z}_2$, the complementary set of size 4, defined by the following four Boolean functions:

$$Q + \sum_{k=0}^{3} c_k x_k + c',$$
$$Q + \sum_{k=0}^{3} c_k x_k + c' + x_0,$$
$$Q + \sum_{k=0}^{3} c_k x_k + c' + x_1,$$
$$Q + \sum_{k=0}^{3} c_k x_k + c' + x_0 + x_1.$$

Recall that each of the functions defines a sequence of length 16 with PMEPR at most 4. □

So far, the considered methods for polyphase sequences allowed construction of sequences with a guaranteed PMEPR of at most 2, and the passage from biphase to multiphase setting just resulted in the increased number of such sequences. However, the lower bound of (4.5) in Lemma 4.1 restricts the PMEPR to be at least 1. The question is whether by increasing the number of allowed phases we can decrease PMEPR below 2. In the limiting case of unrestricted phase modulation, the answer is given in the following theorem.

Theorem 7.37 *There exist MC signals*

$$P_{\mathbf{a}}(t) = \sum_{k=0}^{n-1} a_k e^{2\pi \imath k t},$$

with $|a_k| = 1$ for $k = 0, 1, \ldots, n - 1$, such that

$$(1 - \varepsilon_n)\sqrt{n} \leq |P_{\mathbf{a}}(t)| \leq (1 + \varepsilon_n)\sqrt{n},$$

and $\lim_{n \to \infty} \varepsilon_n = 0$. □

This result guarantees an existence of signals with PMEPR approaching 1 when n grows.

7.6 Trace codes

In the previous section, a construction of sequences with low PMEPR was derived from evaluations of Boolean functions. The next step is to use functions in one variable over finite fields or Galois rings. In this case we obtain, e.g., a polynomial in n number of signals with PMEPR of order $(\log n)^2$. Just to compare, the construction from Boolean functions gives only $n(\log_2 n)!$ BPSK signals, however, this is with a constant PMEPR. Moreover, the constructed codes consisting of code words with a low PMEPR have a high minimum Hamming or Lee distance, and thus can be employed for error correcting. For definitions related to these codes see Section 3.4.5.

We start by considering code words of duals of primitive BCH codes. These codes are defined as follows. Let $\mathbb{F}_q$ be the finite field of size $q = 2^m$, with a primitive element α, and Tr be the trace function from $\mathbb{F}_q$ to $\mathbb{F}_2$. For a $t \geq 1$, consider all polynomials in one variable,

$$f(x) = f_1 x + f_3 x^3 + \ldots + f_{2t-1} x^{2t-1}, \quad f_k \in \mathbb{F}_q, k = 1, 3, \ldots 2t-1.$$

The trace evaluation of f is the vector $\mathbf{c}_f$ of length $n = 2^m - 1$ defined by

$$(\mathbf{c}_f)_k = (-1)^{\mathrm{Tr}(f(\alpha^k))}, \quad k = 0, 1, \ldots, 2^m - 2.$$

Under the condition $2(t-1) < 2^{\lceil \frac{m}{2} \rceil}$, the collection of q^t possible functions define q^t distinct vectors – code words of the dual of t-error correcting BCH code, $\mathcal{BCH}^\perp(t, m)$.

Theorem 7.38 *Any nonconstant code word* $\mathbf{c}$ *of* $\mathcal{BCH}^\perp(t, m)$, $m \geq 3$, *satisfies*

$$\mathrm{PMEPR}(\mathbf{c}) \leq (2t-1)^2 \left(\frac{2 \ln 2}{\pi} \cdot m + 2 \right)^2.$$

Proof A nonconstant word $\mathbf{c}_f$ of $\mathcal{BCH}^\perp(t, m)$ is obtained from a nonzero, nondegenerate polynomial $f(x) = \sum_{k=1}^{t} f_{2k-1} x^{2k-1}$, and we are interested in bounding

$$\max_{t \in [0,1)} \left| F_{\mathbf{c}_f}(t) \right| = \max_{t \in [0,1)} \left| \sum_{k=0}^{n-1} (\mathbf{c}_f)_k \, e^{2\pi \imath k t} \right|.$$

Consider the Nyquist samples of the considered function for $t = \frac{j}{n}$, $j = 0, 1, \ldots, n-1$,

$$\left| F_{\mathbf{c}_f} \left(\frac{j}{n} \right) \right| = \left| \sum_{k=0}^{n-1} (-1)^{\mathrm{Tr}(f(\alpha^k))} e^{2\pi \imath k \frac{j}{n}} \right| = \left| \sum_{\beta \in \mathbb{F}_q^*} \psi(f(\beta)) \chi_j(\beta) \right|,$$

where ψ and χ are correspondingly additive and multiplicative characters of $\mathbb{F}_q$.

For $j = 0$, ξ is the trivial character, and the above expression reduces to

$$\left| \sum_{\beta \in \mathbb{F}_q^*} \psi(f(\beta)) \right|.$$

By the Weil–Carlitz–Uchiyama bound (Theorem 3.36) this can be bounded above by $(2t-2)2^{\frac{m}{2}} + 1$.

For $j \neq 0$, χ_j is a nontrivial multiplicative character, and by Theorem 3.37 with $r = 2t - 1$ and $s = 1$, this yields

$$\left| F_{\mathbf{c}_f} \left(\frac{j}{n} \right) \right| \leq (2t-1)2^{\frac{m}{2}}.$$

In the notations of Chapter 4, this yields

$$M_d\left(F_{\mathbf{c}_f}\right) \le (2t-1)2^{\frac{m}{2}}.$$

Using Theorem 4.2 with a slightly relaxed additive constant, we obtain the claim.

$\square$

Now we consider signals corresponding to M-PSK modulation, with $M \ge 2$. To construct such signals, we employ algebraic codes over Galois rings. Let $M = 2^e$, and $f \in R_{e,m}[x]$ be a polynomial. Let β be a generator for the cyclic subgroup of $\mathcal{T}_{e,m}^*$. With f we associate a length $n = 2^m - 1$ vector $\mathbf{c}_f$ whose components are

$$(\mathbf{c}_f)_k = e^{2\pi i \frac{\mathrm{Tr}(f(\beta^k))}{2^e}}.$$

Setting $e = 2$ (and thus QPSK modulation) we consider the following codes.

Kerdock codes over $\mathbb{Z}_4$:

$$\mathcal{K} = \{\mathbf{c}_f : f(x) = b_0 x, \quad b_0 \in R_{2,m}\},$$

Delsarte–Goethals codes over $\mathbb{Z}_4$:

$$\mathcal{DG}_t = \left\{\mathbf{c}_f : f(x) = b_0 x + 2 \sum_{j=1}^{t} b_j x^{1+2^j}, \quad b_0 \in R_{2,m}, b_j \in \mathcal{T}_{2,m}\right\}.$$

Clearly, the quaternary Kerdock codes correspond to the case $t = 0$ of the Delsarte–Goethals codes. The length and the number of words in the Delsarte–Goethals codes are $2^m - 1$ and $2^{(2+t)m}$, respectively.

Theorem 7.39 *Any nonconstant code word $\mathbf{c}$ of $\mathcal{DG}_t$ satisfies*

$$\mathrm{PMEPR}(\mathbf{c}) \le (2^t + 1)^2 \left(\frac{2\ln 2}{\pi} m + 2\right)^2.$$

Proof Let, for $t \ge 0$,

$$f(x) = b_0 x + 2 \sum_{j=1}^{t} b_j x^{1+2^j}, \quad b_0 \in R_{2,m}, b_j \in \mathcal{T}_{2,m}.$$

Suppose further that there is at least one nonzero b_j. Then f is a nondegenerate polynomial of weighted degree $2^t + 1$ which yields a nonconstant code word of $\mathcal{DG}_t$.

For $\ell = 0, 1, \ldots, n-1$, we have

$$\left|F_{\mathbf{c}_f}\left(\frac{\ell}{n}\right)\right| = \left|\sum_{k=0}^{n-1} e^{2\pi i \frac{\mathrm{Tr}(f(\beta^k))}{2^e}} e^{2\pi i \ell \frac{k}{n}}\right| = \left|\sum_{x \in \mathcal{T}_{2,m}^*} \psi(f(x)) \chi_\ell(x)\right|,$$

where ψ and χ are, respectively, additive and multiplicative characters for $R_{2,m}$. For $\ell = 0$, the expression above reduces to

$$\left| \sum_{x \in \mathcal{T}_{2,m}^*} \psi(f(x)) \right| ,$$

which can be bounded above by $2^t \cdot 2^{\frac{m}{2}} + 1$, using Theorem 3.38. For $\ell \neq 0$, χ_ℓ is a nontrivial multiplicative character and Theorem 3.39 yields

$$\left| F_{\mathbf{c}_f} \left(\frac{\ell}{n} \right) \right| \leq (2^t + 1) \cdot 2^{\frac{m}{2}}, \quad \ell = 1, 2, \ldots, n - 1.$$

Arguing further as in the case of the dual BCH codes yields the claim. $\square$

The final considered family of codes is that of weighted degree trace codes. For $t \geq 1$ satisfying $2t - 1 < 2^{\lceil \frac{m}{2} \rceil} + 1$, the code is defined by

$$\mathcal{WD}_t = \left\{ \mathbf{c}_f : f \in R_{e,m}[x], \quad f = \sum_{j=0}^{t-1} f_{2j+1} x^{2j+1}, \quad D_f \leq 2t - 1 \right\}.$$

The code $\mathcal{WD}_t$ has length $n = 2^m - 1$, and for $e = 1$ coincides with $\mathcal{BCH}_t^\perp$. It can be shown, using 2-adic expansions and simple counting, that when $e = 2$,

$$|\mathcal{WD}_t| = 2^{(2t-1-\lceil \frac{2t-1}{4} \rceil)m}.$$

Theorem 7.40 *Any nonconstant code word* $\mathbf{c}$ *of* $\mathcal{WD}_t$ *satisfies*

$$\mathrm{PMEPR}(\mathbf{c}) \leq (2t - 1)^2 \left(\frac{2 \ln 2}{\pi} m + 2 \right)^2 .$$

Proof This proof is identical to the proof of the previous theorem. $\square$

The presented codes have length $2^m - 1$. It could be beneficial to consider codes whose length is a power of 2. This can be achieved by lengthening the described codes. It is done by adding an extra coordinate corresponding to $f(0)$ (an overall parity check) followed by adding modulo M to every code word multiple of the all-1 code word. The described operation increases the length of the code by 1, and the number of code words becomes M times the one of the initial code. Since the operation converts a trigonometric polynomial $F_f(t)$ into a polynomial $\hat{F}(t) = c + e^{2\pi i t} F_f(t)$ where c is -1 or 1,

$$\max_{t \in [0,1)} |\hat{F}_f(t)| \leq 1 + \max_{t \in [0,1)} |F_f(t)|.$$

Therefore, for the extended codes of those described in this section we have to substitute the additive term 2 by 3 in all the upper bounds for PMEPR.

7.7 *M*-sequences

Let $\mathbb{F}_q$, $q = 2^m$, be the finite field with a primitive element α. For a $\beta \in \mathbb{F}_q^*$, an M-sequence $\mathbf{a}(\beta) = (a_0(\beta), \ldots, a_{n-1}(\beta))$, of length $n = q - 1$, is defined by

$$a_k(\beta) = (-1)^{\text{Tr}(\beta\alpha^k)}, \qquad k = 0, \ldots, n - 1.$$

All M-sequences of length n can be obtained from the initial one, $\mathbf{a}(1)$, by cyclic shifts. The set of M-sequences constitutes a particular case of the duals of BCH codes, $\mathcal{BCH}_1^\perp$.

The MC signal corresponding to the M-sequence $\mathbf{a}(\beta)$ is

$$F_\beta(t) = \sum_{k=0}^{n-1} a_k(\beta) e^{2\pi \imath k t},$$

and the PMEPR of an M-sequence, $P_q(\beta)$ is

$$P_q(\beta) = \max_{t \in [0,1)} \frac{\left| \sum_{k=0}^{n-1} a_k(\beta) e^{2\pi \imath k t} \right|^2}{\sum_{k=0}^{n-1} a_k^2(\beta)} = \frac{1}{n} \max_{t \in [0,1)} \left| F_\beta(t) \right|^2. \tag{7.18}$$

Let

$$P_q = \min \max_{\beta \in F_q^*} P_q(\beta), \tag{7.19}$$

where the minimum is taken over all possible choices of the primitive element.

The following result about sums of additive characters is a particular case of Theorem 3.35. For completeness, I provide it with a proof.

Lemma 7.41

$$\sum_{\gamma \in \mathbb{F}_q} (-1)^{\text{Tr}(\gamma)} = 0.$$

Proof Clearly, there exists an element of $\mathbb{F}_q$, say η, such that $\text{Tr}(\eta) = 1$. Then

$$(-1)^{\text{Tr}(\eta)} \sum_{\gamma \in \mathbb{F}_q} (-1)^{\text{Tr}(\gamma)} = \sum_{\gamma \in \mathbb{F}_q} (-1)^{\text{Tr}(\gamma + \eta)} = \sum_{\gamma \in \mathbb{F}_q} (-1)^{\text{Tr}(\gamma)}.$$

In the last equality we took into account that when γ goes over $\mathbb{F}_q$ the sum $\gamma + \eta$ also passes through all the field elements of $\mathbb{F}_q$. Therefore,

$$(1 - (-1)^{\text{Tr}(\eta)}) \cdot \sum_{\gamma \in \mathbb{F}_q} (-1)^{\text{Tr}(\gamma)} = 2 \cdot \sum_{\gamma \in \mathbb{F}_q} (-1)^{\text{Tr}(\gamma)} = 0,$$

and the claim follows. $\qquad\square$

Theorem 7.42 *For $j = 0, \ldots, n - 1$,*

$$\left| F_\beta \left(\frac{j}{n} \right) \right| = \sqrt{q}.$$

Proof

$$\left| F_\beta \left(\frac{j}{n} \right) \right|^2 = \left| \sum_{k=0}^{n-1} (-1)^{\mathrm{Tr}(\beta\alpha^k)} e^{2\pi i k \frac{j}{n}} \right|^2$$

$$= \sum_{k_1=0}^{n-1} \sum_{k_2=0}^{n-1} (-1)^{\mathrm{Tr}(\beta(\alpha^{k_1}+\alpha^{k_2}))} e^{2\pi i (k_2-k_1)}$$

$$= \sum_{k=0}^{n-1} \sum_{d=0}^{n-1} (-1)^{\mathrm{Tr}(\beta(\alpha^k(1+\alpha^d)))} e^{2\pi i d}$$

$$= \sum_{d=0}^{n-1} e^{2\pi i d} \sum_{k=0}^{n-1} (-1)^{\mathrm{Tr}(\beta\alpha^k(1+\alpha^d))}$$

$$= \sum_{d=0}^{n-1} e^{2\pi i d} \left(\sum_{\gamma \in F_q} (-1)^{\mathrm{Tr}(\gamma(1+\alpha^d))} + 1 \right)$$

$$= \sum_{d=0}^{n-1} e^{2\pi i d} \sum_{\gamma \in F_q} (-1)^{\mathrm{Tr}(\gamma(1+\alpha^d))} + \sum_{d=0}^{n-1} e^{2\pi i d}$$

$$= \sum_{d=0}^{n-1} e^{2\pi i d} \sum_{\gamma \in F_q} (-1)^{\mathrm{Tr}(\gamma(1+\alpha^d))},$$

where in the last equality we used (3.1). By Lemma 7.41 the inner sum is zero if $1 + \alpha^d \neq 0$, and is q otherwise. Since $\alpha^{q-1} = 1$, we obtain

$$\left| F_\beta \left(\frac{j}{n} \right) \right|^2 = e^{2\pi i (q-1)} q = q,$$

and the claim follows.   □

Theorem 7.43   *For $q = 2^m$,*

$$P_q \leq \left(\frac{2 \ln 2}{\pi} \cdot m + 2 \right)^2.$$

Proof   This is analogous to Theorem 7.38.   □

The theorem shows that PMEPR of an M-sequence is at most of order $(\ln q)^2$. However, it does not exclude that it could be constant. In what follows, I will show that indeed P_q is growing with q, namely, that there exists q_0 such that for all $q > q_0$,

$$P_q \geq \frac{1}{2\pi^2} (\ln \ln q)^2. \tag{7.20}$$

Theorem 7.44 *For at least one $\beta \in \mathbb{F}_q^*$,*

$$\max_{t \in [0,1)} \left| F_\beta(t) \right| \geq c \sqrt{q} \ln \ln q$$

where c is a constant independent of q.

Proof We have, see (4.6),

$$F_\beta(t) = \frac{1}{n} \sum_{j=0}^{n-1} F_\beta \left(\frac{j}{n} \right) D_n \left(t - \frac{j}{n} \right), \qquad (7.21)$$

where

$$D_n(t) = \sum_{k=0}^{n-1} e^{2\pi \imath kt} \qquad (7.22)$$

is the Dirichlet kernel. By (4.8),

$$|D_n(t)| = \left| \frac{\sin \pi nt}{\sin \pi t} \right|.$$

Let $\beta = \alpha^s$. Since

$$\sum_{k=0}^{n-1} (-1)^{\mathrm{Tr}(\alpha^{k+s})} e^{2\pi \imath \frac{kj}{n}} = \sum_{h=0}^{n-1} (-1)^{\mathrm{Tr}(\alpha^h)} e^{2\pi \imath \frac{(h-s)j}{n}}$$

$$= \sum_{h=0}^{n-1} (-1)^{\mathrm{Tr}(\alpha^h)} e^{-2\pi \imath \frac{js}{n}} e^{2\pi \imath \frac{hj}{n}},$$

we have

$$F_{\alpha^s} \left(\frac{j}{n} \right) = e^{-2\pi \mathrm{i} \frac{sj}{n}} F_1 \left(\frac{j}{n} \right). \qquad (7.23)$$

In what follows, I omit the lower index in F_1, i.e., $F(x) = F_1(x)$.

Let $\delta(s)$ be a function to be defined later. From (7.21) and (7.23) we have

$$\sum_{s=0}^{n-1} \delta(s) F_{\alpha^s}(t) = \frac{1}{n} \sum_{j=0}^{n-1} \sum_{s=0}^{n-1} \delta(s) F_{\alpha^s} \left(\frac{j}{n} \right) \cdot D_n \left(t - \frac{j}{n} \right)$$

$$= \frac{1}{n} \sum_{j=0}^{n-1} F \left(\frac{j}{n} \right) \cdot D_n \left(t - \frac{j}{n} \right) \cdot \sum_{s=0}^{n-1} \delta(s) e^{-2\pi \imath \frac{js}{n}}. \qquad (7.24)$$

Denoting

$$f(j) = \sum_{s=0}^{n-1} \delta(s) e^{-2\pi \imath \frac{js}{n}}, \qquad (7.25)$$

we have

$$\sum_{s=0}^{n-1} \delta(s) F_{\alpha^s}(t) = \frac{1}{n} \sum_{j=0}^{n-1} F\left(\frac{j}{n}\right) \cdot D_n\left(t - \frac{j}{n}\right) \cdot f(j). \tag{7.26}$$

For some integers x_0 and H, to be chosen later, we define $\delta(s)$ in (7.25) in such a way that

$$f(h) = \begin{cases} 1 - \frac{h - x_0}{H} & h = x_0, x_0 + 1, \dots, x_0 + H; \\ 1 - \frac{x_0 - h}{H} & h = x_0, x_0 - 1, \dots, x_0 - H; \\ 0 & \text{otherwise.} \end{cases}$$

Then (7.26) can be rewritten as

$$\sum_{s=0}^{n-1} \delta(s) F_{\alpha^s}(t) = \frac{1}{n} \sum_{h=x_0-H}^{x_0+H} F\left(\frac{h}{n}\right) \cdot D_n\left(t - \frac{h}{n}\right) \cdot f(h). \tag{7.27}$$

For the described choice of $f(h)$ we can calculate $\delta(s)$, namely,

$$\begin{aligned}
\delta(s) &= \frac{1}{n} \sum_{h=x_0-H}^{x_0+H} f(h) e^{2\pi \imath \frac{sh}{n}} \\
&= \frac{1}{n} \left(\sum_{h=x_0}^{x_0+H} \left(1 - \frac{h - x_0}{H}\right) e^{2\pi \imath \frac{sh}{n}} + \sum_{h=x_0-h}^{x_0-1} \left(1 - \frac{x_0 - h}{H}\right) e^{2\pi \imath \frac{sh}{n}} \right) \\
&= e^{2\pi \imath \frac{sx_0}{n}} \cdot \frac{1}{n} \sum_{|h| \le H} \left(1 - \frac{|h|}{H}\right) e^{2\pi \imath \frac{sh}{n}} \\
&= e^{2\pi \imath \frac{sx_0}{n}} \cdot \frac{1}{n} K_H\left(\frac{sh}{n}\right), \tag{7.28}
\end{aligned}$$

where $K_H(t)$ is the Fejér kernel. By (4.64), $K_H(t) \ge 0$ for all t.

From (7.27) and (7.28) we have

$$\begin{aligned}
\left| \sum_{s=0}^{n-1} \delta(s) F_{\alpha^s}(t) \right| &\le \max_s |F_{\alpha^s}(t)| \sum_{s=0}^{n-1} |\delta(s)| \\
&\le \max_s |F_{\alpha^s}(t)| \cdot \frac{1}{n} \sum_{s=0}^{n-1} K_H\left(\frac{sh}{n}\right) \\
&= \max_s \left| F_{\alpha_s}(t) \right|. \tag{7.29}
\end{aligned}$$

To see that the last equality is valid, notice that

$$\sum_{s=0}^{n-1} K_H\left(\frac{sh}{n}\right) = \sum_{|h| \le H} \left(1 - \frac{|h|}{H}\right) \cdot \sum_{s=0}^{n-1} e^{2\pi \imath \frac{sh}{n}} = n,$$

since the last sum is nonzero only if $h = 0$.

Then, by (7.27) and (7.29), we obtain

$$\frac{1}{n} \left| \sum_{h=x_0-H}^{x_0+H} F\left(\frac{h}{n}\right) \cdot D_n\left(t - \frac{h}{n}\right) f(h) \right| \leq \max_s |F_{\alpha^s}(t)| . \qquad (7.30)$$

Let us address now the left-hand side of (7.30). We know that for $h = 0, 1, \ldots, n-1$,

$$\left| F\left(\frac{h}{n}\right) \right| = \sqrt{q}.$$

Thus

$$\sum_{h=x_0-H}^{x_0+H} F\left(\frac{h}{n}\right) D_n\left(t - \frac{h}{n}\right) f(h)$$

$$= \sqrt{q} \sum_{h=x_0-H}^{x_0+H} e^{2\pi \iota \varphi(h)} \left| D_n\left(t - \frac{h}{n}\right) \right| e^{2\pi \iota \psi(h)} f(h),$$

where

$$2\pi \varphi(h) = \arg F_n\left(\frac{h}{n}\right), \quad 2\pi \psi(h) = \arg D_n\left(t - \frac{h}{n}\right).$$

Then, by (7.30), we have

$$\max_s |F_{\alpha^s}(t)| \geq \frac{\sqrt{q}}{n} \left| \sum_{|h| \leq H} \left(1 - \frac{|h|}{H}\right) \cdot \left| D_n\left(t - \frac{|x_0 + h|}{n}\right) \right| \cdot e^{2\pi \iota (\varphi(x_0+h)+\psi(x_0+h))} \right|$$

$$\geq \frac{\sqrt{q}}{n} \cdot \sum_{|h| \leq H} \left(1 - \frac{|h|}{H}\right) \Re\left(e^{2\pi \iota (\varphi(x_0+h)+\psi(x_0+h))}\right) \cdot \left| D_n\left(t - \frac{|x_0 + h|}{n}\right) \right|.$$

$$(7.31)$$

Now assume that for given x_0 and all $|h| \leq H$ we have

$$|\varphi(x_0 + h) + \psi(x_0 + h)| \leq \frac{1}{8}. \qquad (7.32)$$

The existence of such x_0 and estimates on H will be provided in what follows. Then, for all such h, we obtain

$$\Re\left(e^{2\pi \iota (\varphi(x_0+h)+\psi(x_0+h))}\right) \geq \frac{1}{\sqrt{2}},$$

and by (7.31) we have

$$\max_s |F_{\alpha^s}(t)| \geq \frac{1}{\sqrt{2}n} \sum_{|h| \leq H} \left(1 - \frac{|h|}{H}\right) \left| D_n\left(t - \frac{x_0 + h}{n}\right) \right|.$$

Choosing, in the last inequality,

$$t = \frac{x_0 + \frac{1}{2}}{n},$$

we have

$$\max_s |F_{\alpha^s}(t)| \geq \frac{1}{\sqrt{2n}} \sum_{|h| \leq H} \left(1 - \frac{|h|}{H}\right) \cdot \frac{1}{\left|\sin \pi \frac{h + \frac{1}{2}}{n}\right|}$$

$$\geq \sqrt{\frac{n}{2\pi^2}} \sum_{|h| \leq H} \left(1 - \frac{|h|}{H}\right) \frac{1}{|h + \frac{1}{2}|}$$

$$\geq \frac{\sqrt{2n}}{\pi} \ln H. \tag{7.33}$$

Next, it will be shown that H can be chosen in such a way that it satisfies (7.32) and the right-hand side of (7.33) is at least $c\sqrt{n} \ln \ln n$ for an appropriate positive constant c. Let

$$\mathbf{e}(x_0) = (\varphi(x_0), \varphi((x_0 + 1) \bmod n), \ldots, \varphi((x_0 + H - 1) \bmod n)).$$

For all $x_0 = 0, 1, \ldots, n - 1$, we have $\mathbf{e}(x_0) \in T^H$ where $T = [0, 1)$. Our goal is to prove an equidistribution of $\mathbf{e}(x_0)$ on the torus T^H. For this, we need the following result, which is based on Theorem 9.3 from [195] along with a quantitative estimate for the uniformity of the distribution.

Theorem 7.45 *If $H \leq 0.739\sqrt{\ln q}$ then, for every $\mathbf{y} = (y_0, \ldots, y_{H-1}) \in T^H$, there exists $x_0 \in \{0, \ldots, n - 1\}$ such that*

$$\max_{0 \leq h \leq H-1} |y_h + \varphi((x_0 + h) \bmod n)| \leq \frac{1}{8}.$$

Proof Let, in what follows, $\theta = \frac{1}{8}$. For a $\mathbf{z} = (z_0, \ldots, z_{H-1}) \in T^H$ define

$$\gamma(z_0, \ldots, z_{H-1}) = \begin{cases} 1 & \text{if } |z_j| \leq \frac{\theta}{2} \quad \text{for all } j = 0, 1, \ldots, H - 1, \\ 0 & \text{otherwise.} \end{cases}$$

Let $g(\mathbf{z}) = \gamma * \gamma(\mathbf{z})$ be the convolution of γ with itself,

$$g(\mathbf{z}) = \int_{T^H} \gamma(\mathbf{x}) \gamma(\mathbf{x} + \mathbf{z}) \, d\mathbf{x}.$$

Clearly, if $g(\mathbf{z}) \neq 0$, there exists an $\mathbf{x} \in T^H$ such that both $\mathbf{x}$ and $\mathbf{z} + \mathbf{x}$ belong to the cube with side θ and centered in the origin. This yields that $\mathbf{z}$ is in the cube of side 2θ centered in the origin.

It is known, see, e.g., [457], that $g(\mathbf{z})$ has the following Fourier expansion

$$g(\mathbf{z}) = \sum_{\mathbf{m} \in \mathbf{Z}^H} |c(\mathbf{m})|^2 \cdot e^{2\pi \iota (\mathbf{m}, \mathbf{z})},$$

where $(\mathbf{m}, \mathbf{z}) = m_0 z_0 + m_1 z_1 + \ldots + m_{H-1} z_{H-1}$,

$$c(\mathbf{m}) = \prod_{j=0}^{H-1} \frac{\sin \pi m_j \theta}{\pi m_j}$$

and for $m = 0$,

$$\frac{\sin \pi m \theta}{\pi m} = \theta.$$

The series is absolutely converging, thus

$$\sum_{x_0=0}^{n-1} g(\mathbf{e}(x_0) + \mathbf{y})$$

$$= \sum_{\mathbf{m} \in \mathbf{Z}^H} |c(\mathbf{m})|^2 \cdot e^{2\pi \iota (\mathbf{m}, +\mathbf{y})} \cdot \sum_{x_0=0}^{n-1} e^{2\pi \iota (m_0 \varphi(x_0) + m_1 \varphi(x_0+1) + \ldots + m_{H-1} \varphi(x_0+H-1))}$$

$$= \sum_{\mathbf{m} \in \mathbf{Z}^H} |c(\mathbf{m})|^2 \cdot e^{2\pi \iota (\mathbf{m}, \mathbf{y})} \cdot \sum_{x_0=0}^{n-1} \prod_{h=0}^{H-1} e^{2\pi \iota m_h \varphi(x_0+h)}$$

$$= \sum_{\mathbf{m} \in \mathbf{Z}^H} |c(\mathbf{m})|^2 \cdot e^{2\pi \iota (\mathbf{m}, \mathbf{y})} \cdot S(\mathbf{m}).$$

Recalling that

$$e^{2\pi \iota \varphi(x_0+h)} = \frac{F\left(\frac{x_0+h}{n}\right)}{\sqrt{q}}$$

we have

$$S(\mathbf{m}) = \sum_{x_0=0}^{n-1} \prod_{h=0}^{H-1} \left(\frac{F\left(\frac{x_0+h}{n}\right)}{\sqrt{q}} \right)^{m_h}.$$

It is proved in [195, Theorem 9.6] that for $\mathbf{m} \neq 0$,

$$|S(\mathbf{m})| \leq \frac{\sum_{j=0}^{H-1} |m_j|}{\sqrt{q}} + \frac{2H}{n}. \tag{7.34}$$

Extracting the term corresponding to $\mathbf{m} = 0$ and using the last estimate, we obtain

$$\left| \sum_{x_0=0}^{n-1} g(\mathbf{e}(x_0) + \mathbf{y}) - n |c(0)|^2 \right| \leq \sum_{\mathbf{m} \neq 0} |c(\mathbf{m})|^2 \cdot |S(\mathbf{m})|.$$

We partition the right-hand side sum into two sums, corresponding to the cases when $\mathbf{m}$ satisfies $\max_j |m_j| \leq \mu$ and $\max_j |m_j| > \mu$, μ to be chosen later. Further, we will obtain an upper estimate on the first sum $|S(\mathbf{m})|$ using (7.34), while in the second sum we will use for $|S(\mathbf{m})|$ the trivial upper bound equal to the number of summands. We have

$$
\left| \sum_{x_0=0}^{n-1} g(\mathbf{e}(x_0) + \mathbf{y}) - n|c(0)|^2 \right|
$$

$$
\leq \sum_{\substack{\mathbf{m} \neq \mathbf{0}, \max_j |m_j| \leq \mu}} |c(\mathbf{m})|^2 \cdot |S(\mathbf{m})| + \sum_{\max_j |m_j| > \mu} |c(\mathbf{m})|^2 \cdot |S(\mathbf{m})|
$$

$$
\leq \sum_{\substack{\mathbf{m} \neq \mathbf{0}, \max_j |m_j| \leq \mu}} |c(\mathbf{m})|^2 \cdot \left(\frac{\sum_{j=0}^{H-1} |m_j|}{\sqrt{q}} + \frac{2H}{n} \right) + n \sum_{\max_j |m_j| > \mu} |c(\mathbf{m})|^2
$$

$$
= \Sigma_1 + \Sigma_2.
$$

Since

$$
\left| \frac{\sin \pi \theta m_j}{\pi m_j} \right| \leq \theta,
$$

then

$$
|c(\mathbf{m})|^2 \leq \theta^{2H}
$$

and

$$
\Sigma_1 \leq \frac{\theta^{2H}}{\sqrt{q}} \cdot \sum_{\max_j |m_j| \leq \mu} \sum_{j=0}^{H-1} |m_j| + \frac{\theta^{2H}}{n} \cdot 2H \cdot (2\mu + 1)^H
$$

$$
\leq \frac{\theta^{2H}}{\sqrt{q}} \cdot \frac{(2\mu + 1)^{H-1}(\mu + 1)}{4} + \frac{\theta^{2H}}{n} \cdot 2H \cdot (2\mu + 1)^H
$$

$$
\leq \frac{\theta^{2H}}{\sqrt{q}} \cdot (2\mu + 1)^H.
$$

The last inequality is valid for large enough μ and we took into account that H will later be chosen to be much smaller than $\sqrt{q}$.

Furthermore, in Σ_2 we may use

$$
|c(\mathbf{m})|^2 \leq \left(\prod_{j=0}^{H-1} \min \left(\theta, \frac{1}{\pi |m_j|} \right) \right)^2,
$$

and we obtain

$$\Sigma_2 \leq n \sum_{\max_j |m_j| > \mu} \prod_{j=0}^{H-1} \min\left(\theta^2, \frac{1}{\pi^2 |m_j|^2}\right)$$

$$\leq n \cdot 2H \cdot \sum_{m_0 > \mu} \prod_{j=0}^{H-1} \min\left(\theta^2, \frac{1}{\pi^2 |m_j|^2}\right)$$

$$\leq n \cdot 2H \cdot \sum_{m_0 > \mu} \frac{1}{\pi^2 m_0^2} \prod_{j=1}^{H-1} \min\left(\theta^2, \frac{1}{\pi^2 |m_j|^2}\right)$$

$$\leq n \cdot 2H \cdot \frac{1}{\pi^2(\mu - 1)} \cdot \prod_{j=1}^{H-1} \sum_{m_j \in \mathbf{Z}} \min\left(\theta^2, \frac{1}{\pi^2 |m_j|^2}\right)$$

$$\leq \frac{2Hn}{\pi^2} \cdot \frac{1}{\mu - 1} \left(5\theta^2 + \frac{1}{\pi^2}\right)^{H-1}.$$

Finally,

$$\left| \sum_{x_0=0}^{n-1} g(\mathbf{e}(x_0) + \mathbf{y}) - nc^2(\mathbf{0}) \right| \leq \frac{\theta^{2H}}{\sqrt{q}} \cdot (2\mu + 1)^H + \frac{2Hq}{\pi^2(\mu - 1)} \left(5\theta^2 + \frac{1}{\pi^2}\right)^{H-1}.$$

Choosing μ such that the two terms are (almost) equal we set

$$\mu = \left(\frac{5}{2} + \frac{1}{2\theta^2 \pi^2}\right) \cdot q^{\frac{3}{2H+2}}$$

and then, for large enough q, we have

$$\left| \sum_{x_0=0}^{n-1} g(\mathbf{e}(x_0) + \mathbf{y}) - nc^2(0) \right| \leq \left(5\theta^2 + \frac{1}{\pi^2}\right)^H \frac{4H}{\pi^2} \cdot q^{1 - \frac{3}{2H+2}}.$$

Clearly, $c^2(0) = \theta^{2H}$. Thus if

$$n\theta^{2H} > \left(5\theta^2 + \frac{1}{\pi^2}\right)^H \frac{4H}{\pi^2} \cdot q^{1 - \frac{3}{2H+2}} \tag{7.35}$$

then

$$\left| \sum_{x_0=0}^{n-1} g(\mathbf{e}(x_0) + \mathbf{y}) \right| \neq 0$$

and there exists x_0 such that

$$g(\mathbf{e}(x_0) + \mathbf{y}) > 0.$$

For (7.35) to be valid it is enough that

$$H \leq \sqrt{\frac{3}{2 \ln\left(5 + \frac{1}{\pi^2 \theta^2}\right)}} \cdot \sqrt{\ln q} < 0.783 \cdot \sqrt{\ln q},$$

which accomplishes the proof.   □

Finally, choosing in the last theorem $\mathbf{y} = (\psi(0), \ldots, \psi(H - 1))$ and recalling that by varying β we produce cyclic shifts of the M-sequence, we validate (7.32), and may substitute the estimate on H into (7.33). Thus we obtain the following result:

Theorem 7.46 *For q large enough*

$$\max_t \max_s \left| \sum_{k=0}^{q-2} (-1)^{Tr(\alpha^{k+s})} \cdot e^{2\pi \iota k t} \right| \geq \frac{1}{\pi \sqrt{2}} \cdot \sqrt{q} \ln \ln q$$

$\square$

7.8 Legendre sequences

Another example of signals arising from characters is provided by Legendre sequences.

We will need the following facts about *Gaussian sums*. Let $\ell \in \mathbb{N}$, and g.c.d.$(\ell, n) = 1$. Then

$$S(n) = \sum_{k=0}^{n-1} e^{2\pi \iota \frac{\ell k^2}{n}}$$

is called the Gaussian sum. The absolute value of $|S(n)|$ can be calculated easily.

Theorem 7.47

$$|S(n)| = \begin{cases} \sqrt{n} & \text{if } n = 1 \mod 2 \\ \sqrt{2n} & \text{if } n = 0 \mod 4 \\ 0 & \text{if } n = 2 \mod 4. \end{cases}$$

Proof Indeed,

$$|S(n)|^2 = S^*(n)S(n) = \sum_{m=0}^{n-1} e^{-2\pi \iota \frac{\ell m^2}{n}} \sum_{k=0}^{n-1} e^{2\pi \iota \frac{\ell k^2}{n}}.$$

We use Theorem 3.10 and replace k in the internal sum with $k + m$. Then, changing the order of summation, we get

$$|S(n)|^2 = \sum_{k=0}^{n-1} \sum_{m=0}^{n-1} e^{2\pi \iota \frac{\ell(k+m)^2 - \ell m^2}{n}} = \sum_{k=0}^{n-1} e^{2\pi \iota \frac{\ell k^2}{n}} \sum_{m=0}^{n-1} e^{2\pi \iota \frac{2\ell km}{n}}. \tag{7.36}$$

If n is odd, then by Theorem 3.1 the internal sum is nonzero only if $k = 0$, and therefore we obtain

$$|S(n)|^2 = n.$$

For even n there are two nonzero summands in the right-hand side of (7.36), emerging when $k = 0$ or $k = \frac{n}{2}$. Since g.c.d.$(\ell, n) = 1$ we conclude that ℓ is odd, and using Theorem 3.10 we have

$$|S(n)|^2 = n\left(1 + e^{2\pi \iota \frac{\ell n}{4}}\right) = n\left(1 + e^{2\pi \iota \frac{n}{4}}\right)$$
$$= \begin{cases} 2n & \text{if } n \equiv 0 \bmod 4 \\ 0 & \text{if } n \equiv 2 \bmod 4. \end{cases}$$
□

Now let $n = p$, where $p > 2$ is an odd prime number. We construct BPSK modulated sequences using the notion of the *Legendre symbol* $\left(\frac{k}{p}\right)$. It is defined as follows:

$$\left(\frac{k}{p}\right) = \begin{cases} 0 & \text{if } k = 0 \\ 1 & \text{if there is } x \in \mathbb{Z}, \text{ such that } k = x^2 \bmod p \\ -1 & \text{if there is no } x \in \mathbb{Z}, \text{ such that } k = x^2 \bmod p. \end{cases}$$

According to the value of the Legendre symbol nonzero k are called quadratic residues when $\left(\frac{k}{p}\right) = 1$, and quadratic nonresidues otherwise. It is easy to check that there are exactly $\frac{p-1}{2}$ residues and $\frac{p-1}{2}$ nonresidues.

We analyze the following sequence constructed from the Legendre symbols,

$$\mathbf{a} = (a_0, a_1, \ldots, a_{p-1}),$$

with

$$a_0 = 1, \quad a_k = \left(\frac{k}{p}\right) \text{ for } k \neq 0.$$

Theorem 7.48 *For $\ell \not\equiv 0 \bmod p$,*

$$\sum_{k=1}^{p-1} \left(\frac{k}{p}\right) e^{2\pi \iota \frac{\ell k}{p}} = \sum_{k=0}^{p-1} e^{2\pi \iota \frac{\ell k^2}{p}}.$$

Proof Indeed, if k runs from 1 to $p - 1$ then k^2 takes only the values of the quadratic residues (twice on each one). Since

$$1 + \left(\frac{k}{p}\right) = \begin{cases} 2 & \text{if } k \text{ is a quadratic residue;} \\ 0 & \text{if } k \text{ is a quadratic nonresidue,} \end{cases}$$

we have

$$\sum_{k=0}^{p-1} e^{2\pi \iota \frac{\ell k^2}{p}} = 1 + \sum_{k=1}^{p-1} e^{2\pi \iota \frac{\ell k^2}{p}} = 1 + \sum_{k=1}^{p-1} \left[1 + \left(\frac{k}{p}\right)\right] e^{2\pi \iota \frac{\ell k}{p}}.$$

By Theorem 3.1, we have

$$1 + \sum_{k=1}^{p-1} e^{2\pi \iota \frac{\ell k}{p}} = 0,$$

and the result follows.
□

Corollary 7.49 *Let* **a** *be the Legendre sequence of length p. Then for j =* 0, 1, ..., *p* − 1,

$$F_{\mathbf{a}}\left(\frac{j}{p}\right) = \sqrt{p}.$$

Proof This follows from Theorem 7.47. □

Theorem 7.50 *For the Legendre sequence* **a** *of length p,*

$$\mathrm{PMEPR}(\mathbf{a}) \leq \left(\frac{2\ln 2}{\pi} \cdot \ln p + 2\right)^2.$$

Proof This is analogous to Theorem 7.43. □

7.9 Notes

Section 7.1 Theorem 7.2 is from Tellambura [404]; see also Ermolova and Vainikainen [106]. In what follows, I summarize our knowledge about the maximum of the aperiodic correlation and merit factor.

Maximum of the aperiodic correlation: Let

$$\rho_{\mathbf{a}} = \max_{j=1,\ldots,n-1} |\rho_{\mathbf{a}}(j)|$$

and

$$\rho_n = \min_{\mathbf{a} \in \{-1,1\}^n} \rho_{\mathbf{a}}.$$

The value of ρ_n has been computed up to $n = 70$, and it has been found that: $\rho_n \leq 2$ for $n \leq 21$, see Turyn [415]; $\rho_n \leq 3$ for $n \leq 48$, see Lindner [246] for $n \leq 40$, Cohen *et al.* [73] for the rest; $\rho_n \leq 4$ for $n \leq 70$, see Coxson *et al.* [77] for $n \leq 69$, Coxson and Russo [78] for $n = 70$. Moon and Moser [277] proved that most of the sequences have $\rho_n > o(\sqrt{n})$ and $\rho_n \leq (2 + \varepsilon)\sqrt{n \ln n}$ with ε tending to 0 when n increases. Mercer [264] proved that $\rho_n \leq (\sqrt{2} + \varepsilon)\sqrt{n \ln n}$ for sufficiently large n and any $\varepsilon > 0$.

A sequence, **a**, is called a Barker sequence if $\rho_{\mathbf{a}} = 1$. The longest known Barker sequence has length 13, and it is conjectured that there are no longer Barker sequences. Currently no sequences **a** of growing length are known with $\rho_{\mathbf{a}} = o(\sqrt{n \ln n})$. For some constructions of sequences with low autocorrelation see Golay [137], Koukouvinos [209] and Schroeder [361]. Høholdt *et al.* [166] proved that $\rho_{\mathbf{a}} = O(n^{0.9})$ for **a** being Rudin–Shapiro sequences of length n. For M-sequences, Sarwate [352] proved the upper bound of $1 + \frac{2}{\pi}\sqrt{n+1} \ln \frac{4n}{\pi}$; see also McEliece [262]. Jedwab and Yoshida [179] provided numerical evidence that the maximum of aperiodic correlation of rotated Legendre sequences and M-sequences is likely to be of order $\sqrt{n \ln n}$ rather than $\sqrt{n}$.

Merit factor: Let μ_n stand for the maximum merit factor among all bi-phase sequences of length n. A survey by Jedwab [178] provides a summary of the known results. It was proved by Newman and Byrnes [289] that

$$\sum_{\mathbf{a}\in\{-1,1\}^n} \frac{1}{\mu(\mathbf{a})} = \frac{n-1}{n}.$$

It is known that

$$6 \le \lim_{n\to\infty} \sup \mu_n \le \infty.$$

The value of μ_n has been calculated for $n \le 60$ using exhaustive computation by Lunelli for $n \le 6$, see Turyn [415]; by Swinnerton-Dyer for $7 \le n \le 19$, see Littlewood [254]; by Turyn for $n \le 32$, see Golay [138]; by Mertens [265] for $n \le 48$; and by Mertens and Bauke for $n \le 60$ [266].

Høholdt and Jensen [165] determined the asymptotic factor of cyclically rotated Legendre sequences, see also Golay [139]. If the sequence is rotated by a quarter of the length, it asymptotically achieves 6, the best known asymptotic result. Jensen and Høholdt [180] showed that the asymptotic merit factor of any rotation of an M-sequence is 3. Littlewood [255] proved that the merit factor of any of the Rudin–Shapiro sequences of length 2^m is $3\left(1 - \left(-\frac{1}{2}\right)^m\right)$, and therefore is asymptotically 3. Notice that the merit factor of the Barker sequence of length 13 is 14.08. Borwein et al. [37] and Kristiansen and Parker [218] constructed very long sequences with merit factors of more than 6.3.

Section 7.2   Rudin–Shapiro sequences were introduced by Shapiro [365] in his M.Sc. thesis. They were rediscovered independently by Rudin [344]. Theorems 7.6 and 7.8 are from Brillhart [47]; see also Brillhart and Morton [48]. Theorem 7.9 is by Rudin [344]. Properties of Rudin–Shapiro polynomials were considered by Borwein and Mossinghoff [40], Kervaire et al. [203], and Newman [288].

Section 7.3   Golay sequences were introduced by Golay [133, 134]. Parker et al. [318] provide a comprehensive survey of complementary sequences. The use of complementary sequences for peak power reduction was proposed by Popoviĉ [332], who generalized the work of Boyd [41].

Theorems 7.11 and 7.12 are by Golay [135]. Theorem 7.14 is by Eliahou et al. [104]; see also [105]. Earlier, Griffin [144] excluded lengths of type $n = 2 \cdot 9^t$, and Kounias et al. [210] proved that there are no complementary pairs of length $n = 2 \cdot 7^{2m}$. For small lengths, the value $n = 18$ was excluded by Golay [135], Kruskal [223], and Yang [449]. A complete classification of complementary pairs of modest lengths was started by Andres and Stanton [7], and accomplished for all lengths up to 100 by Borwein and Ferguson [39]. Open cases for lengths $n < 200$ are as follows: $n = 106, 116, 122, 130, 136, 146, 148, 164, 170, 178, 194$. In [39] a system of recursions is introduced, for which there is an extra primitive complementary pair,

namely, $n = 20$

$\mathbf{a} = (1,1,1,1,-1,1,-1,-1,-1,1,1,1,-1,-1,1,1,-1,1,-1,-1,1),$
$\mathbf{b} = (1,1,1,1,-1,1,1,1,1,1,-1,-1,-1,1,1,-1,1,1,-1,1,1,-1)$

Golay [135] points out that the two primitive complementary pairs of length 10 are equivalent under decimation. Specifically, the second pair of length 10 is obtained from the first one by taking successive third sequence elements, cyclically.

The recursive constructions from Theorems 7.15, 7.16 and 7.17 are by Golay [135], the construction of Theorem 7.18 and the result of Theorem 7.19 are by Turyn [416]. For other constructions, see Guangguo [145].

Budiŝin [53] proposed an efficient implementation of a correlator with an incoming data stream having a Golay sequence of length, n, which achieves complexity of $2 \log_2 n$ operations per sample, see also Popoviĉ [333].

Periodic complementary sequences were considered by Arasu and Xiang [8], Dokovic [96], Lüke [256], and Yang [448, 450]. Other papers dealing with complementary sequences and their applications are by Budiŝin [50, 54], Golay [136], Jauregui [172], Jiang and Zhu [184], Seberry et al. [362], Tsen [412], Turyn [414], and Weng and Guangguo [431].

Section 7.4   Golay [133] introduced complementary sets and found complementary sets of size 4. Tseng and Liu [413] were the first to treat the subject of complementary sets. They proved Theorems 7.20, 7.21, and 7.22. Theorem 7.23 is by Turyn [416]. Tseng and Liu [413] proved that the size of complementary sets of odd length n must be a multiple of 4. Dokovic [97] showed that complementary sets of size 4 exist for all even $n < 66$. Turyn [416] presented constructions for complementary sets of size 4 for all odd lengths $n < 33$, and $n = 59$. Theorems 7.24, 7.25, 7.26, 7.27, and 7.29 are by Tseng and Liu [413]. A recursive construction of mates for use in Theorem 7.25 similar to the one used for complementary sequences is presented in [413]. Theorems 7.30 and 7.31 are by Feng et al. [111].

Section 7.5   The first reference to multilevel complementary pairs is by Gutleber [146]. Darnell and Kemp [85] were the first to describe constructions of multilevel sequences; see also Kemp and Darnell [199] and Budiŝin [51, 52]. Ternary complementary sequences were studied by Gavish and Lempel [127]. Other generalizations and specific classes of complementary sequences were considered by: Sivaswamy [382]–subcomplementary sequences, Budiŝin [49]–supercomplementary sequences, Bömer and Antweiler [35], Dokovic [97]–periodic complementary sequences. Sequences over QAM constellations were considered by Rößling and Tarokh [346] and Tarokh and Sadjadpour [396, 397]. For a survey of early results on polyphase complementary sequences, see Fan and Darnell [107, Chapter 13].

Extensive research of multiphase sequences has been undertaken by Sivaswamy [381] and Frank [117]; see also Craigen [79]. Theorems 7.32 and 7.33 are from

Fiedler and Jedwab [115]. The construction of complementary sets from Reed–
Muller codes is by Davis and Jedwab [86, 87], who derived Theorem 7.34 and
Corollary 7.35. Theorem 7.36 is by Paterson [323] see also [322]. A summary of the
coding methods can be found in Davis *et al.* [88]. Further relevant generalization of
Reed–Muller codes yielding new multiphase sequences was proposed by Schmidt
[355] and Schmidt and Finger [356]. Holzmann and Kharaghani [167] found for
quadriphase complementary sequences that the symmetry operations generate an
equivalence class of up to 1024 sequences.

Decoding of nonbinary Reed–Muller codes was considered by Ashikhmin and
Litsyn [16]. Maximum likelihood decoding of Reed–Muller codes in OFDM sys-
tems was considered by Jones and Wilkinson [188]. For the generalized Reed–
Muller codes studied in this section the algorithms developed by Grant and van
Nee [140, 141], Paterson and Jones [327], and Greferath and Vellbinger [143] can
be applied.

A trigonometric polynomial $F(t)$ is called *flat* if, for $t \in [0, 1)$,

$$c_1\sqrt{n} \leq |F(t)| \leq c_2\sqrt{n},$$

for some positive constants c_1 and c_2, and is called *ultra-flat* if, for $t \in [0, 1)$,

$$(1 - o(1))\sqrt{n} \leq |F(t)| \leq (1 + o(1))\sqrt{n}.$$

The study of flat polynomials was initiated by Hardy and Littlewood [150].

The existence of ultra-flat trigonometric polynomials with coefficients of absolute value 1 (Theorem 7.37) is proved by Kahane [192]; see Beller and Newman
[23], Byrnes [57], Körner [206], Littlewood [254, 255] for earlier results. The error
term in the theorem is estimated in [192] as

$$\varepsilon_n = O\left(n^{-\frac{1}{17}}\sqrt{\log n}\right).$$

Beck [21] proved the existence of flat polynomials with the coefficients being 400th
roots of unity.

Complementarity with respect to other transformations was considered by Parker
[317] and Parker and Tellambura [320, 321].

Section 7.6   This section is based on the results of Paterson and Tarokh [328].

Section 7.7   I follow here Alrod *et al.* [3]. The use of M-sequences for power
reduction was considered by Li and Ritcey [240], Jedwab [177], and Tellambura
[405].

Section 7.8   Montgomery [276] considered lower bounds for Legendre sequences.
He derived a bound similar to the one from the previous section (with constant $\frac{2}{\pi}$).

8

Methods to decrease peak power in MC systems

In this chapter, I consider methods of decreasing peak power in MC signals. The simplest method is to clip the MC signal deliberately before amplification. This method is very simple to implement and provides essential PMEPR reduction. However, it suffers some performance degradation, as estimated in Section 8.1. In selective mapping (SLM), discussed in Section 8.2, one favorable signal is selected from a set of different signals that all represent the same information. One possibility for SLM is to choose the best signal from those obtained by inverting any of the coordinates of the coefficient vector. The method of deciding which of the coordinates should be inverted is described in Section 8.3. Further, in Section 8.4 a modification of SLM is analyzed. There the favorable vector is chosen from a coset of a code of given strength. Trellis shaping, where the relevant modification is chosen based on a search on a trellis, is described in Section 8.5. In Section 8.6, the method of tone injection is discussed. Here, instead of using a constellation point its appropriately shifted version can be used. In active constellation extension (ACE), described in Section 8.7, some of the outer constellation points can be extended, yielding PMEPR reduction. In Section 8.8, a method of finding a constellation in the frequency domain is described, such that the resulting region in the time domain has a low PMEPR. In partial transmit sequences (PTS), the transmitted signal is made to have a low PMEPR by partitioning the information-bearing vector to sub-blocks followed by multiplying by a rotating factor the coefficients belonging to the same sub-block. This method is discussed in Section 8.9. In Section 8.10, I discuss the possibility of allocating some redundant carriers used only for peak reduction rather than for information transmission. Finally, in Section 8.11, the described methods are compared.

8.1 Deliberate clipping and filtering

The simplest approach to decreasing the PMEPR of any MC signal to a prescribed level is deliberately to clip it before amplification. However, clipping is a nonlinear process and may cause significant in-band distortion, which degrades the SER or BER performance, and out-of-band noise, which reduces the spectral efficiency. The clipping could be applied to signals at different stages of processing. The possibilities are either to clip the continuous signal at the output of the LPF or to clip discrete samples at the output of IDFT. In what follows, I will analyze both cases.

8.1.1 Clipping continuous signal

Consider a continuous time-unconstrained, baseband MC signal $F(t) = x(t)e^{i\theta(t)}$, input to a soft limiter HPA with output $y(t)e^{i\theta(t)}$ described by

$$y(t) = y = h(x) = \begin{cases} -A & \text{if } x(t) \leq -A, \\ x(t) & \text{if } |x(t)| < A, \\ A & \text{if } x(t) \geq A. \end{cases} \tag{8.1}$$

Let us introduce the clipping ratio γ defined as

$$\gamma = \frac{A}{\sqrt{P_{\text{in}}}}, \tag{8.2}$$

where P_{in} is the input power of the MC signal before clipping. Let $r(t) = |x(t)|$ and $s(t) = |y(t)|$ and, assuming validity of the central limit theorem, we approximate $r(t)$ by a Rayleigh random variable with p.d.f.

$$f(r) = \frac{2r}{P_{\text{in}}} \cdot e^{-\frac{r^2}{P_{\text{in}}}}. \tag{8.3}$$

Then the output power after clipping is

$$P_{\text{out}} = E(h^2(r)) = \int_0^\infty h^2(r)f(r)\,dr = \left(1 - e^{-\gamma^2}\right)P_{\text{in}}. \tag{8.4}$$

Normalizing the clipped signal, we have

$$s(t) = \frac{h(r)}{\sqrt{P_{\text{out}}}} = \frac{h(r)}{\sqrt{\left(1 - e^{-\gamma^2}\right)P_{\text{in}}}}. \tag{8.5}$$

Notice that for $\gamma = 0$, we have to use

$$\lim_{\gamma \to 0} s(t) \approx \lim_{\gamma \to 0} \frac{A}{\sqrt{P_{\text{out}}}} = \lim_{\gamma \to 0} \frac{\gamma}{\sqrt{1 - e^{-\gamma^2}}} = 1,$$

i.e., $s(t)$ is constant (hard limiter).

For a memoryless nonlinearity such as (8.1), the output $y(t)$ may be decomposed into two uncorrelated signal components

$$y(t) = \alpha x(t) + c(t). \tag{8.6}$$

This can be rewritten as

$$y^R(t) = \alpha x^R(t) + c^R(t),$$
$$y^I(t) = \alpha x^I(t) + c^I(t),$$

for, respectively, the real and imaginary parts of the signals. We assume that $x^R(t)$ and $y^R(t)$ are independent Gaussian random variables. To estimate α we apply

$$\alpha = \frac{E(x^R(t)y^R(t))}{E(x^R(t)x^R(t))} = \frac{E(x^I(t)y^I(t))}{E(x^I(t)x^I(t))}.$$

By changing the variables as

$$x^R(t) = r(t)\cos\varphi(t), \quad x^I(t) = r(t)\sin\varphi(t),$$

where $r(t)$ is Rayleigh distributed and $\varphi(t)$ is uniformly distributed, we have

$$\alpha = \frac{2E(r\cos\varphi\, h(r)\sin\varphi)}{P_{\text{in}}}$$
$$= 1 - e^{-\gamma^2} + \gamma\sqrt{\pi}\,Q(\sqrt{2}\gamma), \tag{8.7}$$

where

$$Q(z) = \frac{1}{2\pi}\int_z^\infty e^{-\frac{u^2}{2}}\,du.$$

We start with estimating the error probability using the assumption that the distortion caused by clipping is described by additive Gaussian noise, with a variance equal to the energy of the clipped portion of the signal. Let the signal power be normalized to unity. Then the power of the clipped portion $c(t)$ in (8.6) is

$$\sigma_c^2 = 2\int_A^\infty (z-A)^2 e^{-\frac{z^2}{2}}\,dz = -\sqrt{\frac{2}{\pi}}\,Ae^{-\frac{A^2}{2}} + 2(1+A^2)Q(A). \tag{8.8}$$

To compute the error probability, we will assume that each subcarrier carries a square constellation of L^2 points. Each (real/imaginary) component has L levels, equally spaced and separated by $2d$, with a total power equal to $\frac{1}{2n}$ (since we have normalized the total power to unity). Therefore,

$$d = \sqrt{\frac{3}{2n(L^2 - 1)}}. \tag{8.9}$$

Thus we get the probability of error in any subcarrier as

$$\Pr(\text{error}) = \frac{4(L-1)}{L}Q\left(\frac{\sqrt{3}}{\sigma_c\sqrt{(L^2 - 1)}}\right). \tag{8.10}$$

However, in most cases, particularly when the desired error probability is low, the clipping level is set high enough such that clipping is a rare event, occurring more infrequently than once every symbol interval. Clipping under these conditions then forms a kind of impulsive noise. Since γ is quite large here we conclude from (8.7) that, in this case, $\alpha \approx 1$.

For a description of such noise, I will use a model of a stationary Gaussian process, possessing finite second-order moment, and continuous with probability one. The following three asymptotic properties of the large excursions of such processes will form the basis of this analysis.

- The sequence of upward level crossings of a stationary and ergodic process asymptotically approaches a Poisson process for large levels A. For a Gaussian signal $x(t)$, the rate of this Poisson process is given by

$$\lambda_A = \frac{1}{2\pi} \sqrt{\frac{m_2}{m_0}} \cdot e^{-\frac{A^2}{2m_0}}, \tag{8.11}$$

where

$$m_j = \int \omega^j \, dF(\omega), \quad j = 0, 2, \ldots \tag{8.12}$$

and $dF(\omega)$ is the power spectral density of $x(t)$. Without loss of generality, we may normalize the power m_0 in the signal $x(t)$ to unity. The relevant quantity, m_2, then represents the power in the first derivative of $x(t)$.

- The length of intervals τ during which the signal stays above the high level $|A|$ is (asymptotically) Rayleigh distributed, with density function

$$\rho_\tau(\tau) = \frac{\pi}{2} \frac{\tau}{\tau_m^2} \cdot e^{-\frac{\pi}{4} \left(\frac{\tau}{\tau_m}\right)^2}, \quad \tau \geq 0. \tag{8.13}$$

Here τ_m denotes the expectation of τ. Since $\lambda_A \tau_m = \Pr(x(t) \geq A)$, the expected value of the duration of a clip may be approximated as

$$\tau_m = \frac{Q(A)}{\lambda_A} \approx \frac{\sqrt{2\pi}}{A\sqrt{m_2}}. \tag{8.14}$$

The approximation in (8.14) is valid for large values of A.

- The shape of pulse excursions above level $|A|$ are parabolic arcs of the form

$$p_\tau(t) = \left(-\frac{1}{2}m_2 t^2 + \frac{1}{8}m_2\tau^2\right) A \cdot \text{rect}\left(\frac{t}{\tau}\right), \tag{8.15}$$

where rect($\cdot$) denotes a rectangular window function, and τ, the random duration of the clip, forms the support of the parabolic arc (see Fig. 8.1).

We will also require the instantaneous spectrum of the clipped signal component. This can be obtained by taking the Fourier transform of the parabolic arc (8.15), to get

$$g_\tau(\omega) = m_2 \frac{A\tau}{\omega^2} \left(\text{sinc}\frac{\omega\tau}{2} - \cos\frac{\omega\tau}{2}\right), \tag{8.16}$$

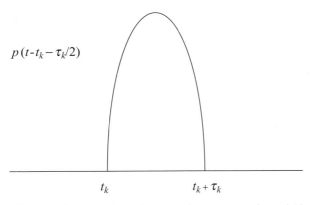

Figure 8.1 Excursion of a Gaussian process above $|A|$

where

$$\operatorname{sinc} \varphi = \frac{\sin \varphi}{\varphi}.$$

Under the assumption of many subcarriers, the power spectrum of $x(t)$ will tend to be rectangular over the total frequency. Hence, the rate of the Poisson process corresponding to the clipping of $x(t)$ can be defined from (8.11) as

$$\lambda_A = \frac{1}{2\pi} \sqrt{m_2} \cdot e^{-\frac{A^2}{2}} = \frac{n}{\sqrt{3}} \cdot e^{-\frac{A^2}{2}}. \tag{8.17}$$

As well, by (8.14) the expected value for the duration of a clip is

$$\tau_m^{-1} \approx \sqrt{\frac{2\pi}{3}} \cdot nA. \tag{8.18}$$

The effect of a clip of duration τ, occurring at time t_0, on the kth subcarrier is given by the Fourier transform of the clipped portion of the pulse,

$$F_k = \frac{1}{\sqrt{n}} \sum_{j=0}^{n-1} f_j e^{-2\pi i j \frac{k}{n}}, \tag{8.19}$$

where f_j are samples of the clipped pulse $p(t)$ in (8.15),

$$f_j = p\left(t - t_0 - \frac{\tau}{2}\right)\Big|_{t=\frac{j}{n}}, \tag{8.20}$$

and the factor $\frac{1}{\sqrt{n}}$ preserves total power. For increasing n we can replace the discrete Fourier transform with the conventional one by substituting

$$f_j = n \int_{\frac{j}{n} + \frac{1}{2n}}^{\frac{j}{n} - \frac{1}{2n}} p(t)\, dt. \tag{8.21}$$

Then

$$F_k = \sqrt{n} \int_{t_0}^{t_0+\tau} p\left(t - t_0 - \frac{\tau}{2}\right) \cdot e^{-2\pi i \frac{kt}{2}} \, dt. \tag{8.22}$$

Substituting $u = t - t_0$ into (8.19) yields

$$F_k = \sqrt{n} e^{-2\pi i k t_0} \int_0^{\tau} p\left(u - \frac{\tau}{2}\right) \cdot e^{-2\pi i k u} \, du$$
$$= \sqrt{n} e^{-2\pi i k t_0} g(2\pi k), \tag{8.23}$$

where $g(\cdot)$ is the pulse spectrum given by (8.16). Thus,

$$F_k = \frac{\sqrt{n} m_2 A \tau}{4\pi^2 k^2} \cdot e^{-2\pi i k t_0} (\text{sinc}\,(\pi k \tau) - \cos \pi k \tau). \tag{8.24}$$

Since $\tau \ll 1$, we can use the expansion

$$\text{sinc}\,\alpha - \cos \alpha \approx \frac{\alpha^2}{3} \tag{8.25}$$

in (8.24) to approximate the response of the kth subcarrier to a clip of duration τ as

$$F_k = \frac{1}{12} \sqrt{n} m_2 A \tau^3 e^{i\theta}, \tag{8.26}$$

where θ is uniformly distributed over $[0, 2\pi]$, and the Rayleigh probability distribution of τ is given by $\rho_{\tau}(\tau)$ in (8.13). Note that the expansion (8.25) will not be valid in general for the higher frequency subcarriers in the MC signal. In fact, (8.23) indicates that the probability of error from a clip will vary across subcarriers, the overall probability being dominated by that of the error probability due to clipping. Rewriting (8.26) as

$$F_k = \eta e^{i\theta}, \tag{8.27}$$

we are interested in obtaining the probability distribution of η:

$$\Pr(\eta > R) = \Pr\left(\tau > \left(\frac{12R}{\sqrt{n} m_2 A}\right)^{\frac{1}{3}}\right). \tag{8.28}$$

Using the distribution of τ in (8.13), and substituting for m_2 from (8.12), we get, after some simplification

$$\Pr(\eta > R) = e^{-\left(\frac{3\pi^2 R^2 A^4 n}{8}\right)^{\frac{1}{3}}}. \tag{8.29}$$

If we warp the complex plane by the mapping $ae^{i\theta} \to a^{\frac{1}{3}} e^{i\theta}$, then (8.29) would represent a Rayleigh distribution in the warped complex plane. The real or imaginary part of F_k therefore has a normal distribution with

$$\Pr(\eta \cos \theta > x) = Q\left(\frac{x^{\frac{1}{3}}}{\sigma}\right), \tag{8.30}$$

where

$$\sigma = \left(\frac{2}{\sqrt{3n} \cdot \pi A^2}\right)^{\frac{1}{3}}.$$

To compute the error probability, we again assume that each subcarrier carries a square constellation of L^2 points. Each (real or imaginary) component has L levels, equally spaced and separated by $2d$, with a total power equal to $\frac{1}{2n}$ (since we have normalized the total power to unity). Therefore, d is defined by (8.9). Combining (8.29) and (8.9), we get the probability of error in any subcarrier, given a clip occurs, as

$$\Pr(\text{error}|\text{clip}) = \frac{4(L-1)}{L} Q\left(\left(\frac{3\pi A^2}{\sqrt{8(L^2-1)}}\right)^{\frac{1}{3}}\right). \tag{8.31}$$

To compute the overall probability of error, we will need to multiply the conditional probability in (8.31) by the probability of occurrence of a clip in one symbol duration. This can be obtained using (8.11), the Poisson rate of two-side clipping events, as

$$\Pr(\text{clip}) \approx 2\lambda_A. \tag{8.32}$$

Thus the overall probability of symbol error is upper bounded by

$$\Pr(\text{error}) = \frac{8n(L-1)}{\sqrt{3}L} \cdot e^{-\frac{A^2}{2}} \cdot Q\left(\left(\frac{3\pi A^2}{\sqrt{8(L^2-1)}}\right)^{\frac{1}{3}}\right). \tag{8.33}$$

In particular, for $L = 2$ (QPSK) the error probability is given by

$$\Pr(\text{error}) = \frac{4n}{\sqrt{3}} \cdot e^{-\frac{A^2}{2}} Q\left(\left(\sqrt{\frac{3}{8}}\pi A^2\right)^{\frac{1}{3}}\right). \tag{8.34}$$

On the other hand, for large constellation sizes, the error probabilility may be approximated as

$$\Pr(\text{error}) = \frac{8n}{\sqrt{3}} \cdot e^{-\frac{A^2}{2}} Q\left(\left(\frac{3\pi A^2}{\sqrt{8}L}\right)^{\frac{1}{3}}\right). \tag{8.35}$$

Notice that the results here are overly pessimistic, since they do not account for the fact that some of the noise power will fall out-of-band, and therefore not cause any in-band distortion.

8.1.2 Clipping on discrete samples

In this section, I will consider clipping on discrete samples of the signal. Clipping of a continuous signal can thus be viewed as clipping of an oversampled signal with an infinitely large oversampling factor.

Let us consider the MC signals sampled at the Nyquist rate followed by clipping of the samples. Let the coefficient vector be $\mathbf{a} = (a_0, \ldots, a_{n-1})$, $a_k \in \mathcal{Q}$. For convenience, we consider here the signals with the power normalized to unity. Therefore, the MC signal is

$$F_{\mathbf{a}}(t) = \frac{1}{\sqrt{n}} \sum_{k=0}^{n-1} a_k e^{2\pi \imath k t}.$$

Picking the samples of the MC signal,

$$x_j e^{\imath \theta_j} = F\left(\frac{j}{n}\right),$$

we clip them as in (8.1) and obtain

$$h(x_j) e^{\imath \theta_j} = y_j e^{\imath \theta_j}.$$

Now, the coefficient vector corresponding to this set of values is $\hat{\mathbf{a}} = (\hat{a}_0, \ldots, \hat{a}_{n-1})$, where

$$\hat{a}_k = \frac{1}{\sqrt{n}} \sum_{j=0}^{n-1} y_j e^{\imath \theta_j} e^{-2\pi \imath k \frac{j}{n}}.$$

The error in the MC signal happens if, for at least one k, the closest to $\hat{a}_k$ point from the constellation $\mathcal{Q}$ differs from a_k. Let us elaborate on several approaches to the estimation of this error probability.

Using the techniques of the previous section we can significantly tighten the error bound in (8.33) by replacing the clip probability with its corresponding discrete time value. Specifically, approximating the probability of a single sample to exceed the level $\pm A$ as $2Q(A)$, and assuming that there is, at most, one clipping at a sample in the Nyquist sampled signal, we conclude that the probability of the clip is approximately $2nQ(A)$ and, therefore, (8.33) becomes

$$\Pr(\text{error}) = \frac{8n(L-1)}{L} \cdot Q(A) \cdot Q\left(\left(\frac{3\pi A^2}{\sqrt{8(L^2-1)}}\right)^{\frac{1}{3}}\right). \tag{8.36}$$

Another approach is based on Gaussian approximations. As with (8.6) we present the output of the soft limiter as

$$y_j = \alpha x_j + c_j, \tag{8.37}$$

where c_j are independent from x_j. This can be rewritten as

$$y_j^R = \alpha x_j^R + c_j^R,$$
$$y_j^I = \alpha x_j^I + c_j^I,$$

for the real and imaginary parts of the samples, respectively. Here α is determined in (8.7). By (8.4) the total output signal power S can be given by

$$S = \alpha^2 P_{\text{in}} = \frac{\alpha^2}{1 - e^{-\gamma^2}} P_{\text{out}}. \tag{8.38}$$

Let

$$C_k = \frac{1}{\sqrt{n}} \sum_{j=0}^{n-1} c_j e^{-2\pi \imath k \frac{j}{n}}.$$

We approximate C_k by complex Gaussian random variables with zero mean. Since all the distortion components fall within the signal bandwidth, the total variance of the distortion D, is given by

$$D = P_{\text{out}} - S = \left(1 - \frac{\alpha^2}{1 - e^{-\gamma^2}}\right) P_{\text{out}}. \tag{8.39}$$

Finally, the error probability is

$$\text{Pr(error)} = Q\left(\sqrt{\frac{S}{D}}\right) = Q\left(\frac{\alpha}{\sqrt{1 - e^{-\gamma^2} - \alpha^2}}\right). \tag{8.40}$$

Simulations show, however, that the reduction of PMEPR is mostly insignificant for the Nyquist sampling. To improve on this method, it is suggested that the oversampled MC signals be clipped followed by filtering of the out-of-band components of DFT. However, the filtering could yield peak regrowth, and therefore the process might be reiterated. The analysis of this method could be based on similar previous ideas. However, it is much more involved and thus I have omitted it.

8.2 Selective mapping

The idea of selective mapping is simple: partition all possible signals to subsets and pick from each subset a representative with the minimum PMEPR.

More formally, let $\mathcal{Q}^n$ be the collection of all vectors of length n with coordinates belonging to a constellation $\mathcal{Q}$, $|\mathcal{Q}| = q$. Assume there exists a partition of $\mathcal{Q}^n$

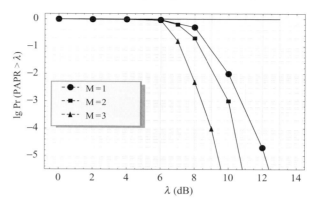

Figure 8.2 PMEPR distribution for $n = 512$ and QPSK modulation for selective mapping method

into M nonintersecting subsets Q_j, $j = 0, 1, \ldots, M - 1$, of equal size $\frac{q^n}{M}$. The information is conveyed by the index of the chosen subset, and is transmitted by picking one of the vectors belonging to the corresponding subset. The channel code, C, consists of the vectors, one per subset, possessing the minimum PMEPR among the vectors in the subset. Thus

$$\text{PMEPR}(C) \leq \max_{j=0,\ldots,M-1} \min_{\mathbf{a} \in Q_j} \text{PMEPR}(\mathbf{a}).$$

The rate of the defined code is $1 - \frac{1}{n} \log_q M$.

There are several simple methods of defining the partitioning. For example, let $M = q^r$, and $g_0, g_1, \ldots, g_{M-1}$, be invertible mappings from Q^{n-r} to itself. Given an information vector $\mathbf{v} \in Q^{n-r}$ we determine the minimum PMEPR of the vectors $g_0(\mathbf{v}), \ldots, g_{M-1}(\mathbf{v})$, and transmit the vector with minimum PMEPR along with the index of the best transform (side information). This will clearly be a vector in Q^n. For instance, one can choose g_0 to be identity, and g_1 to be a pseudo-random (scrambling) transform. In Fig. 8.2, simulation results for a QPSK modulated MC system with 512 subcarriers are presented, where the modifying vectors are chosen at random.

Under the nonrigorous assumption that the M resulting vectors are independent, the CCDF of PMEPR becomes as follows:

$$\Pr \left(\min_{j=0,1,\ldots,M-1} \text{PMEPR}(g_j(\mathbf{v})) > \lambda \right) = (\Pr(\text{PMEPR}(\mathbf{v}) > \lambda))^M. \quad (8.41)$$

A rigorous analysis for a specially defined transform will be given in Section 8.4.

For the following, we need several definitions. Let $|C|$ be the number of possible code words in a channel code, C. The *rate* of the code, C, chosen from a $q$ary constellation, is

$$R = \frac{1}{n} \log_q |\mathcal{C}|.$$

The *rate hit* of the code is $1 - R$.

8.3 Balancing method

Let $\mathcal{Q}$ be a constellation scaled such that $E_{av} = 1$, and let $E_{\max}$ be the maximal power of a point in the constellation. In this section, for any code word $\mathbf{c} = (c_0, \ldots, c_{n-1})$, $c_i \in \mathcal{Q}$, we study the design of optimum signs $\epsilon_k \in \{-1, 1\}, k = 0, 1, \ldots, n - 1$, for each subcarrier, in order to reduce the PMEPR of the resulting code word $\mathbf{c}_\epsilon = (\epsilon_0 c_0, \epsilon_1 c_1, \ldots, \epsilon_{n-1} c_{n-1})$. Thus, we consider the following minimization problem: given a complex vector $\mathbf{c}$ where $|c_k| \le \sqrt{E_{max}}$, find

$$\min_{\epsilon} \max_{t \in [0,1)} \left| \sum_{k=0}^{n-1} \epsilon_k c_k e^{2\pi \imath k t} \right|. \tag{8.42}$$

It is difficult to address this problem in its continuous form. Therefore we deal with its discrete approximation (see Section 5.4). Given $r, r > 1$, the oversampling factor such that rn is integer, $h, h > 1$, the number of projection axes, we are facing joint minimization of $r \cdot h \cdot n$ bounded linear forms, cf. (5.65),

$$L_j(\mathbf{c}_\epsilon) = L_j(\epsilon_0 c_0, \ldots, \epsilon_{n-1} c_{n-1}) = \sum_{k=0}^{n-1} \eta_{j,k} \cdot \epsilon_k, \quad j = 0, 1, \ldots, rhn - 1, \tag{8.43}$$

where $t_j = \frac{j}{rn}$ and

$$\eta_{j,k} = \begin{cases} \Re\left(c_k e^{2\pi \imath (t_j k)}\right), & j = 0, 1, \ldots, rn - 1, \\ \Re\left(c_k e^{2\pi \imath \left(t_j k - \frac{1}{h}\right)}\right), & j = rn+, rn + 1, \ldots, 2rn - 1, \\ \ldots \\ \Re\left(c_k e^{2\pi \imath \left(t_j k - \frac{h-1}{h}\right)}\right), & j = (h - 1)rn, \ldots, rhn - 1. \end{cases} \tag{8.44}$$

Notice that all the linear forms are bounded, $|\eta_{j,k}| \le \sqrt{E_{\max}}$ and the number of forms exceeds the number of variables. For the signal $F_{\mathbf{c}_\epsilon}(t) = \sum_{k=0}^{n-1} \epsilon_k c_k e^{2\pi \imath k t}$,

$$\mathcal{M}_c(F_{\mathbf{c}_\epsilon}) = \max_{t \in [0,1)} |F_{\mathbf{c}_\epsilon}(t)| \le C_r C_h \cdot \max_j |L_j(\mathbf{c}_\epsilon)|, \tag{8.45}$$

where $C_m \le \frac{1}{\cos \frac{\pi}{2m}}$. Denote

$$\alpha = \frac{1}{r \cdot h}.$$

We will need the following version of Lemma 5.13, which is easily derivable from its proof.

Lemma 8.1   *Let $L_j(\epsilon_0, \ldots, \epsilon_{n-1})$ be as defined in (8.43). Then*

$$\max_{j=0,1,\ldots,\frac{n}{\alpha}-1} |L_j(\epsilon_0, \ldots, \epsilon_{n-1})| \leq \frac{K(\alpha)}{\sqrt{\alpha}} \sqrt{n\, E_{\max}}, \qquad (8.46)$$

where $K(\alpha)$ is a constant independent of n and is bounded by

$$K(\alpha) \leq 11 \sqrt{\alpha \ln \frac{2}{\alpha}}. \qquad (8.47)$$

$\square$

Substituting (8.46) and (8.47) into (8.45), and using the definition of PMEPR, we arrive at the following result.

Theorem 8.2   *For any code word, $\mathbf{c} \in \mathcal{Q}^n$, there exists a vector $\epsilon \in \{-1, 1\}^n$, such that for $\mathbf{c}_\epsilon = (\epsilon_0 c_0, \ldots, \epsilon_{n-1} c_{n-1})$,*

$$\mathrm{PMEPR}(\mathbf{c}_\epsilon) \leq E_{\max} \cdot \min_{r,h} \left(\frac{1}{\left(\cos \frac{\pi}{2r}\right)^2} \cdot \frac{1}{\left(\cos \frac{\pi}{2h}\right)^2} \cdot rh \cdot \left(K\left(\frac{1}{rh}\right)\right)^2 \right), \quad (8.48)$$

where the minimum is taken over $r > 1$ such that rn is integer, $h \geq 2$ an integer, and $K(\cdot)$ is upper bounded in (8.47).

$\square$

Notice that the right-hand side of (8.48) does not depend on n, and we achieve a significantly smaller PMEPR than in a randomly picked word, where it is about $\ln n$.

Corollary 8.3   *For any q-ary symmetric constellation $\mathcal{Q}$ containing antipodal points there exists a code of size $\left(\frac{q}{2}\right)^n$ and constant PMEPR.*

$\square$

Now let us pass to an efficient method of deterministically designing the balancing vector, ϵ. Let us consider the set of equiprobable vectors $\epsilon = (\epsilon_0, \ldots, \epsilon_{n-1}) \in \{-1, 1\}^n$. Then, for any code word $\mathbf{c}$, we define A_p^λ as the event that the absolute value of the pth linear form defined in (8.43) is greater than λ. Furthermore, assume that λ is chosen such that $\sum_{j=0}^{rhn-1} \Pr(A_j^\lambda)$ is less than 1, and, therefore, there exists a vector ϵ with the above property. We would like, efficiently, to find the vector ϵ, such that none of the bad events, A_p^λ occur.

We determine the coefficients of ϵ sequentially. Assume that we can compute the conditional probability $\Pr(A_p^\lambda | \epsilon_0, \ldots, \epsilon_{\ell-1})$, the probability of A_p^λ, given we have chosen $\epsilon_0, \ldots, \epsilon_{\ell-1}$. At the ℓth step, given the optimally chosen signs $\epsilon_0^*, \ldots, \epsilon_{\ell-2}^*$, we choose $\epsilon_{\ell-1}^* \in \{-1, 1\}$ such that

$$\sum_{j=0}^{rhn-1} \Pr\left(A_j^\lambda | \epsilon_0^*, \ldots, \epsilon_{\ell-2}^*, \epsilon_{\ell-1}^*\right) \le \sum_{j=0}^{rhn-1} \Pr\left(A_j^\lambda | \epsilon_0^*, \ldots, \epsilon_{\ell-2}^*, -\epsilon_{\ell-1}^*\right).$$

Therefore,

$$\sum_{j=0}^{rhn-1} \Pr\left(A_j^\lambda | \epsilon_0^*, \ldots, \epsilon_{\ell-2}^*\right) = \frac{1}{2}\left(\sum_{j=0}^{rhn-1} \Pr\left(A_j^\lambda | \epsilon_0^*, \ldots, \epsilon_{\ell-2}^*, \epsilon_{\ell-1}^* = 1\right)\right.$$

$$\left. + \sum_{j=0}^{rhn-1} \Pr\left(A_j^\lambda | \epsilon_0^*, \ldots, \epsilon_{\ell-2}^*, \epsilon_{\ell-1}^* = -1\right)\right)$$

$$\ge \sum_{j=0}^{rhn-1} \Pr\left(A_j^\lambda | \epsilon_0^*, \ldots, \epsilon_{\ell-2}^*, \epsilon_{\ell-1}^*\right).$$

Finally,

$$\sum_{j=0}^{rhn-1} \Pr\left(A_j^\lambda | \epsilon_0^*, \ldots, \epsilon_{n-1}^*\right) \le \sum_{j=0}^{rhn-1} \Pr\left(A_j^\lambda\right) < 1.$$

Since there is no randomness in the left-hand side expression, it can be either 0 or 1, and by the inequality we conclude that it is zero, and none of the events A_j^λ occur.

The difficulty here is in the efficient computation of the conditional probabilities. Instead of using the exact expressions, we can use upper bounds defined as

$$\Pr\left(A_j^\lambda | \epsilon_0, \ldots, \epsilon_{\ell-1}\right) \le \tau_j^\lambda\left(\epsilon_0, \ldots, \epsilon_{\ell-1}\right),$$

satisfying the following extra conditions:

i)
$$\sum_{j=0}^{rhn-1} \tau_j^\lambda < 1,$$

$$(8.49)$$

ii)
$$\tau_j^\lambda(\epsilon_0, \ldots, \epsilon_{\ell-1}) \ge \min_{\epsilon_{\ell-1} \in \{-1,1\}} \tau(\epsilon_0, \ldots, \epsilon_{\ell-1}).$$

As will be shown in Theorem 8.4, Chernoff's bound can be used, giving

$$\Pr\left(\left|\sum_{k=0}^{n-1} \epsilon_k \eta_{p,k}\right| > \lambda \mid \epsilon_0, \ldots, \epsilon_{\ell-1}\right) \le \tau_p^\lambda(\epsilon_0, \ldots, \epsilon_{\ell-1}), \qquad (8.50)$$

where

$$\tau_p^\lambda(\epsilon_0, \ldots, \epsilon_{\ell-1}) = 2e^{-\gamma\lambda} \cosh\left(\gamma \sum_{s=0}^{\ell-1} \epsilon_s \eta_{p,s}\right) \cdot \prod_{s=\ell}^{n-1} \cosh \gamma \eta_{p,s},$$

for any $\gamma > 0$ and $p = 0, \ldots, rhn - 1$. The approach is summarized in the following algorithm.

Algorithm   For any $\mathbf{c} \in \mathcal{Q}^n$, let $\eta_{p,k}$ be as in (8.44), and r and h as in Theorem 8.2. Then set $\epsilon_0 = 1$, and determine ϵ_ℓ recursively for $\ell = 1, \ldots, n - 1$, as

$$\epsilon_\ell = - \left(\sum_{p=0}^{rhn-1} \sinh \left(\gamma^* \sum_{s=0}^{\ell-1} \epsilon_s \eta_{p,s} \right) \sinh(\gamma^* \eta_{p,\ell}) \cdot \prod_{s=\ell+1}^{n-1} \cosh(\gamma^* \eta_{p,s}) \right), \quad (8.51)$$

where

$$\gamma^* = \sqrt{\frac{2 \ln(2rhn)}{n E_{\max}}}.$$

$\square$

The following result gives the worst case guarantee on the PMEPR of $\mathbf{c}_\epsilon$.

Theorem 8.4   *Let $\mathbf{c} \in \mathcal{Q}^n$ be given and $\mathbf{c}_\epsilon$ be determined according to the algorithm. Then*

$$\mathrm{PMEPR}\,(\mathbf{c}_\epsilon) < 2 E_{\max} \cdot \min_{r,h} \frac{1}{\cos^2 \frac{\pi}{2r}} \cdot \frac{1}{\cos^2 \frac{\pi}{2h}} \cdot (\ln n + \ln 2rh).$$

Proof   We estimate first the conditional probabilities,

$$\Pr \left(\left| \sum_{k=0}^{n-1} \epsilon_k \eta_{p,k} \right| > \lambda \,|\, \epsilon_0, \ldots, \epsilon_{\ell-1} \right)$$

$$= \Pr \left(\sum_{k=0}^{n-1} \epsilon_k \eta_{p,k} > \lambda \,|\, \epsilon_0, \ldots, \epsilon_{\ell-1} \right)$$

$$+ \Pr \left(\sum_{k=0}^{n-1} \epsilon_k \eta_{p,k} < -\lambda \,|\, \epsilon_0, \ldots, \epsilon_{\ell-1} \right)$$

$$= \Pr \left(\sum_{k=\ell}^{n-1} \epsilon_k \eta_{p,k} > \lambda - \sum_{k=0}^{\ell-1} \epsilon_k \eta_{p,k} \,|\, \epsilon_0, \ldots, \epsilon_{\ell-1} \right)$$

$$+ \Pr \left(- \sum_{k=\ell}^{n-1} \epsilon_k \eta_{p,k} > \lambda + \sum_{k=0}^{\ell-1} \epsilon_k \eta_{p,k} \,|\, \epsilon_0, \ldots, \epsilon_{\ell-1} \right)$$

$$\leq e^{\gamma \sum_{k=0}^{\ell-1} \epsilon_k \eta_{p,k}} \cdot e^{-\gamma \lambda} \cdot E \left(e^{\gamma \sum_{k=\ell}^{n-1} \epsilon_k \eta_{p,k}} \right)$$

$$+ e^{-\gamma \sum_{k=0}^{\ell-1} \epsilon_k \eta_{p,k}} \cdot e^{-\gamma \lambda} \cdot E \left(e^{-\gamma \sum_{k=\ell}^{n-1} \epsilon_k \eta_{p,k}} \right)$$

$$= 2 e^{-\gamma \lambda} \cosh \left(\gamma \sum_{s=0}^{\ell-1} \epsilon_s \eta_{p,s} \right) \cdot \prod_{s=\ell}^{n-1} \cosh \gamma \eta_{p,s} = \tau_p^\lambda(\epsilon_0, \ldots, \epsilon_{\ell-1}).$$

Here we used the Chernoff bound and the fact that $\epsilon_k \in \{-1, 1\}$ are equiprobable. The coefficient γ should be positive and is subject to optimization. We then show that the upper bound satisfies the conditions in (8.49). Using

$$\cosh(a + b) + \cosh(a - b) = 2\cosh a \cdot \cosh b,$$

we obtain

$$\tau_p^\lambda(\epsilon_0, \ldots, \epsilon_{\ell-2}) = e^{-\gamma\lambda} \prod_{s=\ell}^{n-1} \cosh \gamma\eta_{p,s} \cdot \left(\cosh\left(\gamma \sum_{s=0}^{\ell-2} \epsilon_s\eta_{p,s} + \gamma\eta_{p,\ell-1} \right) \right.$$

$$\left. + \cosh\left(\gamma \sum_{s=0}^{\ell-2} \epsilon_s\eta_{p,s} - \gamma\eta_{p,\ell-1} \right) \right)$$

$$= \frac{\tau_p^\lambda(\epsilon_0, \ldots, \epsilon_{\ell-2}, \epsilon_{\ell-1} = 1) + \tau_p^\lambda(\epsilon_0, \ldots, \epsilon_{\ell-2}, \epsilon_{\ell-1} = -1)}{2}$$

$$\geq \min_{\epsilon_{\ell-1} \in \{-1, 1\}} \tau_p^\lambda(\epsilon_0, \ldots, \epsilon_{\ell-2}, \epsilon_{\ell-1}).$$

Thus, the second condition of (8.49) is satisfied. To verify the first condition, we have

$$\sum_{j=0}^{rhn-1} \tau_p^\lambda = \sum_{j=0}^{rhn-1} 2e^{-\gamma\lambda} \prod_{s=0}^{n-1} \cosh \gamma\eta_{p,s} < \sum_{j=0}^{rhn-1} 2e^{-\gamma\lambda} \prod_{s=0}^{n-1} e^{\frac{\gamma^2\eta_{p,s}^2}{2}}$$

$$\leq \sum_{j=0}^{rhn-1} 2e^{-\gamma\lambda} \cdot e^{\frac{\gamma^2}{2}nE_{\max}} \leq 2rhn \cdot e^{-\gamma\lambda + \frac{\gamma^2}{2}nE_{\max}},$$

where we used $\cosh x < e^{\frac{x^2}{2}}$ for $x \neq 0$. Setting $\gamma^* = \frac{\lambda}{nE_{\max}}$ and choosing $\lambda = \sqrt{2nE_{\max} \ln 2rhn}$, we obtain

$$\sum_{j=0}^{rhn} \tau_p^\lambda < 2rhn \cdot e^{-\frac{\lambda^2}{2nE_{\max}}} = 1,$$

thus satisfying the first condition. Use of (8.45) and of the definition of PMEPR accomplishes the proof. $\square$

Although the previous result does not guarantee a restriction on the maximum of PMEPR, it significantly improves its statistics. This is illustrated in Fig. 8.3, which shows the PMEPR distribution of randomly chosen MC QPSK signals against that of the signals balanced by optimized sign vectors. Another advantage of the balancing method is that it does not require transmission of the side information. The receiver just identifies the antipodal points of the constellations and does not need to use any additional processing.

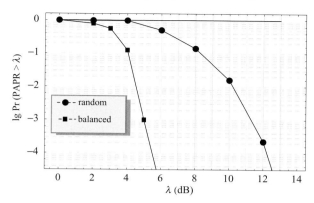

Figure 8.3 PMEPR distribution for $n = 128$ and QPSK using the balancing method (after [373])

8.4 Use of codes of given strength

Although the balancing method provides a very efficient reduction of PMEPR, the rate hit is considerable. In this section, I will analyze a method employing a much smaller collection of balancing vectors taken from a code of given strength. Let D be a binary code of length n and $\mathcal{D}$ be its image under mapping $0 \to 1, 1 \to (-1)$. Recall that the *strength*, t, of $\mathcal{D}$ is the maximal number such that for any fixed set of t positions, as we let the code words vary over $\mathcal{D}$, every possible t-tuple (out of 2^t possibilities) occurs in these positions the same number of times, namely $\frac{|\mathcal{D}|}{2^t}$.

It is known that a code which is dual to a code with the minimum distance $t + 1$ has strength t. An example of codes whose strength is fixed or slowly increasing with length is given by the codes dual to BCH codes. These codes have length $n = 2^m - 1$, the number of information bits ms (i.e., the number of code words is 2^{ms}), and strength $2s$. They are dual to BCH codes having the minimum distance $2s + 1$. For the length $n = 2^m$, we will exploit duals of the extended BCH codes (with extra overall parity check bit), thus obtaining codes of length $n = 2^m$, size 2^{ms+1}, and strength $2s + 1$. These codes can evidently also be considered as being of strength $2s$. Moreover, these codes may easily be encoded using a linear register of size ms with feedback.

I hereby establish a connection between the strength of codes over $\{-1, 1\}$, and their ability to balance linear forms when code vectors are used as the sign vectors.

Theorem 8.5 *Let $\mathcal{D}$ be a code over $\{-1, 1\}$ of length n and strength $2s$, and let M bounded linear forms be*

$$L_j(\mathbf{x}) = L_j(x_0, \ldots, x_{n-1}) = \sum_{k=0}^{n-1} a_{jk} x_j, \quad j = 0, \ldots, M - 1.$$

Then

$$\min_{\mathbf{d}\in\mathcal{D}} \max_{j=0,\dots,M-1} |L_j(\mathbf{d})| \leq \left(\frac{(2s)!}{2^s s!} \cdot \sum_{j=0}^{M-1} \left(\sum_{k=0}^{n-1} a_{jk}^2 \right)^s \right)^{\frac{1}{2s}}. \tag{8.52}$$

Moreover, for any real $\alpha > 1$, a randomly chosen code word $\mathbf{d} \in \mathcal{D}$, and $j = 0, 1, \dots, M - 1$,

$$\Pr_{\mathbf{d}\in\mathcal{D}} \left(|L_j(\mathbf{d})| \geq \left(\alpha \cdot M \cdot \frac{(2s)!}{2^s s!} \cdot \left(\sum_{k=0}^{n-1} a_{jk}^2 \right)^s \right)^{\frac{1}{2s}} \right) \leq \frac{1}{\alpha \cdot M}. \tag{8.53}$$

Proof Define

$$\Gamma_j = \sum_{\mathbf{d}\in\mathcal{D}} (L_j(\mathbf{d}))^{2s} = \sum_{\mathbf{d}\in\mathcal{D}} \left(\sum_{k=0}^{n-1} a_{jk} d_k \right)^{2s}.$$

Rewrite the expression for Γ_j,

$$\Gamma_j = \sum_{\mathbf{d}\in\mathcal{D}} \sum_{\substack{k_0,\dots,k_{2s-1} \\ k_0,\dots,k_{2s-1} \in \{1,\dots,n\}}} \prod_{\ell=0}^{2s-1} a_{jk_\ell} d_{k_\ell}$$

$$= \sum_{\mathbf{k}} \prod_{\ell=0}^{2s-1} a_{jk_\ell} \cdot \sum_{\mathbf{d}\in\mathcal{D}} \prod_{\ell=0}^{2s-1} d_{k_m}$$

$$= \sum_{\mathbf{k}} \prod_{k=0}^{n-1} a_{jk}^{\tau_k(\mathbf{k})} \cdot \sum_{\mathbf{d}\in\mathcal{D}} \prod_{k=0}^{n-1} d_k^{\tau_k(\mathbf{k})},$$

where the summation is over all vectors $\mathbf{k} = (k_0, \dots, k_{2s-1})$, and $\tau_k(\mathbf{k})$ is the number of ℓ, $\ell = 0, \dots, 2s - 1$, such that $k_\ell = k$.

For a given $\mathbf{k}$, if there exists a k, such that $\tau_k(\mathbf{k})$ is odd, then since $\mathcal{D}$ is a strength $2s$ code, we have $\sum_{\mathbf{d}\in\mathcal{D}} \prod_{k=0}^{n-1} d_k^{\tau_k(\mathbf{k})} = 0$. Otherwise,

$$\sum_{\mathbf{d}\in\mathcal{D}} \prod_{k=0}^{n-1} d_k^{\tau_k(\mathbf{k})} = \sum_{\mathbf{d}\in\mathcal{D}} 1 = |\mathcal{D}|.$$

Let $\mathcal{K} = \{\mathbf{k} : \tau_k(\mathbf{k}) \text{ is even for all } k\}$. We thus have

$$\Gamma_j = \left(\sum_{\mathbf{k}\in\mathcal{K}} \prod_{k=0}^{n-1} a_{jk}^{\tau_k(\mathbf{k})} \right) \cdot |\mathcal{D}|. \tag{8.54}$$

It is easily shown that

$$\sum_{\mathbf{k}\in\mathcal{K}} \prod_{k=0}^{n-1} a_{jk}^{\tau_k(\mathbf{k})} \leq \frac{(2s)!}{2^s s!} \cdot \left(\sum_{k=0}^{n-1} a_{jk}^2 \right)^s. \tag{8.55}$$

To see this, note that $\sum_{\mathbf{k}\in\mathcal{K}}\prod_{k=0}^{n-1}a_{jk}^{\tau_k(\mathbf{k})}$ and $\left(\sum_{k=0}^{n-1}a_{jk}^2\right)^s$ contain the same terms, but with different coefficients. Indeed,

$$\sum_{\mathbf{k}\in\mathcal{K}}\prod_{k=0}^{n-1}a_{jk}^{\tau_k(\mathbf{k})} = \sum_{\substack{\{s_0,\ldots,s_{n-1}\}\in\{0,1,2,\ldots,s\}\\ s_0+\ldots+s_{n-1}=s}}\frac{(2s)!}{(2s_0)!(2s_1)!\cdots(2s_{n-1})!}\cdot a_{j,0}^{2s_0}a_{j,1}^{2s_1}\cdots a_{j,n-1}^{2s_{n-1}}$$

$$= \sum_{\substack{\{s_0,\ldots,s_{n-1}\}\in\{0,1,2,\ldots,s\}\\ s_0+\ldots+s_{n-1}=s}}K_{s_0s_1\ldots s_{n-1}}^{(1)}\cdot a_{j,0}^{2s_0}a_{j,1}^{2s_1}\cdots a_{j,n-1}^{2s_{n-1}},$$

and

$$\left(\sum_{k=0}^{n-1}a_{jk}^2\right)^s = \sum_{\substack{\{s_0,\ldots,s_{n-1}\}\in\{0,1,2,\ldots,s\}\\ s_0+\ldots+s_{n-1}=s}}\frac{(s)!}{(s_0)!(s_1)!\cdots(s_{n-1})!}\cdot a_{j,0}^{2s_0}a_{j,1}^{2s_1}\cdots a_{j,n-1}^{2s_{n-1}}$$

$$= \sum_{\substack{\{s_0,\ldots,s_{n-1}\}\in\{0,1,2,\ldots,s\}\\ s_0+\ldots+s_{n-1}=s}}K_{s_0s_1\ldots s_{n-1}}^{(2)}\cdot a_{j,0}^{2s_0}a_{j,1}^{2s_1}\cdots a_{j,n-1}^{2s_{n-1}}.$$

To obtain (8.55), note that

$$\frac{K_{s_0s_1\ldots s_{n-1}}^{(1)}}{K_{s_0s_1\ldots s_{n-1}}^{(2)}} = \frac{(2s)!}{s!}\frac{s_0!}{(2s_0)!}\frac{s_1!}{(2s_1)!}\cdots\frac{s_{n-1}!}{(2s_{n-1})!}\le\frac{(2s)!}{s!}\cdot\frac{1}{2^s}.$$

Consequently,

$$\Gamma_j \le |\mathcal{D}|\frac{(2s)!}{2^ss!}\left(\sum_{k=0}^{n-1}a_{jk}^2\right)^s.$$

Furthermore,

$$\sum_{j=0}^{M-1}\Gamma_j = \sum_{j=0}^{M-1}\sum_{\mathbf{d}\in\mathcal{D}}(L_j(\mathbf{d}))^{2s}$$

$$= \sum_{\mathbf{d}\in\mathcal{D}}\sum_{j=0}^{M-1}\left(L_j(\mathbf{d})\right)^{2s}\le|\mathcal{D}|\frac{(2s)!}{2^ss!}\sum_{j=0}^{M-1}\left(\sum_{k=0}^{n-1}a_{jk}^2\right)^s.$$

Since all $(L_j(\mathbf{d}))^{2s}\ge 0$, from the last inequality it follows that, for some $\mathbf{d}'\in\mathcal{D}$,

$$\sum_{j=0}^{M-1}(L_j(\mathbf{d}'))^{2s}\le\frac{(2s)!}{2^ss!}\sum_{j=0}^{M-1}\left(\sum_{k=0}^{n-1}a_{jk}^2\right)^s.$$

Therefore, for $j=0,1,\ldots,M-1$,

$$|L_j(\mathbf{d}')|\le\left(\frac{(2s)!}{2^ss!}\cdot\sum_{j=0}^{M-1}\left(\sum_{k=0}^{n-1}a_{jk}^2\right)^s\right)^{\frac{1}{2s}},$$

proving (8.52).

Using the Chebyshev inequality, we obtain

$$\Pr_{\mathbf{d}\in\mathcal{D}}\left(|L_j(\mathbf{d})| \geq \left(\alpha \cdot M \cdot \frac{(2s)!}{2^s s!} \cdot \left(\sum_{k=0}^{n-1} a_{jk}^2\right)^s\right)^{\frac{1}{2s}}\right) \leq \frac{E_{\mathbf{d}\in\mathcal{D}}\left(L_j^{2s}(d)\right)}{\alpha \cdot M \cdot \frac{(2s)!}{2^s s!} \cdot \left(\sum_{k=0}^{n-1} a_{jk}^2\right)^s},$$

and thus establish the correctness of (8.53).   □

With no further assumptions about the nature of the coefficients a_{jk}, we have the following result.

Corollary 8.6   *Under the conditions of Theorem 8.5,*

$$\min_{\mathbf{d}\in\mathcal{D}} \max_{j=0,\dots,M-1} |L_j(\mathbf{d})| \leq \left(M \cdot \frac{(2s)!}{2^s s!}\right)^{\frac{1}{2s}} \cdot \sqrt{n} \cdot \max_{\substack{j=0,1,\dots,M-1 \\ k=0,1,\dots,n-1}} |a_{jk}|, \qquad (8.56)$$

$$\Pr_{\mathbf{d}\in\mathcal{D}}\left(\max_{j=0,\dots,M-1} |L_j(\mathbf{d})| \geq \left(\alpha \cdot M \cdot \frac{(2s)!}{2^s s!}\right)^{\frac{1}{2s}} \cdot \sqrt{n} \cdot \max_{\substack{j=0,1,\dots,M-1 \\ k=0,1,\dots,n-1}} |a_{jk}|\right)$$

$$\leq \frac{1}{\alpha}. \qquad (8.57)$$

Moreover, if

$$|a_{jk}| \leq 1, \quad j = 0, 1, \dots, M-1; \; k = 0, 1, \dots, n-1,$$

$M = r \cdot h \cdot n$, $s = \ln n$, $n \geq 2$, *then*

$$\min_{\mathbf{d}\in\mathcal{D}} \max_{j=0,\dots,M-1} |L_j(\mathbf{d})| \leq \sqrt{2n \ln n} \cdot (rh)^{\frac{1}{2\ln n}} \cdot \left(1 + \frac{1}{4\ln n}\right), \qquad (8.58)$$

$$\Pr_{\mathbf{d}\in\mathcal{D}}\left(\max_{j=0,\dots,M-1} |L_j(\mathbf{d})| \geq \sqrt{2\alpha n \ln n} \cdot (rh)^{\frac{1}{2\ln n}} \cdot \left(1 + \frac{1}{4\ln n}\right)\right) \leq \frac{1}{n^{\ln \alpha}}.$$
$$(8.59)$$

Proof   To get (8.56) and (8.57), plug

$$|a_{jk}| \leq \max_{\substack{j=0,1,\dots,M-1 \\ k=0,1,\dots,n-1}} |a_{jk}|$$

into (8.52) and (8.53). For (8.58) and (8.59), use

$$\sqrt{2\pi n} \cdot n^n \cdot e^{-n} \leq n! \leq \sqrt{2\pi n} \cdot n^n \cdot e^{-n+\frac{1}{12n}}, \qquad (8.60)$$

and

$$e^{\frac{\ln 2}{4\ln n} + \frac{1}{48(\ln n)^2}} < 1 + \frac{4}{\ln n}, \quad n \geq 2.$$

□

Now I am ready to describe the method of PMEPR reduction. Let $\mathcal{D}$ be a code from $\{-1, 1\}^n$ of strength $2s$. The vectors $\mathbf{d} \in \mathcal{D}$ are candidates for being chosen as the sign vectors. Let $\xi = (\xi_0, \ldots, \xi_{n-1}) \in \mathcal{Q}^n$. Given r, the oversampling factor, and h, the number of projection axes, we are facing joint minimization of $r \cdot h \cdot n$ bounded linear forms,

$$L_j(\mathbf{d}) = L_j(d_0, \ldots, d_{n-1}) = \sum_{k=0}^{n-1} a_{jk} \cdot d_k,$$

where $t_j = \frac{j}{rn}$ and

$$a_{jk} = \begin{cases} \Re\left(\xi_k e^{2\pi i t_j k}\right), & j = 0, 1, \ldots, rn - 1, \\ \Re\left(\xi_k e^{2\pi i \left(t_j k - \frac{1}{h}\right)}\right), & j = rn, rn + 1, \ldots, 2rn - 1, \\ \ldots \\ \Re\left(\xi_k e^{2\pi i \left(t_j k - \frac{h-1}{h}\right)}\right), & j = (h-1)rn, \ldots, hrn. \end{cases} \tag{8.61}$$

For the case when the linear forms are given by (8.61), a more thorough analysis of the structure of (8.52) allows me to state the following bound, given here without proof.

Theorem 8.7   *Under the conditions of Theorem 8.5, with $M = r \cdot h \cdot n$ linear forms, given by (8.61), we have*

$$\min_{\mathbf{d} \in \mathcal{D}} \max_{j=0, \ldots, M-1} |L_j(\mathbf{d})| \le \mathcal{M}, \tag{8.62}$$

and, for $\alpha > 1$,

$$\Pr_{\mathbf{d} \in \mathcal{D}}\left(\max_j |L_j(\mathbf{d})| \ge \sqrt{\alpha} \mathcal{M}\right) \le \frac{1}{\alpha^s}, \tag{8.63}$$

where, for M-PSK,

$$\mathcal{M} = \left(\frac{(2s)!}{2^s s!}\right)^{\frac{1}{2s}} \cdot \left(r \cdot h \cdot n \left(\left(\frac{n}{2}\right)^s \cdot \left(1 + \frac{n^{-\frac{1-\ln 2}{2}}}{\sqrt{2}}\right)^s + n^{-\ln 2} \cdot n^s\right)\right)^{\frac{1}{2s}}.$$

$\square$

Similar bounds can be derived for other reflection-symmetric constellations, e.g., QAM. I omit the cumbersome details.

Corollary 8.8   *Under the conditions of Theorem 8.7, for $s = \ln n$, $n \ge 2$,*

$$\min_{\mathbf{d} \in \mathcal{D}} \max_{j=0, 1, \ldots, M-1} |L_j(\mathbf{d})| \le \sqrt{n \ln n} \cdot \left(1 + \frac{1}{\ln n}\right),$$

$$\Pr_{\mathbf{d} \in \mathcal{D}}\left(\max_{j=0, 1, \ldots, M-1} |L_j(\mathbf{d})| \ge \sqrt{\alpha n \ln n} \cdot \left(1 + \frac{1}{\ln n}\right)\right) \le \frac{1}{n^{\ln \alpha}}.$$

Proof Use

$$\left(\left(1 + \frac{n^{-\frac{1-\ln 2}{2}}}{\sqrt{2}}\right)^{\ln n} + 1\right)^{\frac{1}{2\ln n}} < 1 + \frac{1}{2\ln n},$$

along with (8.60). □

Theorem 8.9 *Let $\mathcal{D}$ be a code of strength $2s$ from $\{-1, 1\}^n$. For every $\xi \in \mathcal{Q}^n$ there exists $\mathbf{d} \in \mathcal{D}$, such that*

$$\mathrm{PMEPR}(\xi * \mathbf{d}) \leq \Upsilon$$

$$= E_{\max} \cdot \min_{r>1,\, rn \in \mathbb{N}}\ \min_{h>1,\, h \in \mathbb{Z}} \left(\frac{r^{\frac{1}{s}}}{\cos^2 \frac{\pi}{2r}} \cdot \frac{h^{\frac{1}{s}}}{\cos^2 \frac{\pi}{2h}} \cdot \left(\frac{(2s)!n}{2^{2s}s!}\right)^{\frac{1}{s}} \cdot \left(1 + \frac{1}{s}\right)\right), \quad (8.64)$$

$$\xi * \mathbf{d} = (\xi_0 d_0, \dots, \xi_{n-1}d_{n-1}).$$

Since $\mathcal{Q}$ is reflection-symmetric, $\xi * \mathbf{d} \in \mathcal{Q}^n$.

Proof Use Theorem 8.5 combined with Lemma 4.13 and the definition of PMEPR. Also use the inequality

$$\left(\left(1 + \frac{1}{\sqrt{2n^{1-\ln 2}}}\right)^s + 1\right)^{\frac{1}{s}} < 1 + \frac{1}{s}, \quad n \geq 2.$$

□

Corollary 8.10 *Under the conditions of Theorem 8.9, for $s = \ln n$, and for every $n \geq n_0$, we have*

$$\Upsilon \leq E_{\max} \cdot n \ln n \cdot \left(1 + \frac{\sigma_{n_0} \ln \ln n}{\ln n}\right),$$

where $\sigma_{64} = 19$, $\sigma_{128} = 15$, $\sigma_{2048} = 8$, and $\sigma_{n_0} = 1 + \varepsilon$, $\epsilon > 0$, and ε becomes arbitrarily small for large n_0.

Proof Choose $r = h = \sqrt{\ln n}$, and use standard inequalities. □

Now let me describe the implementation of the PMEPR reduction scheme. Let $\mathcal{D}$ be a code of strength $2s$ from $\{-1, 1\}^n$ of size 2^p. The following particular case of selective mapping is used. Let $\xi = (\xi_1, \dots, \xi_n) \in \mathcal{Q}^n$ be the vector we wish to transmit. Compare the PMEPR of 2^p vectors, $\xi * \mathbf{d}$, where $\mathbf{d}$ runs over $\mathcal{D}$, and send the signal corresponding to $\xi' = \xi * \mathbf{d}'$ with the minimum PMEPR, along with the side information of p bits, indicating which balancing vector has been chosen. This allows the receiver to recover $\mathbf{d}'$ by encoding the p information bits into the corresponding word from $\mathcal{D}$, and therefore reconstruct the vector $\xi = \xi' * \mathbf{d}'$. We arrive at the following result.

Theorem 8.11 *Let $\mathcal{D}$ be a binary linear systematic code of strength $2s$ and size 2^p. Then there exists a scheme for PMEPR reduction guaranteeing that PMEPR does not exceed Υ from (8.64) providing the rate hit $\frac{p \log_q 2}{n}$ and implementation complexity proportional to $n2^p$.* □

Using duals of BCH codes we obtain the following corollary.

Corollary 8.12 *The described scheme guarantees the maximum PMEPR of Υ defined in (8.64) with the rate hit $\frac{s \log_q(n+1)}{n}$.* □

Notice that to compute PMEPR in the algorithm it is necessary to calculate the values of $r \cdot n$ complex linear forms; the projection on axes is used only in the proof.

Transmission of the side information is an important issue in implementing the described algorithm. In what follows, I discuss several options. We assume that the signal ξ is obtained as a result of coding that can be distorted by the following multiplication by a balancing vector. A choice at the receiver is that we may either first multiply by the balancing vector followed by decoding, or start from decoding and then multiply by the balancing vector. The simplest situation is when there exist very reliable uncoded bits which can be used to convey the index of the balancing vector (e.g., when only one or two bits from a constellation of size 8 or more are protected by an error-correcting code). If these bits are mapped to antipodal constellation points this does not affect the resulting PMEPR. Another possibility is that we have p reliable subcarriers (this can be achieved, e.g., by decreasing the size of the constellation in these subcarriers). Without loss of generality, assume that these p subcarriers are the first ones, otherwise a permutation of the coordinates in the balancing vectors should be used. Let $\mathcal{D}$ be a systematic code, i.e., having the information bits at its first p positions. Let $\mathcal{Q}^*$ be a half of the constellation $\mathcal{Q}$, in which we pick one out of every pair of antipodal points. Let $\xi = (\xi_0, \dots, \xi_{n-1})$, with $\xi_0, \dots, \xi_{p-1} \in \mathcal{Q}^*$, and $\xi_p, \dots, \xi_{n-1} \in \mathcal{Q}$. Compare the PMEPR of 2^p vectors, $\xi * \mathbf{d}$, where $\mathbf{d}$ runs over $\mathcal{D}$, and send the signal corresponding to $\xi' = \xi * \mathbf{d}'$ with the minimum PMEPR. At the receiving end, one deduces the binary information vector of $\mathbf{d}'$ by checking whether, in the received vector ξ', each of the first p components belongs or does not belong to $\mathcal{Q}^*$.

In this setting it is also possible to compress the information about the chosen code vector to $\lceil s \log_q(n + 1) \rceil$ tones (perhaps reserved). This allows further minimization of the number of the subcarriers affected by the algorithm. In this case the PMEPR is minimized for the signal vector containing the transmitted information. This, however, yields a slight increase by $\lceil s \log_q(n + 1) \rceil$ in the estimate for PMEPR.

Now consider the situation when we prefer to decode first and only then to subtract the balancing vector. Let the transmitted information be protected by some error-correcting code $\mathcal{D}'$, i.e., only vectors $\xi \in \mathcal{D}' \subset \mathcal{Q}^n$ are sent. To start from decoding in $\mathcal{D}'$ we have to guarantee that the modified vector always belongs to $\mathcal{D}'$. For instance, if $q = 2$, i.e., when we use BPSK, and $\mathcal{D}'$ is a linear code, it is sufficient that the code $\mathcal{D}$ we use for balancing is a subcode of $\mathcal{D}'$. Then the modified vector ξ' also belongs to $\mathcal{D}'$ and can be decoded without knowledge of the balancing vector. For higher than BPSK constellations and the use of a linear code, the embedding $\mathcal{D} \subset \mathcal{D}'$ provides a sufficient condition for this scheme to work. This embedding is not very restrictive. For example, if $\mathcal{D}$ is a dual BCH code of fixed strength, it is possible to show that it is nested in BCH codes with a constant minimum distance.

Let us pass to a randomized version of the scheme. Indeed, implementation of the full deterministic scheme for meaningful s is computationally challenging. However, by picking at random at most a fixed number of balancing vectors from the code, we could guarantee that an arbitrary close-to-1 probability of PMEPR, restricted to the derived deterministic bound, is achieved. Possible implementations of the scheme vary according to the chosen method of the balancing vector transmission.

To analyze such a scheme, assume that $s = \ln n$, and the number of balancing vectors used is H. Using Chernoff's bound, for real $\alpha > 1$, and large n, and its tightness for a single linear form, for a random channel code $\mathcal{C}$, we have

$$0.5n^{-\alpha} \leq \Pr\left(\text{PMEPR}(\mathcal{C}) \geq \alpha \ln n\right) \leq 2n^{-\alpha+1}, \tag{8.65}$$

i.e., a polynomial in n decrease. Considering another range of PMEPR, we have for $\beta > 0$,

$$\Pr(\text{PMEPR}(\mathcal{C}) \geq \ln n + \beta \ln \ln n) \leq 2(\ln n)^{-\beta}. \tag{8.66}$$

Theorem 8.13 *For any $\xi \in \mathcal{Q}^n$, let $\mathbf{d}_0, \mathbf{d}_1, \ldots, \mathbf{d}_{H-1}$ be randomly picked from a code $\mathcal{D}$ of strength $2 \ln n$. Then, for all $n \geq n_0$,*

$$\Pr\left(\min_{\ell=0,1,\ldots,H-1} \text{PMEPR}(\xi * \mathbf{d}_\ell) \geq \alpha \ln n + \sigma_{n_0} \ln \ln n\right) \leq n^{-H \ln \alpha}, \tag{8.67}$$

$$\Pr\left(\min_{\ell=0,1,\ldots,H-1} \text{PMEPR}(\xi * \mathbf{d}_\ell) \geq \ln n + (\beta + \sigma_{n_0}) \ln \ln n\right)$$
$$\leq n^{-H \cdot \left(\ln\left(1+\beta \frac{\ln \ln n}{\ln n}\right)\right)}, \tag{8.68}$$

where the constant σ_{n_0} is given by Corollary 8.10.

Proof This is immediate from Corollary 8.10. $\qquad\square$

Indeed we see that the scheme allows to considerably improve the PMEPR statistics, using only $(\ln 2) \cdot (\log_2 n)^2 + 1$ bits of redundancy (i.e., $\left\lceil \frac{(\ln 2) \cdot (\log_2 n)^2 + 1}{\log_2 q} \right\rceil$ redundant subcarriers), and a modest increase in complexity. Moreover, the result is mathematically rigorous, applicable to any reflection-symmetric constellation, and provides the reduction for any information vector. In other words, for *any* information vector, choosing H large enough, we can *provably* make the probability of the large PMEPR arbitrarily small, up to the deterministic bounds, attained at H being equal the code size.

As an example, setting $H = \frac{\rho n}{\ln n \cdot \ln \alpha}$, we have for all $n \geq n_0$,

$$\Pr\left(\min_{\ell = 0, 1, \dots, H-1} \text{PMEPR}(\xi * \mathbf{d}_\ell) \geq \alpha \ln n + \sigma_{n_0} \ln \ln n \right) \leq e^{-\rho n}.$$

As another example, setting $H = \frac{n}{\ln \ln n}$, for all $n \geq n_0$,

$$\Pr\left(\min_{\ell = 0, 1, \dots, H-1} \text{PMEPR}(\xi * \mathbf{d}_\ell) \geq \ln n + (\beta + \sigma_{n_0}) \ln \ln n \right) \leq e^{-\beta n \left(1 - \beta \frac{\ln \ln n}{\ln n}\right)},$$

i.e., comparing with (8.65), we transform the probability from one that decreases polynomially in n into one that decreases *exponentially*.

In the above, I have provided a probabilistic framework for PMEPR reduction towards certain values, depending on the scheme parameters (e.g., balancing code strength, oversampling factor r, number of axes h). Assume for simplicity that the constellation used has $E_{\max} = 1$ (for instance M-PSK).

For any information vector length, n, the balancing code strength, $2s$, prescribes the optimal values of r and h. Denote the PMEPR bound, guaranteed for the channel code $\mathcal{C}$, using the balancing code, $\mathcal{D}_{2s}$, of strength $2s$ (either deterministically, using the whole code, or probabilistically, using a chosen number of candidates, for the wanted peak probability reduction), by $\text{PMEPR}_{\mathcal{D}_{2s}}(\mathcal{C})$. For the balancing code of the least meaningful strength, $2s = 4$, say the dual of the extended BCH code of strength 4 (it is dual to the extended 2-error correcting BCH code), length 2^m, having 2^{2m+1} words, we need $2m + 1$ bits to indicate which specific code word is used. Choosing, e.g., $r = 3$, $h = 3$, we obtain $\text{PMEPR}_{\mathcal{D}_4}(\mathcal{C}) \leq 8\sqrt{\frac{n}{3}}$. Using the optimal strength balancing code, e.g., the dual BCH code of length $n = 2^m$ and strength $2 \ln n$, having $2^{m \ln n + 1}$ words, we need $m \ln n + 1$ bits to indicate which specific code word was used.

Figure 8.4 shows a simulation for $n = 128$, with oversampling $r = 5$, using balancing vectors (BV) randomly chosen from a strength $2s = 10$ dual-BCH code (with only 18 redundant carriers). For example, the peaks higher than 10.8 dB occur with probability 10^{-2}. Using 4 balancing vectors, the probability of such peaks is lowered to 10^{-5}. Looking at it differently, to build a system, for any

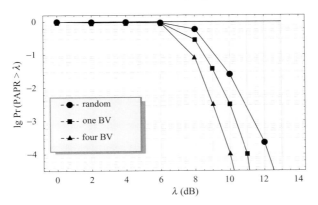

Figure 8.4 PMEPR distribution for $n = 128$ and QPSK using the strength 10 dual BCH code (after [251])

peak probability less than $10^{-2.5}$, we need the dynamic range reduced by 2 dB, at a modest cost of trying 4 balancing vectors. The complexity can thus be traded for PMEPR reduction, up to the theoretical limits provided in the previous sections.

8.5 Trellis shaping

In the two previous sections, I allowed modification of only one (e.g., MSB) bit to determine the point of the used constellation. One of the two choices for the bit corresponded to the selection of one of the two antipodal points, allowing the PMEPR of the corresponding MC signal to be decreased. More generally, one may modify several MSBs to achieve the goal. This method got its name of trellis shaping because of its similarity to the one used in average power reduction.

Let us start with a description of the block code version, which will be followed by the convolutional code one. Let an M-QAM constellation be used, $M = 2^m$, where m is an even integer and $m = m_1 + m_2$. The m_1 bits are MSBs. Let, moreover, C be an $[N = nm_1, K]$ linear code defined by a generating matrix $\mathbf{G}$ of size $K \times N$, and let $\bar{\mathbf{G}}$ of size $(N - K) \times N$ be such that $\begin{pmatrix} \mathbf{G} \\ \bar{\mathbf{G}} \end{pmatrix}$ is a full-rank $N \times N$ matrix.

Denote by $\bar{C}$ the code defined by $\bar{\mathbf{G}}$. Furthermore, let $\bar{\mathbf{H}}$ of size $K \times N$ be the parity check matrix of the code defined by $\bar{\mathbf{G}}$, i.e., $\bar{\mathbf{H}}\bar{\mathbf{G}}^t = 0$. Notice that for any $\bar{\mathbf{c}} \in \bar{C}$, and arbitrary $\mathbf{c} \in C$,

$$\bar{\mathbf{H}}(\mathbf{c} + \bar{\mathbf{c}})^t = \bar{\mathbf{H}}\mathbf{c}^t + \bar{\mathbf{H}}\bar{\mathbf{c}}^t = \bar{\mathbf{H}}\mathbf{c}^t.$$

In other words, we have partitioned the (Hamming) space to cosets of $\bar{C}$ such that the vectors from the same coset have the identical result of multiplication by the matrix $\bar{\mathbf{H}}$. The idea is to transmit information using the index of the coset, and by picking one of the vectors in the coset to minimize the PMEPR of the resulting signal. Notice that the balancing methods correspond to the use of a mapping with only one MSB indicating one of the two antipodal constellation points and $\bar{C}$ of full rank or of a given strength.

The information vector for encoding the MC symbol consists of $K + nm_2$ bits. The first K bits are encoded into a binary vector of length $N = nm_1$ using the generating matrix $\mathbf{G}$, the PMEPR minimizing vector from $\bar{C}$ is added (modulo 2) to the result of the encoding, giving nm_1 bits, and along with nm_2 uncoded LSBs we obtain $n(m_1 + m_2) = nm$ bits determining n constellation points for each of the subcarriers. At the receiver end, the vector undergoes the inverse processing. Namely, the nm_2 are obtained directly from the LSBs of the received constellation points, and the K bits are uniquely determined by the result of multiplication of $\bar{\mathbf{H}}$ on the vector consisting of the nm_2 MSBs.

The suggested scheme is especially simple when we use a convolutional code C_s of rate $\frac{1}{n_s}$. Let $\mathbf{G}$ be the corresponding $1 \times n_s$ generator matrix. Let $\mathbf{H}^t$ and $(\mathbf{H}^{-1})^t$ denote the $n_s \times (n_s - 1)$ parity check matrix and its left inverse, i.e., $(n_s - 1) \times n_s$, matrix for this code respectively. Let $\mathbf{d}$ be a binary information data sequence to be transmitted by each n-subcarrier MC symbol. The information data bits are first divided into the two subsequences, $\mathbf{s}$ and $\mathbf{b}$, where the former is used to choose MSB of the mapping constellation labeling, and the latter chooses its LSB. In choosing the MSB, an $(n_s - 1)$-bit sequence $\mathbf{s}$ is first encoded by the inverse syndrome former to generate an n_s-bit sequence $\mathbf{z}$, i.e.

$$\mathbf{z} = \mathbf{s}(\mathbf{H}^{-1})^t.$$

Any valid code word $\mathbf{y}$ in C_s can be modulo-2 added to the sequence $\mathbf{z}$ without changing the original data sequence $\mathbf{s}$. Indeed,

$$(\mathbf{z} \oplus \mathbf{y})\mathbf{H}^t = \mathbf{s}(\mathbf{H}^{-1})^t \, \mathbf{H}^t \oplus \mathbf{y}\mathbf{H}^t = \mathbf{s} \oplus 0 = \mathbf{s},$$

since $\mathbf{y}\mathbf{H}^t = 0$ for any $\mathbf{y} \in C_s$.

As an example, consider the mapping of 16-QAM depicted in Fig. 8.5. Notice that the points labeled by the same couple of LSB have the same energy, and since a choice of $\mathbf{y}$ affects only MSB, the average power does not depend on $\mathbf{y}$.

What remains is to define a procedure for choosing the best $\mathbf{y}$ for PMEPR minimization. Clearly, going over all possibilities is intractable. Therefore, some suboptimal strategy might be employed. One possible option is to choose the code sequence minimizing the aperiodic autocorrelation. It was shown in Section 7.1 that

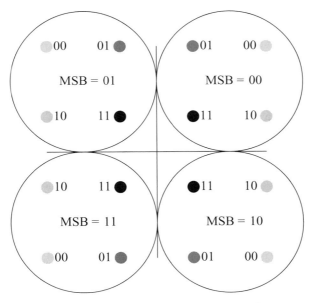

Figure 8.5 Constellation mapping for 16-QAM

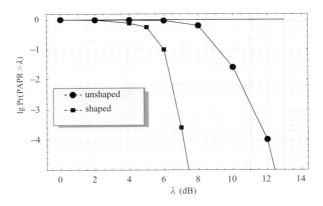

Figure 8.6 Trellis shaping for 256-QAM and $n = 256$ (after [293])

there is a connection between the sum of absolute values of the out-of-phase auto-correlations. To conform with the Viterbi algorithm it is advantageous to calculate the sum of the squares of the absolute values of the out-of-phase correlations. Moreover, it is possible to compute it recursively, choosing, from several possibilities arriving at the same state, the one that minimizes the considered sum for the already chosen subsequence. Figure 8.6 demonstrates simulation results for the trellis shaping in a 256-QAM modulated 256 subcarriers MC system using a convolutional code of rate $\frac{1}{2}$, with the labeling analogous to Fig. 8.5 and the described metrics.

8.6 Tone injection

When using trellis shaping, we have an equally likely choice of each of the subconstellations defined by the MSBs. Notice that the subconstellations are merely shifted versions of one of them. In contrast to this approach, one could concentrate on one centrally placed constellation while exploring the possibility of moving to its shifted version only when the signal exhibits a high value of PMEPR. Since the points of a shifted constellation are spaced from the points of the initial constellation, the error rate will not essentially deteriorate. On the other hand, the substitution could yield the average power growth, and it remains to show that this is compensated by the decrease in PMEPR. This method is called *tone injection*, as substituting the points in the basic constellation for the new points in the larger constellations is equivalent to injecting a tone of the appropriate frequency and phase in the MC symbol.

An example of a 16-QAM constellation with its four shifts is given in Fig. 8.7. The labeling of the points in each of the shifts is identical.

Formally, let the distance between the closest points in the basic constellation Q be d, and let the basic MC signal be $F_\mathbf{a}(t) = \sum_{k=0}^{n-1} a_k e^{2\pi \iota k t}$, $a_k \in Q$. By shifting the coefficients to the neighboring constellations we may choose between the signals

$$\sum_{k=0}^{n-1} (a_k + dp_k + \iota dq_k) e^{2\pi \iota k t},$$

where $p_k, q_k \in \mathbb{Z}$. The simplest choice of $p_k = q_k = 0$ gives the original constellation point. The choice of one from the following options $p_k = q_k = 0$; $p_k = \pm 1, q_k = 0$; $p_k = 0, q_k = \pm 1$; gives points from the constellations in Fig. 8.7. At the receiver, choosing nonzero coefficients p_k and q_k does not cause an ambiguity since the coordinates of the detected constellation point can be reduced, modulo d.

We will search for the coefficients p_k and q_k, $k = 0, 1, \ldots, n - 1$, minimizing the PMEPR of the resulting MC signal. However, even restricting the coefficients to $-1, 0, 1$ will cause an intractable computational problem. Thus a suboptimal algorithm is sought.

One of the options is to update the coefficients in an iterative manner. The algorithm starts by assigning $p_k = q_k = 0$. On each iteration, it is checked whether substituting one of the coefficients p_k by $p_k \pm 1$ or q_k by $q_k \pm 1$ decreases either the maximum of the absolute value of the signal or the PMEPR of the signal. The algorithm is halted, either when no progress in decreasing the optimized function is achieved, or after a fixed number of iterations.

Figure 8.8 presents a simulation of the PMEPR distribution after four iterations of the described algorithm for a 16-QAM modulated MC system with 64 subcarriers. We see that about 5 dB reduction is achieved at a clipping rate of 10^{-5}.

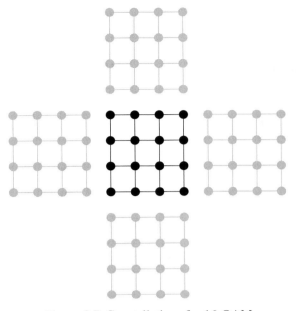

Figure 8.7 Constellations for 16-QAM

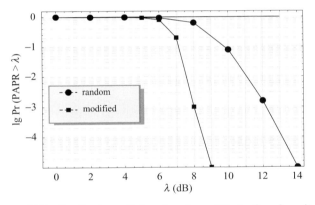

Figure 8.8 PMEPR distribution of the signal modified after four iterations for 16-QAM and $n = 64$ (after [400])

8.7 Active constellation extension

In active constellation extension (ACE), some of the outer constellation points are dynamically extended toward the outside of the original constellation such that the PMEPR of the MC signal is reduced. The main idea of the scheme is easily explained in the case of the MC signal with QPSK modulation. In each subcarrier, there are four possible constellation points that lie in each quadrant in the complex plane

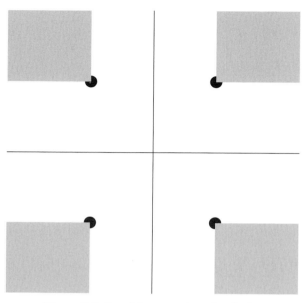

Figure 8.9 Possible extensions for QPSK

(see Fig. 8.9). Assuming AWGN noise, the maximum-likelihood decision regions are the four quadrants bounded by the axes. Thus, any point that is further from the decision boundaries than the nominal constellation point (in the proper quadrant) will offer an increased margin, which guarantees a lower BER. We can, therefore, allow modification of constellation points within the quarter plane outside the nominal constellation points with no degradation in performance. This corresponds to the shaded areas on Fig. 8.9. The idea can be applied to other constellations as well, see, e.g., Fig. 8.10 for the 16-QAM case.

Formally, let for a constellation point $a \in \mathcal{Q}$, $E(a)$ be the extension region of a. For an MC signal $F_{\mathbf{a}}(t)$, we seek a vector $\mathbf{x} = (x_0, \ldots, x_{n-1})$, $x_k \in \mathbb{C}$, such that it delivers

$$\min_{\mathbf{x} \in \mathbb{C}^n} \max_{t \in [0,1)} \left| \sum_{k=0}^{n-1} (a_k + x_k) e^{2\pi \imath k t} \right|,$$

under the conditions

$$a_k + x_k \in E(a_k), \quad k = 0, 1, \ldots, n - 1.$$

If one passes to a discrete version of the problem addressing an oversampled signal, we still arrive at a quadratically constrained quadratic program, and obtaining the optimal solution is computationally challenging. Therefore, the following simplified method can be employed. The algorithm starts with a coefficient vector $\mathbf{a} = (a_0, a_1, \ldots, a_{n-1})$ and assumes the clip level, A, that we wish to attain.

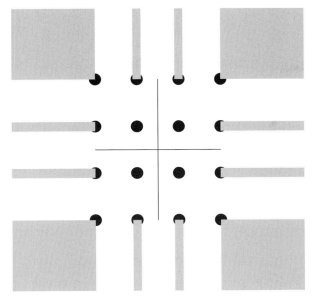

Figure 8.10 Possible extensions for 16-QAM

1. Compute, for $j = 0, 1, \ldots, n - 1$,

$$F_{\mathbf{a}}\left(\frac{j}{n}\right) = \sum_{k=0}^{n-1} a_k e^{2\pi \imath k \frac{j}{n}}.$$

2. Clip the values of $F_{\mathbf{a}}$ with absolute value exceeding A (soft limiter),

$$\hat{F}_{\mathbf{a}}\left(\frac{j}{n}\right) = \begin{cases} F_{\mathbf{a}}\left(\frac{j}{n}\right) & \text{if } \left| F_{\mathbf{a}}\left(\frac{j}{n}\right) \right| \leq A, \\ A e^{\imath \arg F_{\mathbf{a}}\left(\frac{j}{n}\right)} & \text{if } \left| F_{\mathbf{a}}\left(\frac{j}{n}\right) \right| > A. \end{cases}$$

3. Compute for $k = 0, 1, \ldots, n - 1$,

$$\bar{a}_k = \frac{1}{n} \sum_{j=0}^{n-1} \hat{F}_{\mathbf{a}}\left(\frac{j}{n}\right) e^{-2\pi \imath j \frac{k}{n}}.$$

4. Enforce all constraints on $\bar{a}_k$, i.e., find the closest point to $\bar{a}_k$ in $E(a_k)$,

$$\hat{a}_k = \arg \min_{a \in E(a_k)} d_E(a, \bar{a}_k).$$

5. Return to Step 1, and iterate till no points are clipped or the PMEPR is essentially reduced.

Figure 8.11 shows the distribution of the PMEPR of MC signals in an $n = 256$ QPSK-modulated system after application of the algorithm. The parameter A is chosen to be 4.86 dB. Improved convergence of the algorithm can be achieved by employing gradient-like methods.

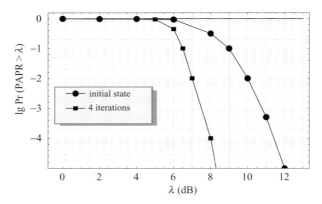

Figure 8.11 PMEPR for QPSK-modulated $n = 256$ MC system with ACE algorithm (after [220])

8.8 Constellation shaping

In constellation shaping, we try to find a constellation in the n-dimensional frequency domain, such that the resulting shaping region in the time domain has low PMEPR. At the same time we would like to have a simple encoding method for the chosen constellation.

Formally, let β be a parameter, and consider all signals $F_{\mathbf{a}}(t) = \sum_{k=0}^{n-1} a_k e^{2\pi \imath k t}$ such that

$$\max_{t \in [0,1)} |F_{\mathbf{a}}(t)| \leq \beta.$$

This restriction translates to the requirement that $\mathbf{a}$ belongs to a parallelotope in $\mathbb{C}^n$. The integer points inside this parallelotope are used as constellation points in transmitting the MC signals. The main challenge in constellation shaping is to find a unique way of mapping the input data to the constellation points such that the mapping (encoding) and its inverse (decoding) can be implemented with reasonable complexity.

Let $\mathbf{y} = (y_0, \ldots, y_{n-1})$ be defined by $y_j = F_{\mathbf{a}}(\frac{j}{n})$. Then

$$\Re(y_j) = \sum_{k=0}^{n-1} \Re(a_k) \cdot \Re\left(e^{2\pi \imath k \frac{j}{n}}\right) - \sum_{k=0}^{n-1} \Im(a_k) \cdot \Im\left(e^{2\pi \imath k \frac{j}{n}}\right),$$

$$\Im(y_j) = \sum_{k=0}^{n-1} \Im(a_k) \cdot \Re\left(e^{2\pi \imath k \frac{j}{n}}\right) + \sum_{k=0}^{n-1} \Re(a_k) \cdot \Im\left(e^{2\pi \imath k \frac{j}{n}}\right).$$

Denoting

$$\mathbf{Y} = (\Re(y_0), \ldots, \Re(y_{n-1}), \Im(y_0), \ldots, \Im(y_{n-1}))^t,$$
$$\mathbf{A} = (\Re(a_0), \ldots, \Re(a_{n-1}), \Im(a_0), \ldots, \Im(a_{n-1}))^t,$$

we may write the corresponding transform as

$$\mathbf{Y} = \mathcal{F}\mathbf{A},$$

with $\mathcal{F}$ being a real $2n \times 2n$ matrix. Let $\mathcal{F}^{-1}$ be its inverse matrix. Define the parallelotope, $\mathcal{P}$, with the bases being the columns of $\mathcal{K} = [\alpha\mathcal{F}^{-1}]$, where $[\cdot]$ is rounding and α is chosen in such a way that the number of integer points within $\mathcal{P}$ is the same as in the signals defined by an unshaped constellation. The following result provides a tool for the encoding procedure of these points.

Theorem 8.14 *The matrix $\mathcal{K}$ can be decomposed into*

$$\mathcal{K} = \mathcal{U}\mathcal{D}\mathcal{V},$$

where $\mathcal{D}$ is diagonal with the integer entries σ_j, $j = 0, \ldots, n-1$, such that $\sigma_0|\sigma_1| \ldots |\sigma_n$, and $\mathcal{U}$ and $\mathcal{V}$ are unimodular matrices. □

This decomposition of an integer matrix is known as the Smith normal form (SNF) decomposition, and the matrix $\mathcal{D}$ is called the SNF of $\mathcal{K}$. Given such decomposition the encoding can be done as follows.

Let $n = 2^m$. For an integer, $\mathcal{J}$, define its canonical representation $\mathbf{j} = (j_0, \ldots, j_{n-1})$ as the result of the following recursive procedure:

$$j_0 = \mathcal{J} \bmod \sigma_0, \quad \mathcal{J}_0 = \frac{\mathcal{J} - j_0}{\sigma_0},$$

$$j_k = \mathcal{J}_{k-1} \bmod \sigma_k, \quad \mathcal{J}_k = \frac{\mathcal{J} - j_k}{\sigma_k}.$$

Then the constellation point $\mathbf{a}$ corresponding to the number $\mathcal{J}$ is

$$\mathbf{a} = \mathcal{U}\mathbf{j} - \mathcal{K}\lfloor\mathcal{K}^{-1}\mathcal{U}\mathbf{j}\rfloor. \tag{8.69}$$

The reverse operation for finding $\mathcal{J}$ from $\mathbf{a}$ is as follows:

$$\mathbf{j} = \mathcal{U}^{-1}\mathbf{a} = (j_0, \ldots, j_{n-1})^t, \quad \tilde{j}_k = j_k \bmod \sigma_k,$$
$$\mathcal{J} = \tilde{j}_0 + \sigma_0(\tilde{j}_1 + \sigma_2(\ldots(\tilde{j}_{n-2} + \sigma_{n-2}\tilde{j}_{n-1})\ldots)). \tag{8.70}$$

The resulting coefficient vector $\mathbf{a}$ defines the corresponding MC signal. The complexity of the encoding procedure is proportional to n^2, which is quite high. Substitute the matrix $\mathcal{K}$ by the Hadamard matrix of the corresponding size.

The Hadamard matrix of size $2^m \times 2^m$ is defined recursively:

$$\mathcal{H}_{2^m} = \begin{pmatrix} \mathcal{H}_{2^{m-1}} & \mathcal{H}_{2^{m-1}} \\ \mathcal{H}_{2^{m-1}} & \mathcal{H}_{2^{m-1}} \end{pmatrix}, \quad \mathcal{H}_1 = (1).$$

The SNF decomposition of $\mathcal{H}_{2^m}$ can easily be computed as $\mathcal{H}_{2^m} = \mathcal{U}_{2^m} \mathcal{D}_{2^m} \mathcal{V}_{2^m}$, where

$$
\mathcal{U}_{2^m} = \begin{pmatrix} \mathcal{U}_{2^{m-1}} & 0 \\ \mathcal{U}_{2^{m-1}} & \mathcal{U}_{2^{m-1}} \end{pmatrix}, \quad \mathcal{D}_{2^m} = \begin{pmatrix} \mathcal{D}_{2^{m-1}} & 0 \\ 0 & 2\mathcal{D}_{2^{m-1}} \end{pmatrix}
$$

$$
\mathcal{V}_{2^m} = \begin{pmatrix} \mathcal{V}_{2^{m-1}} & \mathcal{V}_{2^{m-1}} \\ 0 & -\mathcal{V}_{2^{m-1}} \end{pmatrix}, \quad \mathcal{U}_{2^m}^{-1} = \begin{pmatrix} \mathcal{U}_{2^{m-1}} & 0 \\ -\mathcal{U}_{2^{m-1}} & \mathcal{U}_{2^{m-1}} \end{pmatrix},
$$

where $\mathcal{U}_1 = \mathcal{U}_1^{-1} = \mathcal{D}_1 = \mathcal{V}_1 = (1)$. Therefore, the encoding algorithm for this constellation can be implemented as follows:

$$
\mathbf{a} = \mathcal{U}_n \mathbf{j} - \mathcal{H}_n \left\lfloor \frac{\mathcal{H}_n^t \mathcal{U}_n \mathbf{j}}{n} \right\rfloor. \tag{8.71}
$$

The reverse operation is just as in (8.70) with $\mathcal{U}_n$ replacing $\mathcal{U}$. Notice that the encoding and decoding can be implemented by a butterfly structure using only bit shifting and logical AND, and is much simpler than in the case of $\mathcal{K}$. Although the constellation defined by Hadamard matrices guarantees a slightly less favorable PMEPR than the one based on a real Fourier matrix, the advantageous simplicity of encoding and decoding justifies its application. It remains to determine the number of points in the constellation. It equals the determinant $\det(\mathcal{H}_{2^m})$.

Theorem 8.15 *The constellation size for $n = 2^m$ is*

$$
\det(\mathcal{H}_{2^m}) = 2^{m2^{m-1}} = n^{\frac{n}{2}}.
$$

Proof By the SNF decomposition, $\det(\mathcal{H}_{2^m}) = \det(\mathcal{D}_{2^m})$, because the matrices $\mathcal{U}_{2^m}$ and $\mathcal{V}_{2^m}$ are unimodular. We use induction. Indeed, $\det(\mathcal{H}_2) = 2$. Let the claim hold for $\mathcal{H}_{2^k}$. Then

$$
\det(\mathcal{D}_{2^{k+1}}) = \det(\mathcal{D}_{2^k}) \cdot 2^{2^k} \cdot \det(\mathcal{D}_{2^k}) = 2^{(k+1)2^k}.
$$

$\square$

Simulation results for $n = 128$ and a shaped constellation with the same number of points as an MC system with 16-QAM modulation are presented in Fig. 8.12.

8.9 Partial transmit sequences

In this method, which is similar to selective mapping, modification of phases of subcarriers is allowed. Let $J = \{0, 1, \ldots, n - 1\}$ be the index set, and $J_0, J_1, \ldots, J_{V-1}$, be a partition of J to V pairwise disjoint subsets of the same size, i.e.,

$$
J_0 \cup \ldots \cup J_{V-1} = J, \quad J_k \cap J_m = \emptyset \text{ for } k, m = 0, 1, \ldots, V - 1; k \neq m.
$$

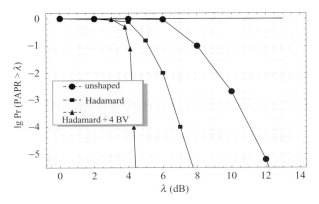

Figure 8.12 PMEPR distribution for $n = 128$ and the number of constellation points equal to the one in 16-QAM (after [274])

For each of the subsets, we introduce a rotation factor $b_v = e^{\iota \theta_v}$, $\theta_v \in [0, 2\pi)$. Let the information bearing vector be $\mathbf{a} = (a_0, \ldots, a_{n-1})$. The vectors

$$\mathbf{a}^{(v)} = \left(a_0^{(v)}, \ldots, a_{n-1}^{(v)} \right), \quad v = 0, \ldots, V - 1,$$

where

$$a_k^{(v)} = \begin{cases} a_k & \text{if } k \in J_v \\ 0 & \text{otherwise} \end{cases}$$

are called *partial transmit sequences* (PTS). Then, the corresponding MC signal is

$$F_{\mathbf{a}}(t) = \sum_{v=0}^{V-1} \sum_{k \in J_v} a_k b_v e^{2\pi \iota k t}$$

$$= \sum_{v=0}^{V-1} b_v \left(\sum_{k=0}^{n-1} a_k^{(v)} e^{2\pi \iota k t} \right) = \sum_{k=0}^{n-1} a_k^{(v)} F_{\mathbf{a}}^{(v)}(t). \tag{8.72}$$

The values of b_v are chosen from a restricted set of possibilities and minimize the PMEPR of $F_{\mathbf{a}}(t)$. These values are delivered to the receiver as side information. Having been given $\mathbf{b} = (b_0, \ldots, b_{V-1})$, the receiver easily reconstructs $\mathbf{a}$ by applying DFT to the received signal followed by multiplication of the results by the corresponding factors b_v^*.

It is possible to refrain from explicitly transmitting side information if differentially encoded modulation across the subcarriers in each block is used. In this case, only the block partitioning must be known to the receiver and one subcarrier in each subblock must be left unmodulated as reference carrier.

Figure 8.13 presents results of simulations for $n = 512$ QPSK modulated MC signals and factors chosen from $\{\pm 1, \pm \iota\}$.

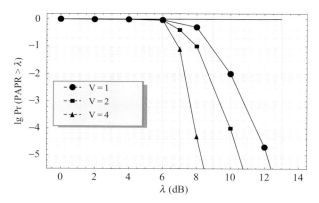

Figure 8.13 PMEPR distribution for $n = 512$ and QPSK modulation for partial transmit sequences with factors chosen from $\{\pm 1, \pm \iota\}$

Let the rotating factors be chosen from the set $\{e^{2\pi \iota \frac{j}{q}}, j = 0, 1, \ldots, q - 1\}$. As can be seen from (8.72) the decision about the optimal values of factors requires analysis of q^{V-1} possibilities (the first factor may be always set to 1). Since increasing V results in improved PMEPR reduction capability, the complexity of the method might become prohibitively high. Therefore, several suboptimal strategies have been proposed.

Iterative flipping. In this method the rotating factors are determined one by one. In the beginning, all the factors are set to 1. On the jth step of the algorithm, $j = 1, \ldots, V - 1$, the factor b_j is chosen in such a way that the PMEPR of the signal defined by $\mathbf{b} = (b_0 = 1, b_1, \ldots, b_j, 1, \ldots, 1)$ is minimal over the q possible choices. This procedure requires computation of PMEPR of the signal defined by the all-one vector $\mathbf{b}$ followed by checking $q - 1$ possibilities ($b_j \neq 1$) on each step, and thus has complexity proportional to $1 + (q - 1)(V - 1)$.

Neighborhood search. The algorithm starts with a predetermined vector $\mathbf{b}$ of phase factors. Next, it finds an updated vector $\mathbf{b}'$ in its neighborhood that results in the largest reduction in PMEPR. The neighborhood of radius r is defined as the set of vectors with Hamming distance equal to or less than r from the initial vector. The procedure continues with $\mathbf{b}'$ as the new initial vector. The search procedure is stopped after a predetermined number of steps, say M. The complexity of the procedure is proportional to $M \sum_{k=0}^{r} \binom{V}{k}(q - 1)^k$. When $r = V$ and $M = 1$ we arrive at the standard search over all possible phase factors.

Dual-layered phase sequencing. In this method the vector $\mathbf{b}$ is partitioned into d subsets. The values of the phase factors in each subset are set as in the iterative flipping, while, within each of the subsets, all possible combinations of the factors are checked. If $d = V$, we have the iterative flipping, while, for $d = 1$, we implement the exhaustive search. The complexity of the method is proportional to $q^{\frac{V}{d}}(d-1)$.

Orthogonal projections. From (8.72) the discrete version of the problem for an oversampled signal with factor r is to find $\mathbf{b}$ that delivers

$$\min_{\mathbf{b}} \max_{j=0,1,\ldots,rn} \left| \sum_{v=0}^{V-1} b_v F_{\mathbf{a}}^{(v)} \left(\frac{j}{rn} \right) \right| = \min_{\mathbf{b}} \max_{j=0,1,\ldots,rn} \left| \sum_{v=0}^{V-1} b_v f_j^{(v)} \right|. \tag{8.73}$$

The idea is to search for the best $\mathbf{b}$ among rn vectors, where each of them is close to being orthogonal to $\mathbf{f}_j = (f_j^{(0)}, f_j^{(0)}, \ldots, f_j^{(0)})$, $j = 0, 1, \ldots, n-1$. This is done as follows. Some predetermined vector (e.g., the all-one vector, $\mathbf{1}$) is projected onto the orthogonal space to $\mathbf{f}_j$, i.e., $\mathbf{y}_j$ is calculated,

$$\mathbf{y}_j = \left(I - \mathbf{f}_j (\mathbf{f}_j^t \mathbf{f}_j)^{-1} \mathbf{f}_j^t \right) \mathbf{1}.$$

This is followed by modification of the elements of $\mathbf{y}_j$ to the closest allowed phase. Finally, the best PMEPR reducing vector is chosen among the rn candidates. The complexity of the algorithm is proportional to rn.

Sphere decoding. The idea of this method is to reduce the number of considered combinations of $\mathbf{b}$ by restriction to a sphere-like area in the total space of possibilities. Assume that $\mathbf{b}$ is a column vector. Let

$$x_j = \sum_{v=0}^{V-1} b_v f_j^{(v)}.$$

Using $*$ for the conjugate transpose, we have

$$|x_j|^2 = \mathbf{b}^* \mathbf{f}_j^* \mathbf{f}_j \mathbf{b} = \mathbf{b}^* \left(\mathbf{f}_j^* \mathbf{f}_j + \alpha^2 I \right) \mathbf{b} - \alpha^2 \mathbf{b}^* \mathbf{b} = \mathbf{b}^* A_j \mathbf{b} - \alpha^2 V.$$

Here, α is an arbitrary nonzero number. The resulting $V \times V$ matrix, A_j, is positive definite due to the addition of $\alpha^2 I$, and therefore can be Cholesky factorized as

$$A_j = Q_j^* Q_j,$$

where Q_j is an upper-triangular matrix. Therefore,

$$|x_j|^2 = \mathbf{b}^* Q_j^* Q_j \mathbf{b} - \alpha^2 V = \| Q_j \mathbf{b} \|^2 - \alpha^2 V.$$

Choosing a goal value, μ, we aim to satisfy

$$\| Q_j \mathbf{b} \|^2 < \mu + \alpha^2 V.$$

This is equivalent to the following set of V inequalities, which has to be satisfied for each $j = 0, 1, \ldots, rn - 1$:

$$\sum_{v=V-m}^{V-1} \left| \sum_{u=v}^{V-1} (Q_j)_{v,u} b_u \right| < \mu + \alpha^2 V, \quad m = 0, 1, \ldots, V-1. \tag{8.74}$$

Notice that, for $m = 0$, the inequality contains only b_{V-1}, for $m = 1$ it comprises b_{V-1} and b_{V-2}, etc. This allows the following iterative choice of the candidate phase rotating vectors. First, find all b_{V-1} satisfying (8.74) for $m = 0$. Then, for each of the chosen b_{V-1}, determine possible b_{V-2} satisfying (8.74) for $m = 1$, etc. For accurately chosen parameters μ and α, the procedure essentially reduces the search space with modest penalty on achievable PMEPR reduction.

8.10 Peak reduction carriers

It is assumed here that a set of subcarriers is not used to transmit information, but rather to reduce peaks. These subcarriers are called *peak reduction carriers* (PRC). The modulation of the allocated subcarriers is chosen such that the PMEPR of the resulting signal is minimized. The receiver ignores the contents of the PRC. Formally, let $m < n$ be the number of PRC, and $K = (k_0, k_1, \ldots, k_{n-m-1})$ be the index set of the information bearing subcarriers, while $J = (j_0, j_1, \ldots, j_{m-1})$ is the index set of PRC. Then the coefficient vector $\mathbf{a} = (a_0, \ldots, a_{n-m-1}) \in \mathcal{Q}^{n-m}$ corresponds to the MC signal

$$F_{\mathbf{a},\mathbf{x}}(t) = \left(\sum_{\ell=0}^{n-m-1} a_\ell \, e^{2\pi i k_\ell t} \right) + \left(\sum_{\ell=0}^{m-1} x_\ell \, e^{2\pi i j_\ell t} \right),$$

where $\mathbf{x} = (x_0, \ldots, x_{m-1})$ is to be determined. We may restrict $\mathbf{x}$ either to belong to $\mathcal{Q}^m$ or not. Our goal is to find $\mathbf{x}$ such that $\mathrm{PMEPR}(\mathbf{a}, \mathbf{x})$ is minimized. By considering only samples of the r-times oversampled MC signal, we may reduce the problem to minimization of A such that

$$\max_{j=0,1,\ldots,rn-1} \left| F_{\mathbf{a},\mathbf{x}} \left(\frac{j}{rn} \right) \right| \leq A.$$

This is a quadratically constrained quadratic programming problem and its solution is computationally challenging. Thus suboptimal methods are called for.

The following is a simple gradient-type algorithm. At each step the algorithm updates the vector $\mathbf{x}$ by adding to it the result of DFT on the positions of PRC of the vector consisting of the differences between the samples of the MC signal and its clipped version.

More formally, we set a goal level A, the PRC set J, oversampling factor r, $r > 1$, an initial vector $\mathbf{x}^{(1)}$, and parameter μ. Let for $j = 0, 1, \ldots, rn - 1$,

$$\hat{F}_{\mathbf{a},\mathbf{x}} \left(\frac{j}{n} \right) = \begin{cases} F_{\mathbf{a},\mathbf{x}} \left(\frac{j}{n} \right) & \text{if } \left| F_{\mathbf{a},\mathbf{x}} \left(\frac{j}{n} \right) \right| \leq A, \\ A e^{i \arg F_{\mathbf{a},\mathbf{x}} \left(\frac{j}{n} \right)} & \text{if } \left| F_{\mathbf{a},\mathbf{x}} \left(\frac{j}{n} \right) \right| > A, \end{cases}$$

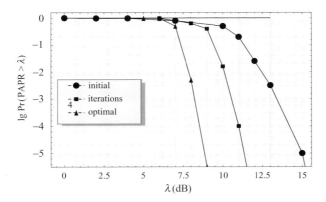

Figure 8.14 PMEPR distribution for $n = 512$ and QPSK modulation with PRC algorithm (5% of PRC) applied with four iterations (after [400])

and

$$\Delta^{(\ell)}(j) = F_{\mathbf{a},\mathbf{x}^{(\ell)}}\left(\frac{j}{n}\right) - \hat{F}_{\mathbf{a},\mathbf{x}^{(\ell)}}\left(\frac{j}{n}\right).$$

Then, for an index $p \in J$, the update on step $\ell + 1$ of the algorithm is

$$x_p^{(\ell+1)} = x_p^{(\ell)} + \mu \sum_{j=0}^{rn-1} \Delta^{(\ell)}(j) e^{2\pi \imath p \frac{j}{rn}}.$$

Another possible algorithm is based on approximation of the vector $\delta_j = (0, 0, \ldots, 1, \ldots, 0) \in \mathbb{C}^{rn}$ with the 1 in the jth position by a vector having nonzero frequency components only at J. For example, if $r = 1$, one may take

$$P_j(t) = \frac{1}{m} \sum_{k \in J} e^{2\pi \imath k \left(-\frac{j}{n} + t\right)}.$$

Clearly,

$$P_j\left(\frac{j}{n}\right) = 1,$$

and is less than 1 otherwise.

Then, in the iterative procedure at each step we find the maximum absolute value, say $\left|F_{\mathbf{a},\mathbf{x}}\left(\frac{j}{rn}\right)\right|$, and modify $F_{\mathbf{a},\mathbf{x}}(t)$ by subtracting $\mu e^{\imath \theta} P_j(t)$ where θ is chosen so that $\arg F_{\mathbf{a},\mathbf{x}}(t) = \arg \mu e^{\imath \theta} P_j(t)$.

Experiments with the described algorithms showed that contiguous PRC provide worse results in comparison with randomly distributed PRC. In Fig. 8.14, results of simulations of the gradient-type algorithm are compared with the optimal (computationally intractable in real-time applications) solution.

Table 8.1 *Comparison of methods*

Method	PMEPR reduction	Distortion	Rate hit	Side information	Complexity
Coding	H	N	H	N	H
Clipping	H	Y	L	N	L
SLM	M	N	L–H	Y	L–H
Balancing	M	N	H	N	L
Codes of strength (CS)	M	N	L	Y	H
Trellis shaping (TS)	M	N	L–H	N	H
Tone injection (TI)	M	N	H	N	L–H
ACE	L	N	L	N	H
Constellation shaping	H	N	H	N	H
PTS	M	N	L–H	Y	L–H
Reduction carriers (PRC)	L	N	L–H	N	H

L, low; M, moderate; H, high; Y, yes; N, no.

8.11 Comparison

Several methods for PMEPR reduction have been described in this book. Since, for most of the approaches, there is currently no theoretical method predicting their PMEPR reduction capability, one should gain intuition from particular simulation results. The main characteristics of reduction methods are the capability of PMEPR reduction, distortion in the signal the method yields, the rate hit, whether the method requires transmission of side information, and the complexity of implementation of the method. In Table 8.1, the mentioned characteristics are summarized for the described methods.

Clearly the table provides only a general picture of features of the described algorithms. In what follows, I will elaborate on it.

PMEPR reduction. Notice that we know (see Chapter 5) that the typical PMEPR is close to $\ln n$. Not only the number of the signals with PMEPR essentially greater than this, but also the number of the signals with a PMEPR essentially less than this, is small. The coding methods described in the previous chapter are directly intended for constructing signals having PMEPRs less than the typical ratio. Most of the methods described in this chapter, like SLM and PTS, for a small number of modifications or iterations exclude only the signals with high peaks, while achieving a higher probability of generating a typical vector. This is the reason that these methods usually achieve the typical PMEPR using relatively little effort, and then require much higher complexity to decrease it further. Another distinction between the methods is that most of them improve the statistics of the PMEPR, while in many situations it is required to restrict the PMEPR (and therefore the out-of-band radiation) to some prescribed value. These can be done using coding methods, balancing and cosets of codes of given strength.

Distortion. Clipping introduces in-band and out-of-band distortion, thus increasing the error probability. Filtering removes the out-of-band radiation, but at the same time yields peak regrowth. Other described methods are distortionless.

Rate hit. The price to be paid for PMEPR reduction is a loss in the number of possible transmit sequences. This loss is essential when coding is used and none when clipping or ACE is employed. The rate hit is low when either SLM, PTS, CS, or TS is used for PMEPR reduction to the typical values, while it is high if we want to further decrease it. In TI and constellation shaping, the rate hit is high, since the total number of sequences chosen from the extended constellation is large. The rate loss in using reduction carriers depends on the choice of the number of such carriers. However, according to simulations, the method becomes efficient if the percentage of the reserved tones is high.

Side information. Transmission of side information may be problematic, and requires special attention. Such methods as SLM, PTS, and CS require transmission of side information, although for each of them there exist modifications allowing this to be avoided.

Complexity. Such methods as SLM, PTS, and CS require a comparison of the PMEPR of several sequences, which in turn yields a necessity of several DFTs. This may essentially increase the complexity of the transmitter. On the other hand, TI, ACE, and RC use iterative algorithm implementation, which could be challenging. As well as this, some of the methods, for instance TI and ACE, may lead to power increase in the transmit signal.

8.12 Notes

Section 8.1   O'Neill and Lopes [305], Li and Cimini [239], and Dinis *et al.* [93, 94] studied the effects of clipping and filtering by extensive simulation. Armstrong [9, 10, 11], Armstrong and Feramez [12] and Chen and Haimovich [60] considered iterative methods of clipping and filtering. Chow, Bingham, and Flowers [65], Wulich [433], and Wulich and Goldfeld [437] considered the effect of amplitude limiting and scaling on performance. Panta and Armstrong [311, 312] considered the effect of clipping on performance in fading channels. Ochiai *et al.* [294] and Ochiai and Imai [295, 296, 297, 298] considered the effects of combining block error-correcting coding with clipping.

Section 8.1.1   I follow here Bahai *et al.* [18]. Properties of large excursions of Gaussian properties were studied by Rice [341]; see also van Vleck and Middleton

[420], Kac and Slepian [191], Mazo [261], Leadbetter *et al.* [233], and Blachman [28]. The relation (8.11) appears in [233], (8.13) is from [191], and (8.15) is from [341, 261]. The parabolic shape of the signal's crossings above high levels follows from a Taylor's series approximation, see [28].

Section 8.1.2   I follow here Ochiai and Imai [300, 302, 304]. Other papers dealing with analysis of nonlinearly transformed MC signals are Banelli and Cacopardi [19], Costa *et al.* [75], Costa and Pupolin [76], Dardari *et al.* [83, 84], Feig and Nadas [109], Friese [123], Kim and Stuber [204], Mestdagh *et al.* [267, 268], Mestdagh and Spryut [269], Ochiai [290, 291], Ochiai and Imai [301], O'Neill and Lopes [306], Pauli and Kuchenbecker [329], Rinne and Renfors [343], Saltsberg [350], Tellado *et al.* [403] and Wulich *et al.* [436].

Section 8.2   Selective mapping was proposed by van Eetvelt *et al.* [101] and Bäuml *et al.* [20], see also Müller *et al.* [280, 283]. Avoiding transmission of side information in SLM was considered in Breiling *et al.* [44, 45, 46], Cho *et al.* [62] and Han and Lee [153]. Other results on SLM are by Laroia *et al.* [229], Lim *et al.* [243] and Wang and Ouyang [426]. The use of interleaving transforms in SLM is considered by Hill *et al.* [161, 162], van Eetvelt *et al.* [101], and Jayalath and Tellambura [173, 175].

Section 8.3   This section follows Sharif and Hassibi [372, 373], see also [376]. The suboptimal balancing algorithm for balancing linear forms was suggested by Spencer [388]. The factor 2 in Theorem 8.4 may be removed using a more accurate analysis. The use of simultaneous amplitude and sign adjustment in balancing algorithms is considered by Sharif *et al.* [367, 368].

Section 8.4   I follow here Litsyn and Shpunt [251]. Theorem 8.5 is reminiscent to a result by Honkala and Klapper [168].

Section 8.5   This section is based on Ochiai [292, 293] and Henkel and Wagner [158, 159].

Section 8.6   This method was proposed by Tellado and Cioffi [401, 402, 400], see also Börjesson *et al.* [36], Kou *et al.* [207, 208] and Sumasu *et al.* [390].

Section 8.7   The active constellation extension was proposed by Krongold and Jones [220, 221], see also [190, 219]. Smart gradient-like algorithms for ACE are discussed in [220]. For accurately chosen parameters, they provide a faster convergence than the described method.

Section 8.8   The constellation shaping was proposed by Kwok and Jones [226, 227]. For details on SNF decomposition see [72]. SNF decomposition was introduced by Smith [384]. Kwok [225] proposed a computational method for SNF decomposition of $\mathcal{K}$. The encoding (8.69) and decoding (8.70) are from Kwok [225]. The algorithm for Hadamard matrix decomposition was also proposed by Kwok. The use of Hadamard shaping for PMEPR reduction was proposed by Mobasher and Khandani [272, 273, 274]. Another use of Hadamard transforms for PAPR reduction was suggested by Park *et al.* [316]. For other lattice-based techniques see Collings and Clarkson [74].

Section 8.9   The PTS method was proposed Müller and Huber [282]. Tellambura [410] presented an efficient method for phase factor computation. The iterative flipping algorithm was proposed by Cimini and Sollenberger [68, 69, 70]. The use of cyclically shifted PTS was considered by Hill *et al.* [161, 162]. The neighborhood search was proposed by Han and Lee [152]. Ho *et al.* [164] introduced dual-layered phase sequencing to reduce complexity, at the price of performance degradation. Chen and Pottie [59] proposed an orthogonal-projection based approach for computing the phase factors. Alavi *et al.* [2] suggested using sphere decoding for phase factor optimization. Other papers on efficient implementation of PTS method are Greenstein and Fitzgerald [142], Han and Lee [151], Jyalath and Tellambura [174], Jyalath *et al.* [176], Kang, Kim and Joo [194], Müller and Huber [281], Narahashi and Nojima [284], Sathananthan and Tellambura [353], Tan and Bar-Ness [394], Tellambura [406], and Verma and Arvind [417].

Section 8.10   This approach was proposed by Tellado and Cioffi [401, 400] and Lawrey and Kikkert [232]. The method is also known as *tone reservation*.

A version of the gradient-like algorithm was proposed by Gatherer and Polley [126]. Krongold and Jones [222] suggested a method for allowing convergence that is much faster than the described gradient-type methods; this is called an *active-set* method. Here the strategy is first to decrease the biggest peak to the level of the second one. Then continue decreasing both biggest peaks simultaneously to the level of the third one, etc. Thus, at each step, we increase by one the active set of equal-size peaks. At some stages, several of the equal peaks can be excluded from the active set, and the procedure can be continued with a smaller number of peaks. For alternative strategies see Schmidt and Kammeyer [357] and Tan and Bar-Ness [394].

Section 8.11   Coding methods yielding signals with a constant or almost constant PMEPR were considered in the previous chapter. Along with these methods, several simple strategies of adding redundancy for PMEPR reduction have been considered. The first such scheme was proposed by Jones *et al.* [189]. Wulich [434] suggested the use of a simple cyclic code of rate $^3/_4$ for PMEPR reduction. Other schemes are by Chong and Tarokh [63], Clarkson and Collings [71], Fragicomo *et al.* [116], Friese [120, 121, 122], Goeckel [130, 131], Goeckel and Ananthaswamy [132], Hyo *et al.* [170], Jiang and Zhu [184], Jones and Wilkinson [186, 187], Kamerman and Krishnakumar [193], Pingyi and Xiang [330], Sathananthan and Tellambura [354], Shepherd *et al.* [378], Smith, Cruz and Pinckley [385], Tellambura [407, 408], and Yunjun *et al.* [454, 455].

Here, I did not describe several methods that did not become popular, but probably deserve more attention. For example, I omitted pulse superposition techniques proposed by Farnese *et al.* [108], analog coding suggested by Henkel [157], an additive algorithm designed by Hentali and Schrader [160], companding techniques proposed by Mattson *et al.* [259], Jiang and Zhu [183] and Xianbin *et al.* [446, 447], the use of optimized pilot sequences suggested by Fernández-Getino Garcia *et al.* [112, 113, 114] and Yunjun *et al.* [453], frequency domain swapping, considered by Ouderaa *et al.* [310], pulse shaping, discussed by Slimane [383], and the use of artificial signals suggested by Yang *et al.* [451].

Bibliography

1. V. Ahirvar and B. S. Rajan, Low PAPR full-diversity space-frequency codes for MIMO-OFDM systems, in *Proc. GlobeCom'05*, (2005), 1476–1480.
2. A. Alavi, C. Tellambura, and I. Fair, PAPR reduction of OFDM signals using partial transmit sequences: an optimal approach using sphere decoding, *IEEE Communications Letters*, **9**: 11, (2005), 982–984.
3. I. Alrod, S. Litsyn, and A. Yudin, On the peak-to-average power ratio of *M*-sequences, *Finite Fields and their Applications*, **12**: 1, (2006), 139–150.
4. M. Anderson, *A First Course in Abstract Algebra: Rings, Groups and Fields*, second edition, McGraw-Hill, (2005).
5. A. N. D'Andrea, V. Lottici, and R. Reggiannini, RF power amplifier linearization through amplitude and phase predistortion, *IEEE Transactions in Communications*, **44**: 11, (1996), 1477–1484.
6. S. Andreoli, H. G. McClure, P. Banelli, and S. Cacopardi, Digital linearizer for RF amplifiers, *IEEE Transactions in Broadcasting*, **43**: 1, (1997), 12–19.
7. T. H. Andres and R. G. Stanton, Golay sequences, in *Combinatorial Mathematics V (Melbourne 1976), Lecture Notes in Mathematics*, **622**, (1977), 44–54.
8. K. T. Arasu and Q. Xiang, On the existence of periodic complementary binary sequences, *Designs, Codes and Cryptography*, **2**: 3, (1992), 257–262.
9. J. Armstrong, New OFDM peak-to-average power reduction scheme, in *Proc. VTC'01*, **1**, (2001), 756–760.
10. J. Armstrong, Peak-to-average power reduction for OFDM by repeated clipping and frequency domain filtering, *Electronic Letters*, **38**: 8, (2002), 246–247.
11. J. Armstrong, Peak-to-average power reduction in digital television transmitters, in *Proc. DICTA'02*, Melbourne, (2002), 19–24.
12. J. Armstrong and M. Feramez, Computationally efficient implementation of OFDM peak-to-average power reduction technique, in *Proc. Third International Symposium on Communication Systems, Networks and Digital Signal Processing*, Stafford, (2002), 236–239.
13. P. M. Asbeck, H. Kobayashi, M. Iwamoto, *et al.*, Augmented behavioral characterization for modeling the nonlinear response of power amplifiers, in *IEEE MTT-S Digest*, **1**, (2002), 135–137.
14. A. Ashikhmin, A. Barg, and S. Litsyn, Estimates of the distance distribution of codes and designs, *IEEE Transactions in Information Theory*, **47**: 3, (2001), 1050–1061.

15. A. Ashikhmin and S. Litsyn, A fast search for the maximum element of the Fourier spectrum, in *Algebraic Coding, Lecture Notes in Computer Science*, Springer, **573**, (1992), 134–141.

16. A. Ashikhmin and S. Litsyn, Fast decoding of non-binary first order Reed–Muller codes, *Applicable Algebra in Engineering Communication and Computing*, **7**, (1996), 299–308.

17. A. R. S. Bahai, B. R. Saltzberg, and M. Ergen, *Multi-carrier Digital Communications: Theory and Applications of OFDM*, Springer, (2004).

18. A. R. S. Bahai, M. Singh, A. J. Goldsmith, and B. R. Saltsberg, A new approach for evaluating clipping distortion in multicarrier systems, *IEEE Journal on Selected Areas in Communication*, **20**: 5, (2002), 1037–1046.

19. P. Banelli and S. Cacopardi, Theoretical analysis and performance of OFDM signals in nonlinear AWGN channels, *IEEE Transactions on Communication*, **48**: 3, (2000), 430–441.

20. R. W. Bäuml, R. F. H. Fischer, and J. B. Huber, Reducing the peak-to-average power ratio of multicarrier modulation by selected mapping, *Electronics Letters*, **32**: 22, (1996), 2056–2057.

21. J. Beck, Flat polynomials on the unit circle – note on a problem of Littlewood, *Bulletin of the London Mathematical Society*, **23**, (1991), 269–277.

22. E. F. Beckenbach and R. Bellman, *Inequalities*, Springer, (1983).

23. E. Beller and D. J. Newman, An ℓ^1 extremal problem for polynomials, *Proceedings of the American Mathematical Society*, **29**, (1971), 474–481.

24. R. Bellman, Almost orthogonal series, *Bulletin of the American Mathematical Society*, **50**: 2, (1944), 517–519.

25. S. N. Bernstein, Sur la valeur asymptotique de la mellieure approximation des fonctions analytiques, *Comptes Rendus de l'Academie des Sciences*, **155**, (1912), 1062–1065.

26. J. A. C. Bingham, Multicarrier modulation for data transmission: an idea whose time has come, *IEEE Communications Magazine*, **28**, (1990), 5–14.

27. J. A. C. Bingham, *ADSL, VDSL, and Multicarrier Modulation*, Wiley-Interscience, (2000).

28. N. M. Blachman, The distribution of local extrema of Gaussian noise and of its envelope, *IEEE Transactions in Information Theory*, **45**: 6, (1999), 2115–2121.

29. R. Blahut, *Fast Algorithms for Digital Signal Processing*, Addison-Wesley, (1985).

30. R. P. Boas, A general moment problem, *American Journal of Mathematics*, **63**, (1941), 361–370.

31. H. Boche and G. Wunder, Analytical methods for estimating the statistical distribution of the crest-factor in OFDM systems, in *Proc. 10th Aachen Symposium on Signal Theory*, (2001), 223–228.

32. H. Boche and G. Wunder, Über eine Verallgemeinerung eines Resultats von Riesz über trigonometrische Polynome auf allgemeine bandbegrenzte Funktionen, *Zeitschrift für angewandte Mathematik und Mechanik*, **82**: 5, (2002), 347–351.

33. H. Boche and G. Wunder, Abschätzung der Normen von Operatoren analytischer Funktionen, *Complex Variables*, **48**: 3, (2003), 211–219.

34. E. Bombieri, A note on the large sieve, *Acta Arithmetica*, **18**, (1971), 401–404.

35. L. Bömer and M. Antweiler, Periodic complementary binary sequences, *IEEE Transactions in Information Theory*, **36**, (1990), 1487–1494.

36. P. O. Börjesson, H. G. Feichtinger, N. Grip, *et al.*, A low-complexity PAR-reduction method for DMT-VDSL, in *Proc. 5th International Symposium DSPCS'99*, (1999), 164–169.

37. P. Borwein, K.-K. S. Choi, and J. Jedwab, Binary sequences with merit factor greater than 6.34, *IEEE Transactions in Information Theory*, **50**: 12, (2004), 3234–3249.

38. P. Borwein and T. Erdélyi, *Polynomials and Polynomial Inequalities*, Springer, (1995).

39. P. B. Borwein and R. A. Ferguson, A complete description of Golay pairs for lengths up to 100, *Mathematics of Computation*, **73**: 246, (2004), 967–985.

40. P. Borwein and M. Mossinghoff, Rudin–Shapiro like polynomials in L_4, *Mathematics of Computation*, **69**, (2000), 1157–1166.

41. S. Boyd, Multitone signals with low crest factor, *IEEE Transactions on Circuits and Systems*, **33**, (1986), 1018–1022.

42. R. N. Bracewell, *The Fourier Transform and its Applications*, McGraw-Hill, (1999).

43. R. N. Braithwaite, Using Walsh code selection to reduce the power variance of band-limited forward-link CDMA waveforms, *IEEE Journal on Selected Areas in Communications*, **18**: 11, (2000), 2260–2269.

44. M. Breiling, S. H. Müller-Weinfurtner, and J. B. Huber, Distortionless peak-power reduction without explicit side information, in *Proc. GlobeCom'00*, (2000), 1494–1498.

45. M. Breiling, S. H. Müller-Weinfurtner, and J. B. Huber, Distortionless peak-power reduction without explicit side information, in *Proc. 5th OFDM Workshop*, Hamburg, Germany, (2000), 28-1–28-4.

46. M. Breiling, S. H. Müller-Weinfurtner, and J. B. Huber, SLM peak-power reduction without explicit side information, *IEEE Communications Letters*, **5**: 6, (2001), 239–241.

47. J. Brillhart, On the Rudin–Shapiro polynomials, *Duke Mathematical Journal*, **40**, (1973), 335–353.

48. J. Brillhart and P. Morton, A case study in mathematical research: the Golay–Rudin–Shapiro sequence, *American Mathematical Monthly*, **103**: 10, (1996), 854–859.

49. S. Z. Budišin, Supercomplementary sets of sequences, *Electronics Letters*, **23**: 10, (1986), 504–506.

50. S. Z. Budišin, Complementary Huffman sequences, *Electronics Letters*, **26**, (1990), 533–534.

51. S. Z. Budišin, New complementary pairs of sequences, *Electronics Letters*, **26**, (1990), 881–883.

52. S. Z. Budišin, New multilevel complementary pairs of sequences, *Electronics Letters*, **26**, (1990), 1861–1863.

53. S. Z. Budišin, Efficient pulse compressor for Golay complementary sequences, *Electronics Letters*, **27**: 3, (1991), 219–220.

54. S. Z. Budišin, Golay complementary sequences are superior to PN sequences, *Proc. IEEE International Conference on Systems Engineering*, (1992), 101–104.

55. S. Z. Budišin, B. M. Popovic, and I. M. Indjin, Designing radar signals using complementary sequences, in *Proc. RADAR'87*, (1987), 593–597.

56. D. Burshtein and G. Miller, Asymptotic enumeration methods for analyzing LDPC codes, *IEEE Transactions on Information Theory*, **50**: 6, (2004), 1115–1131.

57. J. S. Byrnes, On polynomials with coefficients of modulus one, *Bulletin of the London Mathematical Society*, **9**, (1977), 171–176.

58. D. E. Cartwright and M. S. Longuet-Higgins, The statistical distribution of the maximum of a random function, *Proceedings of the Royal Society*, **237**, (1956), 212–232.

59. H. Chen and G. Pottie, An orthogonal projection-based approach for PAR reduction in OFDM, *IEEE Communications Letters*, **6**: 5, (2002), 169–171.
60. H. Chen and M. Haimovich, Iterative estimation and cancellation of clipping noise for OFDM signals, *IEEE Communications Letters*, **7**: 7, (2003), 305–307.
61. C. H. Cheng and E. J. Powers, Optimal Volterra kernel estimation algorithms for a nonlinear communication system for PSK and QAM inputs, *IEEE Transactions on Signal Processing*, **49**: 1, (2001), 147–163.
62. Y. C. Cho, S. H. Han, and J. H. Lee, Selected mapping technique with novel phase sequences for PAPR reduction of an OFDM signal, in *Proc. VTC'04*, **7**, (2004), 4781–4785.
63. C. V. Chong and V. Tarokh, A simple encodable/decodable OFDM QPSK code wih low peak-to-mean envelope power ratio, *IEEE Transactions on Information Theory*, **47**: 7, (2001), 3025–3029.
64. B. Choi, E. Kuan, and L. Hanzo, Crest-factor study of MC-CDMA and OFDM, in *Proc. VTC'99*, Amsterdam, the Netherlands, (1999), 233–237.
65. J. S. Chow, J. A. C. Bingham, and M. S. Flowers, Mitigating clipping noise in multi-carrier systems, in *Proc. ICC'97*, Montreal, Canada, (1997), 715–719.
66. L. J. Cimini, Jr., B. Daneshrad, and N. R. Sollenberger, Clustered OFDM with transmitter diversity and coding, in *Proc. GlobeCom'96*, (1996), 703–707.
67. L. J. Cimini, Jr. and N. R. Sollenberger, OFDM with diversity and coding for advanced cellular Internet services, in *Proc. GlobeCom'97*, (1997), 305–309.
68. L. J. Cimini, Jr. and N. R. Sollenberger, Peak-to-average power ratio reduction of an OFDM signal using partial transmit sequences, in *Proc. ICC'99*, Vancouver, Canada, **1**, (1999), 511–515.
69. L. J. Cimini, Jr. and N. R. Sollenberger, Peak-to-average power ratio reduction of an OFDM signal using partial transmit sequences, *IEEE Communications Letters*, **4**, (2000), 86–88.
70. L. J. Cimini, Jr. and N. R. Sollenberger, Peak-to-average power ratio reduction of an OFDM signal using partial transmit sequences with embedded side information, *Electronics Letters*, **4**: 3, (2000), 86–88; also in *Proc. GlobeCom'00*, **2**, (2000), 746–750.
71. I. V. L. Clarkson and I. B. Collings, A new joint coding and modulation scheme for channels with clipping, *Digital Signal Processing*, **12**, (2002), 223–241.
72. H. Cohen, *A Course in Computational Algebraic Number Theory*, Springer-Verlag, (1993).
73. M. N. Cohen, M. R. Fox, and J. M. Baden, Minimum peak sidelobe pulse compression codes, in *Proc. International Radar Conference*, (1990), 633–638.
74. I. B. Collings and I. V. L. Clarkson, A low complexity lattice-based low-PAR transmission scheme for DSL channels, *IEEE Transactions on Communications*, **52**, (2004), 755–764.
75. R. Costa, M. Midrio, and S. Pupolin, Impact of amplifier nonlinearities on OFDM transmission system performance, *IEEE Communications Letters*, **3**: 2, (1999), 37–39.
76. E. Costa and S. Pupolin, M-QAM-OFDM system performance in the presence of a nonlinear amplifier and phase noise, *IEEE Transactions on Communications*, **50**: 3, (2002), 462–472.
77. G. E. Coxson, A. Hirschel, and M. N. Cohen, New results on minimum-PSL binary codes, in *Proc. IEEE Radar Conference*, (2001), 153–156.

78. G. E. Coxson and J. Russo, Efficient exhaustive search for optimal-peak-sidelobe binary codes, *IEEE Transactions on Aerospace and Electronic Systems*, **41**, (2005), 302–308.
79. R. Craigen, Complex Golay sequences, *Journal of Combinatorial Mathematics and Combinatorial Computing*, **15**, (1994), 1251–1253.
80. P. Crama and J. Schoukens, Hammerstein-Wiener system estimation initialization, in *Proc. ISMA'02*, (2002), 1169–1176.
81. S. Cripps, *RF Power Amplifiers for Wireless Communications*, Artech House Publishers, (1999).
82. S. Cripps, *Advanced Techniques in RF Power Amplifier Design*, Artech House Publishers, (2002).
83. D. Dardari, V. Tralli, and A. Vaccari, A theoretical approach to error probability evaluation for OFDM systems with memoryless nonlinearities, in *Proc. Communications MiniConf'98*, Sydney, (1998), 207–212.
84. D. Dardari, V. Tralli, and A. Vaccari, A theoretical characterization of nonlinear distortion effects in OFDM systems, *IEEE Transactions on Communications*, **48**: 10, (2000), 1764.
85. M. Darnell and A. H. Kemp, Synthesis of multi-level complementary sequences, *Electronics Letters*, **24**, (1988), 1251–1252.
86. J. Davis and J. Jedwab, Peak-to-mean power control and error correction for OFDM transmission using Golay sequences and Reed–Muller codes, *Electronics Letters*, **33**: 4, (1997), 267–268.
87. J. Davis and J. Jedwab, Peak-to-mean power control in OFDM, Golay complementary sequences and Reed–Muller codes, *IEEE Transactions on Information Theory*, **45**: 7, (1999), 2397–2417.
88. J. Davis, J. Jedwab, and K. G. Paterson, Codes, correlations and power control in OFDM, in *Difference Sets, Sequences and their Correlation Properties*, A. Pott, *et al.*, eds., NATO Science Series C, Kluwer, **542**, (1999), 113–132.
89. A. Deitmar, *A First Course in Harmonic Analysis*, Springer, (2002).
90. C. Di, A. Montanari, and R. Urbanke, Weight distributions of LDPC code ensembles: combinatorics meets statistical physics, in *Proc. ISIT'04*, (2004), 102.
91. C. Di, R. Urbanke, and T. Richardson, Weight distributions: how deviant can you be?, in *Proc. ISIT'01*, (2001), 50.
92. J. I. Diaz, C. Pantaleon, and I. Santamaria, Nonlinearity estimation in power amplifiers based on subsampled temporal data, *IEEE Transactions on Instrument and Measurement*, **50**: 4, (2001), 882–887.
93. R. Dinis, P. Montezuma, and A. Gusmâo, Performance trade-offs with quasi-linearly amplified OFDM through a two-branch combining technique, in *Proc. VTC'96*, (1996), 899–903.
94. R. Dinis and A. Gusmâo, On the preformance evaluation of OFDM transmission using clipping techniques, in *Proc. VTC'99*, (1999), 2923–2928.
95. N. Dinur and D. Wulich, Peak-to-average power ratio in high-order OFDM, *IEEE Transactions on Communications*, **49**: 6, (2001), 1063–1072.
96. D. Z. Dokovic, Equivalence classes and representatives of Golay sequences, *Discrete Mathematics*, **189**, (1998), 79–93.
97. D. Z. Dokovic, Note on periodic complementary sets of binary sequences, *Designs, Codes and Cryptography*, **13**, (1998), 251–256.
98. D. S. Dummit and R. M. Foote, *Abstract Algebra*, Wiley, (2003).

99. P. Ebert and S. Weinstein, Data transmission by frequency division multiplexing using the discrete Fourier transform, *IEEE Transactions on Communications Technology*, **19**: 5, (1971), 628–634.

100. P. W. J. van Eetvelt, S. J. Shepherd, and S. K. Barton, The distribution of peak-factor in QPSK multi-carrier modulation, *Wireless Personal Communications*, **2**, (1995), 87–96.

101. P. W. J. van Eetvelt, G. Wade, and M. Tomlinson, Peak to average power reduction for OFDM schemes by selective scrambling, *Electronics Letters*, **32**, (1996), 1963–1964.

102. H. Ehlich and K. Zeller, Schwankung von Polynomen zwischen den Gitterpunkten, *Mathematische Zeitschrift*, **86**, (1964), 41–44.

103. H. Ehlich and K. Zeller, Auswertung der Normen von Interpolationsoperatoren, *Mathematische Annalen*, **164**, (1966), 105–112.

104. S. Eliahou, M. Kervaire, and B. Saffari, A new restriction on the lengths of Golay complementary sequences, *Journal of Combinatorial Theory, Series A*, **55**, (1990), 49–59.

105. S. Eliahou, M. Kervaire, and B. Saffari, On Golay polynomial pairs, *Advances in Applied Mathematics*, **12**: 3, (1991), 235–292.

106. N. Y. Ermolova and P. Vainikainen, On the relationship between peak factor of a multicarrier signal and aperiodic correlation of the generating sequence, *IEEE Communications Letters*, **7**: 3, (2003), 107–108.

107. P. Fan and M. Darnell, *Sequence Design for Communication Applications*, Wiley, (1996).

108. D. Farnese, A. Leva, G. Paltenghi, and A. Spalvieri, Pulse superposition: a technique for peak-to-average power ratio reduction in OFDM modulation, in *Proc. ICC'2002*, **3**, (2002), 1682–1685.

109. E. Feig and A. Nadas, The performance of Fourier transform division multiplexing schemes on peak limited channel, in *Proc. GlobeCom'88*, **2**, (1988), 1141–1144.

110. W. Feller, *An Introduction to Probability Theory and its Applications*, 3rd edn., John Wiley & Sons, Inc., (1968).

111. K. Feng, P. J.-S. Shiue, and Q. Xiang, On aperiodic and periodic complementary binary sequences, *IEEE Transactions on Information Theory*, **45**: 1, (1999), 296–303.

112. M. J. Fernández-Getino Garcia, O. Edfors, and J. M. Páez-Borrallo, Joint channel estimation and peak-to-average power reduction in coherent OFDM: a novel approach, in *Proc. VTC'01*, **2**, (2001), 815–819.

113. M. J. Fernández-Getino Garcia, O. Edfors, and J. M. Páez-Borrallo, Peak power reduction for OFDM systems with orthogonal pilot sequences, *IEEE Transactions on Wireless Communications*, **5**: 1, (2006), 47–51.

114. M. J. Fernández-Getino Garcia, J. M. Páez-Borrallo, and O. Edfors, Orthogonal pilot sequences for peak-to-average power ratio reduction in OFDM, in *Proc. VTC'01*, (2001), 650–654.

115. F. Fiedler and J. Jedwab, How do more Golay sequences arise?, *IEEE Transactions on Information Theory*, (2005) (in press), see also www.math.sfu.ca/~jed/publications.html.

116. S. Fragiacomo, C. Matrakidis, and J. J. O'Reilly, Multicarrier transmission peak-to-average power reduction using simple block code, *Electronics Letters*, **34**: 10, (1998), 953–954.

117. R. L. Frank, Polyphase complementary codes, *IEEE Transactions on Information Theory*, **26**: 6, (1980), 641–647.

118. L. Freiberg, A. Annamalai, and V. K. Bhargava, Crest factor reduction using orthogonal spreading codes in multi-carrier CDMA systems, in *Proc. PIMRC'97*, **1**, (1997), 120–124.

119. G. Freiman, S. Litsyn, and A. Yudin, A method to suppress high peaks in BPSK modulated OFDM signal, *IEEE Transactions on Communications*, **52**: 9, (2004), 1440–1443.

120. M. Friese, Multicarrier modulation with low peak-to-mean average power ratio, *Electronics Letters*, **32**, (1996), 713–714.

121. M. Friese, Multitone signals with low crest factors, *IEEE Transactions on Communications*, **45**: 10, (1997), 1338–1344.

122. M. Friese, OFDM signals with low crest factor, in *Proc. GlobeCom'97*, **1**, (1997), 290–294.

123. M. Friese, On the degradation of OFDM signals due to peak-clipping in optimally pre-distorted power amplifiers, in *Proc. GlobeCom'98*, Sydney, Australia, **2**, (1998), 939–942.

124. M. Friese, On the achievable information rate with peak-power limited OFDM, *IEEE Transactions on Information Theory*, **46**: 7, (2000), 2579–2587.

125. R. G. Gallager, *Low Density Parity Check Codes*, M.I.T. Press, (1963).

126. A. Gatherer and M. Polley, Controlling clipping probability in DMT transmission, in *Proc. 31st Asilomar Conference on Signals, Systems, and Computers*, (1997), 578–584.

127. A. Gavish and A. Lempel, On ternary complementary sequences, *IEEE Transactions on Information Theory*, **40**: 2, (1994), 522–526.

128. A. Gersho, B. Gopinpath, and A. Odlyzko, Coefficient inaccuracy in transversal filtering, *The Bell System Technical Journal*, **58**: 10, (1979), 2301–2317.

129. A. Ghorbani and M. Sheikhan, The effect of solid state power amplifiers (SSPAs) nonlinearities on MPSK and M-QAM signal transmission, in *Proc. Sixth International Conference on Digital Processing of Signals in Communications*, (1991), 193–197.

130. D. L. Goeckel, Coded modulation for peak power constrained wireless OFDM systems, in *Proc. 36th Allerton Conference on Communication, Control, and Computing*, (1998), 126–135.

131. D. L. Goeckel, Coded modulation with non-standard signal sets for OFDM systems, in *Proc. ICC'99*, **2**, (1999), 791–795.

132. D. L. Goeckel and G. Ananthaswamy, On the design of multidimensional signal set for OFDM systems, *IEEE Transactions on Communications*, **50**: 3, (2002), 442–452.

133. M. J. E. Golay, Multislit spectroscopy, *Journal of the Optical Society of America*, **39**, (1949), 437–444.

134. M. J. E. Golay, Static multislit spectrometry and its application to the panoramic display of infrared spectra, *Journal of the Optical Society of America*, **41**, (1951), 468–472.

135. M. J. E. Golay, Complementary series, *IRE Transactions on Information Theory*, **7**: 2, (1961), 82–87.

136. M. J. E. Golay, Note on 'Complementary series', *Proc. IRE*, (1962), 84.

137. M. J. E. Golay, Sieves for low autocorrelation binary sequences, *IEEE Transactions on Information Theory*, **23**, (1977), 43–51.

138. M. J. E. Golay, The merit factor of long low autocorrelation binary sequences, *IEEE Transactions on Information Theory*, **28**, (1982), 543–549.

139. M. J. E. Golay, The merit factor of Legendre sequences, *IEEE Transactions on Information Theory*, **29**, (1983), 934–936.

140. A. J. Grant and R. D. J. van Nee, Efficient maximum-likelihood decoding of peak power limiting codes for OFDM, in *Proc. VTC'98*, (1998), 2081–2084.

141. A. J. Grant and R. D. J. van Nee, Efficient maximum likelihood decoding of Q-ary modulated Reed–Muller codes, *IEEE Communications Letters*, **2**, (1998), 134–136.

142. L. J. Greenstein and P. J. Fitzgerald, Phasing multitone signals to minimize peak factors, *IEEE Transactions on Communications*, **29**: 7, (1981), 1072–1074.

143. M. Greferath and U. Vellbinger, Efficient decoding of $\mathbb{Z}_{p^k}$-linear codes, *IEEE Transactions on Information Theory*, **44**, (1998), 1288–1291.

144. M. Griffin, There are no Golay sequences of length $2 \cdot 9^t$, *Aequationes Mathematicae*, **15**, (1977), 73–77.

145. B. Guangguo, Methods of constructing orthogonal complementary pairs and sets of sequences, *ICC'85*, (1985), 839–843.

146. F. S. Gutleber, Spread spectrum multiplexed noise codes, in *Proc. MILCOM'82*, (1982), 15.1.1–15.1.10.

147. G. Halász, On the average order of magnitude of Dirichlet series, *Acta Mathematica Academiae Scientiarum Hungaricae*, **21**, (1970), 227–233.

148. G. Halász, On a result of Salem and Zygmund concerning random polynomials, *Studia Scientiarum Mathematicarum Hungarica*, **8**, (1973), 369–377.

149. A. R. Hammons Jr., P. V. Kumar, A. R. Calderbank, N. J. A. Sloane, and P. Solé, The Z_4-linearity of Kerdock, Preparata, Goethals, and related codes, *IEEE Transactions on Information Theory*, **40**, (1994), 301–319.

150. G. H. Hardy and J. E. Littlewood, Some problems of Diophantine approximation: a remarkable trigonometric series, *Proceedings of the National Academy of Sciences of the USA*, **2**, (1916), 583–586.

151. S. H. Han and J. H. Lee, Reduction of PAPR of an OFDM signal by partial transmit sequence technique with reduced complexity, in *Proc. GlobeCom'03*, (2003), 1326–1329.

152. S. H. Han and J. H. Lee, PAPR reduction of OFDM signals using a reduced complexity PTS technique, *IEEE Signal Processing Letters*, **11**: 11, (2004), 887–890.

153. S. H. Han and J. H. Lee, Modified selected mapping technique for PAPR reduction of coded OFDM signal, *IEEE Transactions on Broadcasting*, **50**: 3, (2004), 335–341.

154. S. H. Han and J. H. Lee, An overview of peak-to-average power ratio reduction techniques for multicarrier transmission, *IEEE Wireless Communications*, **12**: 2, (2005), 56–65.

155. L. Hanzo, M. Münster, B. J. Choi, and T. Keller, *OFDM and MC-CDMA for Broadband Multi-User Communications, WLANs and Broadcasting*, John Wiley & Sons, (2003).

156. T. Helleseth, P. V. Kumar, O. Moreno, and A. G. Shanbhag, Improved estimates via exponential sums for the minimum distance of Z_4-linear trace codes, *IEEE Transactions on Information Theory*, **42**, (1996), 1212–1216.

157. W. Henkel, Analog codes for peak-to-average ratio reduction, in *Proc. 3rd ITG Conference on Source and Channel Coding*, Munich, Germany, (2000), 151–155.

158. W. Henkel and B. Wagner, Trellis shaping for reducing the peak-to-average ratio of multitone signals, in *Proc. 6th Benelux-Japan Workshop*, Essen, Germany, (1996), 4.1–4.2.

159. W. Henkel and B. Wagner, Trellis shaping for reducing the peak-to-average ratio of multitone signals, in *Proc. ISIT'97*, Ulm, Germany, (1997), 519.

160. N. Hentati and M. Schrader, Additive algorithm for reduction of crest factor (AARC), in *Proc. 5th International OFDM Workshop*, (2000), 27-1–27-5.
161. G. R. Hill, M. Faulkner, and J. Singh, Cyclic shifting and time inversion of partial transmit sequences to reduce the peak-to-average power ratio in OFDM, in *Proc. PIMRC'00*, **2**, (2000), 1256–1259.
162. G. R. Hill, M. Faulkner, and J. Singh, Reducing the peak-to-average power ratio in OFDM by cyclically shifting partial transmit sequences, *Electronics Letters*, **36**: 6, (2000), 560–561.
163. T. F. Ho and V. K. Wei, Synthesis of low-crest waveforms for multicarrier CDMA systems, in *Proc. GlobeCom'95*, (1995), 131–135.
164. W. S. Ho, A. S. Madhukumar, and F. Chin, Peak-to-average power reduction using partial transmit sequences: a suboptimal approach based on dual layered phase sequencing, *IEEE Transactions on Broadcasting*, **49**: 2, (2003), 225–231.
165. T. Høholdt and H. E. Jensen, Determination of the merit factor of Legendre sequences, *IEEE Transactions on Information Theory*, **34**, (1988), 161–164.
166. T. Høholdt, H. E. Jensen, and J. Justesen, Aperiodic correlations and merit factor of a class of binary sequences, *IEEE Transactions on Information Theory*, **31**, (1985), 549–552.
167. W. H. Holzmann and H. Kharaghani, A computer search for complex Golay sequences, *Australasian Journal of Combinatorics*, **10**, (1994), 251–258.
168. I. Honkala and A. Klapper, Bounds for the multicovering radii of Reed–Muller codes with applications to stream ciphers, *Designs, Codes and Cryptography*, **23**, (2001), 131–145.
169. W. C. Huffman and V. Pless, *Fundamentals of Error-Correcting Codes*, Cambridge University Press, (2003).
170. J. A. Hyo, S. Yoan, and I. Sungbin, A block coding scheme for peak-to-average power ratio reduction in an orthogonal frequency division multiplexing system, in *Proc. VTC'00*, **1**, (2000), 56–60.
171. M. Ibnkhala, J. Sombrin, and F. Castaine, Neural networks for modeling nonlinear memoryless communication channels, *IEEE Transactions on Communications*, **45**: 7, (1997), 768–771.
172. S. Jauregui, Jr., Complementary series of length 26, *IRE Transactions on Information Theory*, **7**, (1962), 323.
173. A. Jayalath and C. Tellambura, Reducing the peak-to-average power ratio of an OFDM signal through bit or symbol interleaving, *Electronics Letters*, **36**: 13, (2000), 1161–1163.
174. A. D. S. Jayalath and C. Tellambura, Adaptive PTS approach for reduction of peak-to-average power ratio of OFDM signal, *Electronics Letters*, **36**: 14, (2000), 1226–1228.
175. A. D. S. Jayalath and C. Tellambura, Use of data permutation to reduce the peak-to-average power ratio of an OFDM signal, *Wireless Communications and Mobile Computing*, **2**, (2002), 187–203.
176. A. D. S. Jayalath, C. Tellambura, and H. Wu, Reduced complexity PTS and new phase sequences for SLM to reduce PAP of an OFDM signal, in *Proc. VTC'00*, (2000), 1914–1917.
177. J. Jedwab, Comment: *M*-sequences for OFDM peak-to-average power ratio reduction and error-correction, *Electronics Letters*, **33**, (1997), 1293–1294.
178. J. Jedwab, A survey of the merit factor problem for binary sequences, in *Sequences and their Applications-SETA'04, LNCS*, T. Helleseth, *et al.*, eds., **3486**, (2005), 30–55.

179. J. Jedwab and K. Yoshida, The peak sidelobe level of families of binary sequences, *IEEE Transactions on Information Theory*, **52**: 5, (2006), 2247–2254.

180. H. E. Jensen and T. Høholdt, Binary sequences with good correlation properties, in *Proc. AAECC-5, LNCS*, Springer-Verlag, **356**, (1989), 306–320.

181. W. G. Jeon, K. H. Chang, and Y. S. Cho, An adaptive data predistorter for compensation of nonlinear distortion in OFDM systems, *IEEE Transactions on Communications*, **45**: 10, (1997), 1167–1171.

182. K. Jetter, G. Pfander, and G. Zimmerman, The crest-factor for trigonometric polynomials. Part I: approximation theoretical estimates, *Revue d'Analyse Numerique et de Theorie de l'Approximation*, **30**, (2001), 41–56.

183. T. Jiang and G. Zhu, Nonlinear companding transform for reducing peak-to-average power ratio of OFDM signals, *IEEE Transactions on Broadcasting*, **50**, (2004), 342–346.

184. T. Jiang and G. Zhu, Complement block coding for reduction in peak-to-average power ratio of OFDM signals, *IEEE Radio Communications*, **43**: 9, (2005), S17–S22.

185. R. Johannesson and K. Sh. Zigangirov, *Fundamentals of Convolutional Coding*, IEEE Press, (1999).

186. A. E. Jones and T. A. Wilkinson, Minimization of the peak-to-mean envelope power ratio of multicarrier transmission schemes by block coding, in *Proc. VTC'95*, (1995), 825–829.

187. A. E. Jones and T. A. Wilkinson, Combined coding error control and increased robustness to system nonlinearities in OFDM, in *Proc. VTC'96*, (1996), 904–908.

188. A. E. Jones and T. A. Wilkinson, Performance of Reed–Muller codes and maximum likelihood decoding algorithm for OFDM, *IEEE Transactions on Communications*, **47**: 7, (1999), 949–952.

189. A. E. Jones, T. A. Wilkinson, and S. K. Barton, Block coding scheme for reduction of peak-to-mean envelope power ratio of multicarrier transmission schemes, *Electronics Letters*, **30**, (1994), 2098–2099.

190. D. L. Jones, Peak power reduction in OFDM and DMT via active channel modification, in *33rd Asilomar Conference on Signals, Systems, and Computers*, **2**, (1999), 1076–1079.

191. M. Kac and D. Slepian, Large excursions of Gaussian processes, *Annals of Mathematical Statistics*, **30**, (1959), 1215–1228.

192. J. P. Kahane, Sur les polynomes a coefficients unimodulares, *Bulletin of the London Mathematical Society*, **12**, (1980), 321–342.

193. A. Kamerman and A. S. Krishnakumar, OFDM encoding with reduced crest-factor, in *Proc. IEEE Second Symposium on Communications and Vehicular Technology in Benelux*, (1994), 182–186.

194. S. G. Kang, J. G. Kim, and E. K. Joo, A novel subblock partition scheme for partial transmit sequence OFDM, *IEEE Transactions on Broadcasting*, **45**: 3, (1999), 333–338.

195. N. M. Katz, *Gauss Sums, Kloosterman Sums, and Monodromy Groups, Annals of Mathematics Studies*, Princeton University Press, (1988).

196. T. Kasami, T. Fujiwara, and S. Lin, An approximation to the weight distribution of binary linear codes, *IEEE Transactions on Information Theory*, **31**: 6, (1985), 769–780.

197. T. Kasami, T. Fujiwara, and S. Lin, An approximation to the weight distribution of binary primitive BCH codes with designed distances 9 and 11, *IEEE Transactions on Information Theory*, **32**: 5, (1986), 706–709.

198. Y. Katznelson, *An Introduction to Harmonic Analysis*, 3rd edn., Cambridge University Press, (2004).
199. A. H. Kemp and M. Darnell, Synthesis of uncorrelated and nonsquare sets of multilevel complementary sequences, *Electronics Letters*, **25**: 12, (1989), 791–792.
200. P. B. Kennington, *High-linearity RF Amplifier Design*, Artech House, (2000).
201. O. Keren and S. Litsyn, More on the distance distribution of BCH codes, *IEEE Transactions on Information Theory*, **45**: 1, (1999), 251–255.
202. O. Keren and S. Litsyn, The number of solutions to a system of equations and spectra of codes, in *Finite Fields: Theory, Applications, and Algorithms, Contemporary Mathematics*, AMS, **225**, (1999), 177–184.
203. M. Kervaire, B. Saffari, and R. Vaillancourt, Une méthode de détection de nouveaux polynômes vérifiant l'identité de Rudin–Shapiro, *Comptes Rendus de l'Academie des Sciences–Paris*, **I302**: 3, (1986), 95–98.
204. D. Kim and S. L. Stuber, Clipping noise mitigation for OFDM by decision-aided reconstruction, *IEEE Communications Letters*: **3**: 1, (1999), 4–6.
205. D. J. Kleitman, On a combinatorial conjecture of Erdös, *Journal of Combinatorial Theory*, **1**, (1966), 209–214.
206. T. W. Körner, On a polynomial of Byrnes, *Bulletin of the London Mathematical Society*, **12**, (1980), 219–224.
207. Y. Kou, W.-S. Lu, and A. Antoniou, New peak-to-average power-ratio reduction algorithms for multicarrier communications, *IEEE Transactions on Circuits and Systems*, **51**: 9, (2004), 1790–1800.
208. Y. Kou, W.-S. Lu, and A. Antoniou, New peak-to-average power ratio reduction algorithms for OFDM systems using constellation extension, in *Proc. PACRIM'05*, (2005), 514–517.
209. C. Koukouvinos, Sequences with zero autocorrelation, in *CRC Handbook of Combinatorial Designs*, C. J. Colbourn and J. H. Dinitz, eds., CRC Press, (1996), 452–456.
210. S. Kounias, C. Koukouvinos, and K. Sotirakoglou, On Golay sequences, *Discrete Mathematics*, **92**: 1–3, (1991), 177–185.
211. I. Krasikov and S. Litsyn, On spectra of BCH codes, *IEEE Transactions on Information Theory*, **41**: 3, (1995), 786–788.
212. I. Krasikov and S. Litsyn, On the accuracy of the binomial approximation to the distance distribution of codes, *IEEE Transactions on Information Theory*, **41**: 5, (1995), 1472–1474.
213. I. Krasikov and S. Litsyn, Estimates for the range of binomiality in codes' spectra, *IEEE Transactions on Information Theory*, **43**: 3, (1997), 987–991.
214. I. Krasikov and S. Litsyn, Bounds on spectra of codes with known dual distance, *Designs, Codes, and Cryptography*, **13**: 3, (1998), 285–298.
215. I. Krasikov and S. Litsyn, On the distance distributions of duals of BCH codes, *IEEE Transactions on Information Theory*, **45**: 1, (1999), 247–250.
216. I. Krasikov and S. Litsyn, On the distance distribution of BCH codes and their duals, *Designs, Codes, and Cryptography*, **23**, (2001), 223–231.
217. I. Krasikov and S. Litsyn, Survey of binary Krawtchouk polynomials, in *Codes and Association Schemes*, A. Barg and S. Litsyn, eds., *DIMACS*, **56**, (2001).
218. R. Kristiansen and M. G. Parker, Binary sequences with merit factor > 6.3, *IEEE Transactions on Information Theory*, **50**, (2004), 3385–3389.
219. B. S. Krongold and D. L. Jones, A new method for PAR reduction in baseband DMT systems, in *Proc. 35th Asilomar Conference on Signals, Systems and Computing*, (2001), 502–506.

220. B. S. Krongold and D. L. Jones, PAR reduction in OFDM via active constellation extension, *IEEE Transactions on Broadcasting*, **49**: 3, (2003), 258–268.
221. B. S. Krongold and D. L. Jones, PAR reduction in OFDM via active constellation extension, in *Proc. ICASSP'03*, **4**, (2003), IV-525-8.
222. B. S. Krongold and D. L. Jones, An active set approach for OFDM PAR reduction via tone reservation, *IEEE Transactions on Signal Processing*, **52**: 2, (2004), 495–509.
223. J. Kruskal, Golay complementary series, *IRE Transactions on Information Theory*, (1961), 273–276.
224. P. V. Kumar, T. Helleseth, and A. R. Calderbank, An upper bound for Weil exponential sums over Galois rings and applications, *IEEE Transactions on Information Theory*, **41**, (1995), 456–468.
225. H. K. Kwok, *Shape Up: Peak-Power Reduction via Constellation Shaping*, Ph.D. Thesis, University of Illinois, (2001).
226. H. K. Kwok and D. L. Jones, PAR reduction for Hadamard transform-based OFDM, in *Proc. 34th Conference on Signals, Systems, and Computers*, Princeton, (2000), FP4-18–23.
227. H. K. Kwok and D. L. Jones, PAR reduction via constellation shaping, in *Proc. ISIT'00*, Sorrento, (2000), 166.
228. K. Laird, N. Whinnett, and S. Buljore, A peak-to-average power reduction method for third generation CDMA reverse links, in *Proc. VTC'99*, **1**, (1999), 551–555.
229. R. Laroia, T. Richardson, and R. Urbanke, Reduced peak power requirements in FDM and related systems, postscript file available through http://lthcwww.epfl.ch/publications/publist-bib.html#LRU.
230. V. K. N. Lau, On the analysis of peak-to-average ratio (PAR) for IS95 and CDMA2000 systems, *IEEE Transactions on Vehicular Technology*, **49**: 6, (2000), 2174–2188.
231. V. K. N. Lau, Peak-to-average ratio (PAR) reduction by Walsh code selection for IS-95 and CDMA2000 systems, *IEEE Proceedings on Communications*, **147**: 6, (2000), 361–364.
232. E. Lawrey and C. J. Kikkert, Peak to average power ratio reduction of OFDM signals using peak reduction carriers, in *Proc. ISSPA'99*, Brisbane, Australia, (1999), 737–740.
233. M. R. Leadbetter, G. Lindgren, and H. Rootzén, *Extremes and Related Properties of Random Sequences and Processes*, Springer-Verlag, (1983).
234. H. Lee, D. N. Liu, W. Zhu and M. P. Fitz, Peak power reduction using a unitary rotation in multiple transmit antennas, in *Proc. ICC'05*, **4**, (2005), 2407–2011.
235. Y.-L. Lee, Y.-H. You, W.-G. Jeon, J.-H. Paik, and H.-K. Song, Peak-to-average power ratio reduction in MIMO-OFDM systems using selective mapping, *IEEE Communications Letters*, **7**, (2003), 575–577.
236. N. Levanon, Multifrequency complementary phase-coded radar signals, *IEE Proc. Radar, Sonar and Navigation*, **147**: 6, (2000), 276–284.
237. N. Levanon and E. Mozeson, Multicarrier radar signal – pulse train and CW, *IEEE Transactions on Aerospace Electronic Systems*, **38**: 2, (2002), 707–720.
238. N. Levanon and E. Mozeson, *Radar Signals*, J. Wiley & Sons, (2004).
239. X. Li and L. Cimini, Effects of clipping and filtering on the performance of OFDM, *IEEE Communications Letters*, **2**: 5, (1998), 131–133; also in *Proc. VTC'97*, (1997), 1634–1638.
240. X. Li and J. A. Ritcey, *M*-sequences for OFDM peak-to-average power ratio reduction and error correction, *Electronics Letters*, **33**, (1997), 554–555.

241. T.-J. Liang and G. Fettweis, MIMO preamble design with a subset of subcarriers in OFDM-based WLAN, in *Proc. VTC'05*, **2**, (2005), 1032–1036.

242. R. Lidl and H. Niederreiter, *Finite Fields*, Cambridge University Press, (1996).

243. D.-W. Lim, S.-J. Heo, J.-S. No, and H. Chung, A new SLM OFDM scheme with low complexity for PAPR reduction, *IEEE Signal Processing Letters*, **12**: 2, (2005), 93–96.

244. S. Lin and D. J. Costello, *Error Control Coding*, 2nd edn., Prentice Hall, (2004).

245. S. Ling and F. Ozbudak, An improvement on the bounds of Weil exponential sums over Galois rings with some applications, *IEEE Transactions on Information Theory*, **50**: 10, (2004), 2529–2543.

246. J. Lindner, Binary sequences up to length 40 with best possible autocorrelation function, *Electronics Letters*, **11**, (1975), 507.

247. J. H. van Lint, *Introduction to Coding Theory*, Springer-Verlag, (1992).

248. S. Litsyn and A. Yudin, Discrete and continuous maxima in multicarrier communications, *IEEE Transactions on Information Theory*, **51**: 3, (2005), 919–928.

249. S. Litsyn and V. Shevelev, On ensembles of low-density parity check codes: asymptotic distance distributions, *IEEE Transactions on Information Theory*, **48**: 4, (2002), 887–908.

250. S. Litsyn and V. Shevelev, Distance distributions in ensembles of irregular low-density parity-check codes, *IEEE Transactions on Information Theory*, **49**: 12, 3140–3159.

251. S. Litsyn and A. Shpunt, A method to decrease peak-to-average power ratio in OFDM signals, www.eng.tau.ac.il/~litsyn/publications.html.

252. S. Litsyn and G. Wunder, Generalized bounds on the crest-factor distribution of OFDM signals with applications to code design, *IEEE Transactions on Information Theory*, **52**: 3, (2006), 992–1006.

253. S. Litsyn and A. Yudin, On the continuous maximum of oversampled OFDM signals, www.eng.tau.ac.il/~litsyn/publications.html.

254. J. E. Littlewood, On polynomials $\sum \pm z^m$, $\sum \exp(\alpha_m) z^m$, $z = e^{i\theta}$, *Journal of the London Mathematical Society*, **41**, (1966), 367–376.

255. J. E. Littlewood, *Some Problems in Real and Complex Analysis*, D. C. Heath and Co., (1968).

256. H. D. Lüke, Binary odd-periodic complementary sequences, *IEEE Transactions on Information Theory*, **43**: 1, (1997), 365–367.

257. F. J. MacWilliams and N. J. A. Sloane, *The Theory of Error-Correcting Codes*, North Holland, 2nd edn., (1986).

258. J. C. Mason, *Chebyshev Polynomials*, Chapman & Hall/CRC, (2002).

259. A. Mattson, G. Mendenhall, and T. Dittmer, Comments on reduction of peak-to-average power ratio of OFDM system using a companding technique, *IEEE Transactions on Broadcasting*, **45**: 4, (1999), 418–419.

260. T. May and H. Rohling, Reducing the peak-to-average power ratio in OFDM radio transmission systems, in *Proc. VTC'98*, **3**, (1998), 2474–2478.

261. J. E. Mazo, Asymptotic distortion spectrum of clipped, DC-biased, Gaussian noise, *IEEE Transactions on Communications*, **40**: 8, (1992), 1339–1344.

262. R. J. McEliece, Correlation properties of sets of sequences derived from irreducible cyclic codes, *Information and Control*, **45**, (1980), 18–25.

263. R. J. McEliece, *Finite Fields for Computer Scientists and Engineers*, Kluwer, (1987).

264. I. D. Mercer, Autocorrelations of random binary sequences, *Combinatorics, Probability, and Computing*, (in press).

265. S. Mertens, Exhaustive search for low-autocorrelation binary sequences, *Journal of Physics A*, **29**, (1996), L473–L478.

266. S. Mertens and H. Bauke, Ground states of the Bernasconi model with open boundary conditions, www.cecm.sfu.ca/~jknauer/labs/records.html.

267. D. J. G. Mestdagh, P. M. P. Spryut, and B. Biran, Effect of amplitude clipping in DMT-ADSL transceivers, *Electronics Letters*, **29**: 15, (1993), 1354–1355.

268. D. J. G. Mestdagh, P. M. P. Spryut, and B. Biran, Analysis of clipping effect in DMT-based ADSL systems, in *Proc. ICC'94*, **1**, (1994), 293–300.

269. D. J. G. Mestdagh and P. M. P. Spryut, A method to reduce the probability of clipping in DMT-based transceivers, *IEEE Transactions on Communications*, **44**: 10, (1996).

270. S. L. Miller and R. J. O'Dea, Peak power and bandwidth efficient linear modulation, *IEEE Transactions on Communications*, **46**, (1998), 1639–1648.

271. H. Minn, C. Tellambura, and V. K. Bhargava, On the peak factors of sampled and continuous signals, *IEEE Communications Letters*, **5**: 4, (2001), 129–131.

272. A. Mobasher and A. K. Khandani, PAPR reduction using integer structures in OFDM systems, in *Proc. VTC'04*, (2004), 650–654.

273. A. Mobasher and A. K. Khandani, PAPR reduction in OFDM systems using constellation shaping, in *Proc. 22nd Biennial Symposium on Communications*, Kingston, ON, (2004), 258–260.

274. A. Mobasher and A. K. Khandani, Integer-based constellation shaping method for PAPR reduction in OFDM systems, *IEEE Transactions on Communications*, **54**: 1, (2006), 119–127.

275. H. L. Montgomery, *Topics in Multiplicative Number Theory*, Springer, (1971).

276. H. L. Montgomery, An exponential polynomial formed with the Legendre symbol, *Acta Arithmetica*, **37**, (1980), 375–380.

277. J. W. Moon and L. Moser, On the correlation function of random binary sequences, *SIAM Journal on Applied Mathematics*, **16**, (1968), 340–343.

278. J. H. Moon, Y. H. You, W. G. Jeon, K.-W. Kwon, and H. K. Song, Peak-to-average power control for multiple-antenna HIPERLAN/2 and IEEE802.11a systems, *IEEE Transactions on Consumer Electronics*, **49**: 4, (2003), 1078–1083.

279. E. Mozeson and N. Levanon, Multicarrier radar signals with low peak-to-mean envelope power ratio, *IEE Proceedings – Radar, Sonar and Navigation*, **150**: 2, (2003), 71–77.

280. S. H. Müller, R. W. Bäuml, R. F. H. Fischer, and J. B. Huber, OFDM with reduced peak-to-average power ratio by multiple signal representation, *Annals of Telecommunications*, **52**: 1–2, (1997), 58–67.

281. S. H. Müller and J. Huber, A comparison of peak power reduction schemes for OFDM, in *Proc. GlobeCom'97*, Phoenix, (1997), 1–5.

282. S. H. Müller and J. Huber, OFDM with reduced peak-to-average power ratio by optimum combination of partial transmit sequences, *Electronics Letters*, **33**: 5, (1997), 368–369.

283. S. H. Müller and J. Huber, A novel peak power reduction scheme for OFDM, in *Proc. PIMRC'97*, Helsinki, Finland, **3**, (1997), 1090–1094.

284. S. Narahashi and T. Nojima, New phasing scheme of N-multiple carriers for reducing peak-to-average power ratio, *Electronics Letters*, **30**: 17, (1994), 1382–1383.

285. R. D. J. van Nee, OFDM codes for peak-to-average power reduction and error correction, in *Proc. GlobeCom'96*, London, (1996), 740–744.

286. R. van Nee and R. Prasad, *OFDM for Wireless Multimedia Communications*, Artech House Publishers, (2000).

287. R. D. J. van Nee and A. de Wild, Reducing the peak-to-average power ratio of OFDM, in *Proc. VTC'98*, (1998), 2072–2076.

288. D. J. Newman, An L_1 extremal problem for polynomials, *Proceedings of the American Mathematical Society*, **16**, (1965), 1287–1290.

289. D. J. Newman and J. S. Byrnes, The L_4 norm of a polynomial with coefficients ± 1, *American Mathematical Monthly*, **97**, (1990), 42–45.

290. H. Ochiai, Performance analysis of peak power and band-limited OFDM system with linear scaling, *IEEE Transactions on Wireless Communications*, **2**: 5, (2003), 1055–1065.

291. H. Ochiai, Performance of optimal and suboptimal detection for uncoded OFDM system with deliberate clipping and filtering, in *Proc. GlobeCom'03*, **3**, (2003), 1618–1622.

292. H. Ochiai, Trellis shaped coded OFDM systems with high power efficiency, in *Proc. ISIT'03*, (2003), 6.

293. H. Ochiai, A novel trellis-shaping design with both peak and average power reduction for OFDM systems, *IEEE Transactions on Communications*, **52**: 11, (2004), 1916–1926.

294. H. Ochiai, M. P. C. Fossorier, and H. Imai, On decoding of block codes with peak-power reduction in OFDM systems, *IEEE Communications Letters*, **4**: 7, (2000), 226–228.

295. H. Ochiai and H. Imai, Block coding scheme based on complementary sequences for multicarrier signals, *IEICE Transactions on Fundamentals*, **E80-A**, (1997), 2136–2143.

296. H. Ochiai and H. Imai, Forward error correction with reduction of peak to average power ratio of QPSK multicarrier signals, in *Proc. ISIT'97*, (1997), 120.

297. H. Ochiai and H. Imai, Block codes for frequency diversity and peak power reduction in multicarrier systems, *Proc. ISIT'98*, (1998), 192.

298. H. Ochiai and H. Imai, Performance of block codes with peak power reduction for indoor multicarrier systems, *Proc. VTC'98*, **1**, (1998), 338–342.

299. H. Ochiai and H. Imai, OFDM-CDMA with peak power reduction based on the spreading sequences, *Proc. ICC'98*, **3**, (1998), 1299–1303.

300. H. Ochiai and H. Imai, Performance of OFDM-CDMA with simple peak power reduction, *European Transactions on Telecommunications*, **10**: 4, (1999), 391–398.

301. H. Ochiai and H. Imai, On clipping for peak power reduction of OFDM signals, in *Proc. GlobeCom'00*, **2**, (2000), 731–735.

302. H. Ochiai and H. Imai, Performance of the deliberate clipping with adaptive symbol selection for strictly band-limited OFDM systems, *IEEE Journal on Selected Areas in Communications*, **18**: 11, (2000), 2270–2277.

303. H. Ochiai and H. Imai, On the distribution of the peak-to-average power ratio in OFDM signals, *IEEE Transactions on Communications*, **49**: 2, (2001), 282–289.

304. H. Ochiai and H. Imai, Performance analysis of deliberately clipped OFDM signals, *IEEE Transactions on Communications*, **50**: 1, (2002), 89–101.

305. R. O'Neill and L. B. Lopes, Performance of amplitude limited multitone signals, in *Proc. VTC'94*, (1994), 1675–1679.

306. R. O'Neill and L. B. Lopes, Envelope variations and spectral splatter in clipped multicarrier signals, in *Proc. PIMRC'95*, (1995), 71–75.

307. A. V. Oppenheim, R. W. Schafer, and J. R. Buck, *Discrete-Time Signal Processing*, 2nd edn., Prentice Hall, (1999).

308. T. Ottoson, Precoding in multicode DS-CDMA systems, in *Proc. ISIT'97*, (1997), 351.

309. T. Ottoson, Precoding for minimization of envelope variations in multicode DS-CDMA systems, *Wireless Personal Communications*, **13**, (2000), 57–78.

310. E. van der Ouderaa, J. Schoukens, and J. Renneboog, Peak factor minimization using a time frequency domain swapping algorithm, *IEEE Transactions on Instrumentation and Measurement*, **37**: 1, (1988), 207–212.

311. K. R. Panta and J. Armstrong, Use of peak-to-average power reduction technique in HIPERLAN2 and its performance in a fading channel, in *Proc. DSPCS'02*, (2002), 113–117.

312. K. R. Panta and J. Armstrong, Effects of clipping on the error performance of OFDM in frequency selective fading channels, *IEEE Transactions on Wireless Communications*, **3**: 2, (2004), 668–671.

313. A. Papoulis and S. U. Pillai, *Probability, Random Variables and Stochastic Processes*, McGraw-Hill, (2001).

314. I.-S. Park and E. J. Powers, Compensation of nonlinear distortion in OFDM systems using a new predistorter, in *Proc. PIMRC'98*, (1998), 811–815.

315. H.-S. Park, Y.-O. Park, and C.-S. Kim, A design and performance analysis of OFDMA modulator based on IEEE802.16a standard, in *Proc. of 38th Asilomar Conference on Signals, Systems and Computers*, (2004), 536–539.

316. M. Park, H. Jun, J. Cho, *et al.*, PAPR reduction in OFDM transmission using Hadamard transform, in *Proc. ICC'00*, (2000), **1**, 430–433.

317. M. J. Parker, The constabent properties of Golay-Davis-Jedwab sequences, in *Proc. ISIT'00*, Sorrento, Italy, (2000), 302.

318. M. G. Parker, K. G. Paterson, and C. Tellambura, Golay complementary sequences, in *Wiley Encyclopedia of Telecommunications*, J. G. Proakis, ed., Wiley, (2003).

319. M. G. Parker and C. Tellambura, Golay–Davis–Jedwab complementary sequences and Rudin–Shapiro constructions, (2001), www.ii.uib.no/~matthew/ConstaBent2.pdf.

320. M. G. Parker and C. Tellambura, Generalized Rudin–Shapiro constructions, in *Electronic Notes in Discrete Mathematics, Proc. WCC'01*, Augot, Daniel *et al.*, eds. **6**, (2001).

321. M. G. Parker and C. Tellambura, A construction for binary sequence sets with low peak-to-average power ratio, in *Proc. ISIT'02*, (2002), 239.

322. K. G. Paterson, Coding techniques for power controlled OFDM, in *Proc. PIMRC'98*, **2**, (1998), 801–805.

323. K. G. Paterson, Generalized Reed–Muller codes and power control in OFDM modulation, *IEEE Transactions on Information Theory*, **46**: 1, (2000), 104–120.

324. K. G. Paterson, Sequences for OFDM and multi-code CDMA: two problems in algebraic coding theory, in *Proc. of Sequences and their Applications-SETA'01*, T. Helleseth, P. V. Kumar, and K. Yang, eds., Discrete Mathematics and Theoretical Computer Science Series, Springer, (2002), 46–71.

325. K. G. Paterson, On codes with low peak-to-average power ratio for multi-code CDMA, in *Proc. ISIT'02*, (2002), 49.

326. K. G. Paterson, On codes with low peak-to-average power ratio for multi-code CDMA, *IEEE Transactions on Information Theory*, **50**: 3, (2004), 550–559.

327. K. G. Paterson and A. E. Jones, Efficient decoding algorithms for generalized Reed–Muller codes, *IEEE Transactions on Communications*, **48**: 8, (2000), 1272–1285.

328. K. G. Paterson and V. Tarokh, On the existence and construction of good codes with low peak-to-average power ratios, *IEEE Transactions on Information Theory*, **46**: 6, (2000), 1974–1987.

329. M. Pauli and H. P. Kuchenbecker, Minimization of the intermodulation distortion of a nonlinearly amplified OFDM signal, *Wireless Personal Communications*, **4**, (1996), 93–101.

330. F. Pingyi and G. X. Xiang, Block coded modulation for the reduction of the peak-to-average power ratio in OFDM systems, *IEEE Trans. Consumer Electronics*, **45**: 4, (1999), 1025–1029; also in *Proc. IEEE Wireless Communications and Networking Conference*, **3**, (1999), 1095–1099; also in *Proc. SPIE'99*, **3708**, (1999), 34–43.

331. G. Pólya and G. Szegö, *Problems and Theorems in Analysis*, **I**, Springer, (2004).

332. B. M. Popoviĉ, Synthesis of power efficient multitone signals with flat amplitude spectrum, *IEEE Transactions on Communications*, **39**, (1991), 1031–1033.

333. B. M. Popoviĉ, Efficient Golay correlator, *Electronics Letters*, **35**, (1999), 1427–1428.

334. R. Prasad, *OFDM for Wireless Communication Systems*, Artech House, (2004).

335. J. G. Proakis, *Digital Communications*, 4th ed., McGraw-Hill, (2000).

336. J. G. Proakis and M. Salehi, *Fundamentals of Communication Systems*, Prentice Hall, (2004).

337. C. Rapp, Effects of HPA-nonlinearity on a 4-DPSK/OFDM-signal for a digital sound broadcasting system, in *Proc. Second European Conference on Satellite Communications*, (1991), 179–184.

338. H. Reddy and T. M. Duman, Space-time coded OFDM with low PAPR, in *Proc. GlobeCom'03*, **2**, (2003), 799–803.

339. S. O. Rice, Mathematical analysis of random noise, *Bell System Technical Journal*, **23**: 3, (1944), 282–332; **24**: 1, (1945), 45–156.

340. S. O. Rice, Statistical properties of a sine wave plus random noise, *Bell System Technical Journal*, **27**: 1, (1948), 109–157.

341. S. O. Rice, Distribution of the duration of fades in radio transmission, *Bell System Technical Journal*, **37**, (1958), 581–635.

342. M. Riesz, Eine trigonometrische Interpolationsformel und einige Ungleichungen für Polynome, *Jahresbericht der Deutschen Mathematiker Vereinigung*, **23**, (1914), 354–368.

343. J. Rinne and M. Renfors, The behavior of orthogonal frequency division multiplexing signals in an amplitude limiting channel, in *Proc. ICC'94*, (1994), 381–385.

344. W. Rudin, Some theorems on Fourier coefficients, *Proceedings of the American Mathematical Society*, **10**, (1959), 855–859.

345. S. M. Ross, *Stochastic Processes*, Wiley, (1995).

346. C. Rößling and V. Tarokh, A construction of OFDM 16-QAM sequences having low peak power, *IEEE Transactions on Information Theory*, **47**: 5, (2001), 2091–2094.

347. H. E. Rowe, Memoryless nonlinearities with Gaussian inputs: elementary results, *Bell System Technical Journal*, **61**: 7, (1982), 1519–1525.

348. D. J. Ryan, I. B. Collings, and I. V. L. Clarkson, Low-complexity low-PAR transmission for MIMO-DSL, *IEEE Communications Letters*, **9**: 10, (2005), 868–870.

349. A. A. M. Saleh, Frequency-independent and frequency-dependent nonlinear models of TWT amplifiers, *IEEE Transactions on Communications*, **29**: 11, (1981), 1715–1720.

350. B. R. Saltsberg, Intersymbol interference error bounds with application to ideal bandlimited signaling, *IEEE Transactions on Information Theory*, **14**: 4, (1968), 563–568.

351. G. Santella and F. Mazzenga, A hybrid analytical-simulation procedure for performance evaluation in M-QAM-OFDM schemes in presence of nonlinear distortions, *IEEE Transactions on Vehicular Technology*, **47**: 1, (1998), 142–151.

352. D. V. Sarwate, An upper bound on the aperiodic correlation function for a maximal-length sequence, *IEEE Transactions on Information Theory*, **30**, (1984), 685–687.

353. K. Sathananthan and C. Tellambura, Partial transmit sequence and selected mapping schemes to reduce ICI in OFDM systems, *IEEE Communications Letters*, **6**: 8, (2002), 313–315.

354. K. Sathananthan and C. Tellambura, Coding to reduce both PAR and PICR of an OFDM signal, *IEEE Communications Letters*, **6**: 8, (2002), 316–318.

355. K.-U. Schmidt, On cosets of the generalized first-order Reed–Muller code with low PMEPR, *IEEE Transactions on Information Theory*, (2006), (in press).

356. K.-U. Schmidt and A. Finger, New codes for OFDM with low PMEPR, in *Proc. ISIT'05*, (2005), 1136–1140, see also http://arxiv.org/PS_cache/cs/pdf/0508/0508024.pdf.

357. H. Schmidt and K. D. Kammeyer, Reducing the peak to average power ratio of multicarrier signals by adaptive subcarrier selection, in *Proc. ICUPC'98*, Florence, Italy, **2**, (1998), 933–937.

358. W. Schmidt, *Equations Over Finite Fields – An Elementary Approach*, Springer, (1976).

359. A. Schönhage, Fehlerfortpflanzung bei Interpolation, *Numerische Mathematik*, **3**, (1961), 62–71.

360. A. Schönhage, *Approximationstheorie*, De Gruyter Lehrbuch, (1971).

361. M. R. Schroeder, Synthesis of low-peak-factor signals and binary sequences with low autocorrelation, *IEEE Transactions on Information Theory*, **16**: 1, (1970), 85–89.

362. J. Seberry, B. J. Wysocki, and T. A. Wysocki, Golay sequences for DS CDMA applications, in *Proc. 6th Int. Symp. on DSP for Communication Systems, DSPCS2002*, (2002), 103–108.

363. A. G. Shanbhag and E. G. Tiedemann, Peak-to-average reduction via optimal Walsh code allocation in third generation CDMA systems, in *Proc. ISSTA'00*, **2**, (2000) 560–564.

364. C. E. Shannon, Probability of error for optimal codes in a Gaussian channel, in *Claude Shannon. Collected Papers*, IEEE Press, (1993), 279–324.

365. H. S. Shapiro, Extremal problems for polynomials, M.Sc.Thesis, MIT, (1951).

366. M. Sharif, M. Gharavi-Alkhansari, and B. H. Khalaj, On the peak-to-average power of OFDM signals based on oversampling, *IEEE Transactions on Communications*, **51**: 1, (2003), 72–78.

367. M. Sharif, C. Florens, M. Fazel, and B. Hassibi, Peak to average power reduction using amplitude and sign adjustment, in *Proc. ICC'04*, **2**, (2004), 837–841.

368. M. Sharif, C. Florens, M. Fazel, and B. Hassibi, Amplitude and sign adjustment for peak-to-average-power reduction, *IEEE Transactions on Communications*, **53**: 8, (2005), 1243–1247.

369. M. Sharif and B. Hassibi, On the asymptotic peak distribution of multicarrier signals: symmetric QAM/PSK constellations and spherical codes, in *Proc. of the 40'th Annual Allerton Conf.*, (2002), 1657–1666.

370. M. Sharif and B. Hassibi, Asymptotic probability bounds on the peak distribution of complex multicarrier signals without Gaussian assumption, in *Thirty-Sixth Asilomar Conference on Signals, Systems and Computers*, **1**, (2002), 171–175.

371. M. Sharif and B. Hassibi, On the average power of multiple subcarrier intensity modulated optical signals: Nehari's problem and coding bounds, in *Proc. ICC'03*, **4**, (2003), 2969–2973.

372. M. Sharif and B. Hassibi, A deterministic algorithm that achieves the PMEPR of $c \log n$ for multicarrier signals, in *Proc. ICASSP'03*, **4**, (2003), IV-540-3.

373. M. Sharif and B. Hassibi, Existence of codes with constant PMEPR and related design, *IEEE Transactions on Signal Processing*, **52**: 10, part 1, (2004), 2836–2846.

374. M. Sharif and B. Hassibi, On multicarrier signals where the PMEPR of a random codeword is asymptotically $\log n$, *IEEE Transactions on Information Theory*, **50**: 5, (2004), 895–903.

375. M. Sharif and B. Hassibi, On the achievable average power reduction of MSM optical signals, *Communications Letters*, **8**: 2, (2004), 84–86.

376. M. Sharif and B. Hassibi, Towards reducing the gap between PMEPR of multicarrier and single carrier signals, in *Proc. IEEE 6th Workshop Signal Processing Advances in Wireless Communications*, (2005), 380–384.

377. M. Sharif and B. H. Khalaj, Peak-to-mean envelope power ratio of oversampled OFDM signals: an analytic approach, in *Proc. ICC'01*, Helsinki, Finland, (2001), 1476–1480.

378. S. J. Shepherd, P. W. J. van Eetvelet, C. W. Wyatt–Millington, and S. K. Barton, Simple coding scheme to reduce peak factor in QPSK multicarrier modulation, *Electronics Letters*, **31**, (1995), 1131–1132.

379. S. J. Shepherd, J. Orris, and S. K. Barton, Asymptotic limits in peak envelope power reduction by redundant coding in orthogonal frequency-division multiplex modulation, *IEEE Transactions on Communications*, **46**: 1, (1998), 5–10.

380. V. M. Sidel'nikov, Weight spectrum of binary Bose–Chaudhuri–Hoquengham codes, *Problemy Peredachi Informatsii*, **7**: 1, (1971), 14–22.

381. R. Sivaswamy, Multiphase complementary codes, *IEEE Transactions on Information Theory*, **24**, (1978), 546–552.

382. R. Sivaswamy, Self-clutter cancellation and ambiguity properties of subcomplementary sequences, *IEEE Transactions on Aerospace and Electronic Systems*, **18**, 163–187.

383. S. B. Slimane, Peak-to-average power ratio reduction of OFDM signals using pulse shaping, in *Proc. GlobeCom'00*, **3**, (2000), 1412–1416.

384. H. J. S. Smith, On systems of linear indeterminate equations and congruences, *Philosophical Transactions of the Royal Society of London*, **151**, (1861), 293–326.

385. J. A. Smith, J. R. Cruz, and D. Pinckley, Method for reducing the peak-to-average of a multi-carrier waveform, in *Proc. VTC'00*, **1**, (2000), 542–546.

386. P. Solé, A limit law on the distance distribution of binary codes, *IEEE Transactions on Information Theory*, **36**, (1990), 229–232.

387. J. Spencer, Six standard deviations suffice, *Transactions of the American Mathematical Society*, **289**: 2, (1985), 679–706.

388. J. Spencer, *Ten Lectures on the Probabilistic Method*, SIAM CBMS-NSF Regional Conference Series in Applied Mathematics, (1994).

389. C. Suh, C.-S. Hwang, and H. Choi, Preamble design for channel estimation in MIMO-OFDM systems, in *Proc. GlobeCom'03*, **1**, (2003), 317–321.

390. A. Sumasu, T. Ue, M. Uesugi, O. Kato, and K. Homma, A method to reduce the peak power with signal space expansion (ESPAR) for OFDM system, in *Proc. VTC'00*, **1**, (2000), 405–409.

391. T. Starr, M. Sorbara, J. M. Cioffi, *et al.*, *DSL Advances*, Prentice Hall, (2002).

392. G. Szegö, *Orthogonal Polynomials*, AMS Colloq. Publ., Providence, **23**, (1975).

393. C. E. Tan and I. J. Wassel, Sub-optimum peak-reduction carriers for OFDM systems, in *Proc. VTC'03*, (2003), 1183–1187.

394. M. Tan and Y. Bar-Ness, OFDM peak-to-average power ratio reduction by combined symbol rotation and inversion with limited complexity, in *Proc. GlobeCom'03*, (2003), 605–610.

395. M. Tan, Z. Latinovic, and Y. Bar-Ness, STBC MIMO-OFDM peak-to-average power ratio reduction by cross-antenna rotation and inversion, *IEEE Communications Letters*, **9**: 7, (2005), 592–594.

396. B. Tarokh and H. R. Sadjadpour, Construction of OFDM M-QAM sequences with low peak-to-average power ratio, in *Proc. CISS'01*, Baltimore, USA, (2001), 724–729.

397. B. Tarokh and H. R. Sadjadpour, Construction of OFDM M-QAM sequences with low peak-to-average power ratio, *IEEE Transactions on Communications*, **51**: 1, (2003), 25–28.

398. V. Tarokh and H. Jafarkhani, An algorithm for reducing the peak-to-average power ratio in a multicarrier communications system, in *Proc. VTC'99*, (1999), **1**, 680–684.

399. V. Tarokh, and H. Jafarkhani, On the computation and reduction of the peak-to-average power ratio in multicarrier communications, *IEEE Transactions on Communications*, **48**: 1, (2000), 37–44.

400. J. Tellado, *Multicarrier Modulation with Low PAR: Applications to DSL and Wireless*, Kluwer, (2000).

401. J. Tellado and J. M. Cioffi, Peak power reduction for multicarrier transmission, in *Proc. IEEE GlobeCom Communications Theory MiniConf'98*, Sydney, (1998), 219–224.

402. J. Tellado and J. M. Cioffi, Efficient algorithms for reducing PAR in multicarrier systems, in *Proc. ISIT'98*, Boston, (1998), 191.

403. J. Tellado, L. Hoo, and J. M. Cioffi, Maximum likelihood detection of nonlinearly distorted multicarrier symbols by iterative decoding, in *Proc. GlobeCom'99*, (1999), 2493–2497.

404. C. Tellambura, Upper bound on peak factor of N-multiple carriers, *Electronics Letters*, **33**, (1997), 1608–1609.

405. C. Tellambura, Use of m-sequences for OFDM peak-to-average power ratio reduction, *Electronics Letters*, **33**, (1997), 1300–1301.

406. C. Tellambura, Phase optimization criterion for reducing peak-to-average power ratio in OFDM, *Electronics Letters*, **34**, (1998), 169–170.

407. C. Tellambura, A coding technique for reducing peak-to-average power ratio in OFDM, in *Proc. GlobeCom'98*, Sydney, Australia, **5**, (1998), 2783–2787.

408. C. Tellambura, Comment on Multicarrier transmission peak-to-average power reduction using simple block code, *Electronics Letters*, **34**: 17, (1998), 1646.

409. C. Tellambura, Computation of the continuous-time PAR of an OFDM signal with BPSK subcarriers, *IEEE Communications Letters*, **5**: 5, (2001), 185–187.

410. C. Tellambura, Improved phase factor computation for the PAR reduction of an OFDM signal using PTS, *IEEE Communications Letters*, **5**, (2001), 135–137.

411. N. T. T. Trang, T. Han, and N. Kim, Power efficiency improvement by PAPR reduction and predistorter in MIMO-OFDM system, in *Proc. ICACT'05*, **2**, (2005), 1381–1386.

412. C.-C. Tsen, Signal multiplexing in surface-wave delay lines using orthogonal pairs of Golay's complementary sequences, *IEEE Transactions on Sonics and Ultrasonics*, **18**, (1971), 103–107.

413. C.-C. Tseng and C. L. Liu, Complementary sets of sequences, *IEEE Transactions on Information Theory*, **18**: 5, (1972), 644–651.

414. R. Turyn, Ambiguity functions of complementary sequences, *IEEE Transactions on Information Theory*, **19**, (1963), 46–47.

415. R. Turyn, Sequences with small correlation, in *Error Correcting Codes*, H. B. Mann, ed., Wiley, (1968), 195–228.

416. R. Turyn, Hadamard matrices, Baumert–Hall units, four-symbol sequences, pulse compression, and surface wave encodings, *Journal of Combinatorial Theory, Series A*, **16**, (1974), 313–333.

417. A. Verma and M. T. Arvind, Peak-to-average power ratio reduction in multicarrier communication systems, in *Proc. IEEE Conference on Personal Wireless Communications*, (1999), 204–206.

418. S. Vláduts and A. Skorobogatov, On spectra of binary cyclic codes, in *Proc. of the 9th All-Union Conference on Coding Theory and Information Transmission*, Odessa, (1988), 72–74 (in Russian).

419. S. Vláduts and A. Skorobogatov, On spectra of subcodes over subfields of algebraic-geometric codes, *Problemy Peredachi Informatsii*, **27**: 1, (1991), 24–36.

420. J. H. van Vleck and D. Middleton, The spectrum of clipped noise, *Proceedings of the IEEE*, **54**: 1, (1966), 2–19.

421. T. Wada, Characteristic of bit sequences applicable to constant amplitude orthogonal multicode systems, *IEICE Transactions on Fundamentals of Electronics, Communications and Computer Sciences*, **E83-A**: 11, (2000), 2160–2164.

422. T. Wada, T. Yamazato, M. Katayama, and A. Ogawa, A constant amplitude coding for orthogonal multi-code CDMA systems, *IEICE Transactions on Fundamentals of Electronics, Communications and Computer Sciences*, **E80-A**: 12, (1997), 2477–2484.

423. T. Wada, T. Yamazato, M. Katayama, and A. Ogawa, Error correcting capability of constant amplitude coding for orthogonal multi-code CDMA systems, *IEICE Transactions on Fundamentals of Electronics, Communications and Computer Sciences*, **E81-A**: 10, (1998), 2166–2169.

424. Z.-X. Wan, *Lectures on Finite Fields and Galois Rings*, World Scientific, (2003).

425. Z.-X. Wan and C.-H. Wan, *Quaternary Codes*, World Scientific, (1997).

426. C.-L. Wang and Y. Ouyang, Low-complexity selected mapping schemes for peak-to-average power ratio reduction in OFDM systems, *IEEE Transactions on Signal Processing*, **53**: 12, (2005), 4652–4660.

427. H. Wang and B. Chen, Asymptotic distributions and peak power analysis for uplink OFDMA signals, in *Proc. ICASSP'04*, **4**, (2004), 1085–1088.

428. Z. Wang and G. B. Giannakis, Wireless multicarrier communications, *IEEE Signal Processing Magazine*, **17**: 3, (2000), 29–48.

429. S. Wei, D. L. Goeckel, and P. E. Kelly, A modern extreme value theory approach to calculating the distribution of the PAPR in OFDM systems, in *Proc. ICC'02*, (2002), 1686–1690.

430. G. R. Welti, Quaternary codes for pulsed radar, *IRE Transactions on Information Theory*, **6**, (1960), 400–408.

431. T. Weng and B. Guangguo, Application of orthogonal complementary pair of sequences to CDMA-QAM communication system, in *ICCT'87*, (1987), 826–829.

432. E. Wong and B. Hajek, *Stochastic Processes in Engineering Systems*, Springer, (1985).

433. D. Wulich, Peak factor in orthogonal multicarrier modulation with variable levels, *Electronics Letters*, **32**, (1996), 1859–1860.

434. D. Wulich, Reduction of peak to mean ratio of multicarrier modulation by cyclic coding, *Electronics Letters*, **32**, (1996), 432–433.
435. D. Wulich, Comments on the peak factor of sampled and continuous signals, *IEEE Communications Letters*, **4**: 7, (2000), 213–214.
436. D. Wulich, N. Dinur, and A. Glinowiecki, Level clipped high-order OFDM, *IEEE Transactions on Communications*, **49**, (2000), 928–930.
437. D. Wulich and L. Goldfeld, Reduction of peak factor in orthogonal multicarrier modulation by amplitude limiting and coding, *IEEE Transactions on Communications*, **47**: 1, (1999), 18–21.
438. G. Wunder and H. Boche, A baseband model for computing the PAPR in OFDM systems, in *Proc. 4th International ITG Conference on Source and Channel Coding*, Berlin, (2002), 273–280.
439. G. Wunder and H. Boche, Evaluating the SER in OFDM transmission with nonlinear distortion: an analytic approach, in *Proc. 7th Int. OFDM Workshop*, Hamburg, (2002), 193–197.
440. G. Wunder and H. Boche, Peak value estimation of band-limited signals from its samples with application to the peak-to-average power problem in OFDM, in *Proc. ISIT'02*, Lausanne, Switserland, (2002), 17.
441. G. Wunder and H. Boche, Peak magnitude of oversampled trigonometric polynomials, *Frequenz*, **56**: 5–6, (2002), 102–109.
442. G. Wunder and H. Boche, Peak value estimation of band-limited signals from its samples, noise enhancement and a local characterization in the neighborhood of an extremum, *IEEE Transactions on Signal Processing*, **51**: 3, (2003), 771–780.
443. G. Wunder and H. Boche, Upper bounds on the statistical distribution of the crest-factor in OFDM transmission, *IEEE Transactions on Information Theory*, **49**: 2, (2003), 488–494.
444. G. Wunder and S. Litsyn, Generalized bounds on the crest-factor of codes in OFDM transmission, in *Proc. ITW'03*, Paris, (2003), 191–194.
445. G. Wunder and K. G. Paterson, Crest-factor analysis of carrier interferometry MC-CDMA and OFDM systems, in *Proc. ISIT'04*, (2004), 426.
446. W. Xianbin, T. T. Tjhung, and C. S. Ng, Reduction of peak-to-average power ratio of OFDM system using a companding technique, *IEEE Transactions on Broadcasting*, **45**: 3, (1999), 303–307.
447. W. Xianbin, T. T. Tjhung, and C. S. Ng, Reply to the comments on Reduction of peak-to-average power ratio of OFDM system using a companding technique, *IEEE Transactions on Broadcasting*, **45**: 4, (1999), 420–422.
448. C. H. Yang, On Hadamard matrices constructible by circulant submatrices, *Mathematics of Computation*, **25**, (1971), 181–186.
449. C. H. Yang, Maximal binary matrices and sum of two squares, *Mathematics of Computation*, **30**, (1976), 148–153.
450. C. H. Yang, Hadamard matrices, finite sequences, and polynomials defined on the unit circle, *Mathematics of Computation*, **33**: 146, (1979), 688–693.
451. J. Yang, J. Yang, and J. Li, Reduction of the peak-to-average power ratio of the multicarrier signal via artificial signals, in *Proc. ICCT'2000*, **1**, (2000), 581–585.
452. D.-W. Yue and E.-H. Yang, Asymptotically Gaussian weight distribution and performance of multicomponent turbo block codes and product codes, *IEEE Transactions on Communications*, **52**: 5, (2004), 728–736.
453. Z. Yunjun, A. Yongacogiu, and J.-Y. Chouinard, Orthogonal frequency division multiple access peak-to-average power ratio reduction using optimized pilot symbols, in *Proc. ICCT'2000*, **1**, (2000), 574–577.

454. Z. Yunjun, A. Yongacoglu, and J. Y. Chouinard, Reducing multicarrier transmission peak power with a modified simple block code, in *Proc. ICCT'2000*, **1**, (2000), 578–580.
455. Z. Yunjun, A. Yongacoglu, J. Y. Chouinard, and L. Zhang, OFDM peak power reduction by sub-block coding and its extended versions, in *Proc. VTC'99*, **1**, (1999), 695–699.
456. Z. Zhang and J. How, Frequency domain equalization for MIMO space-time transmissions with single-carrier signaling, in *Proc. PIMRC'03*, **3**, (2003), 2262–2266.
457. A. Zygmund, *Trigonometric Series*, 2nd edn., Cambridge University Press, (1959).

Index